D1307033

UNDERSTANDING CLIMATE CHANGE

UNDERSTANDING CLIMATE CHANGE

Climate Variability, Predictability, and Change
in the Midwestern United States

Edited by S. C. Pryor

INDIANA UNIVERSITY PRESS

Bloomington and Indianapolis

This book is a publication of

Indiana University Press
601 North Morton Street
Bloomington, IN 47404-3797 USA

http://iupress.indiana.edu

Telephone orders	800-842-6796
Fax orders	812-855-7931
Orders by e-mail	iuporder@indiana.edu

LIBRARY OF CONGRESS
CATALOGING-IN-PUBLICATION DATA

Understanding climate change : climate variability, predictability, and change in the midwestern United States
/ edited by S. C. Pryor.
　　p.　cm.
　Includes bibliographical references and index.
　ISBN 978-0-253-35344-3 (cloth : alk. paper)
1. Climatic changes—Middle West. 2. Middle West—
Climate. I. Pryor, S. C., date
　QC984.M53U53 2009
　551.6978—dc22　　　　　　　　　2009051664

　　　1　2　3　4　5　14　13　12　11　10　09

As the global climate system evolves
there is an increasing need to understand
how those changes have been and will be manifest
at the regional and local scale.
Herein we present such analyses based on the Midwestern USA.

Contents

Preface

There is now unequivocal evidence for anthropogenic forcing of climate change, that our climate system has evolved and is evolving principally (though not exclusively) due to human activities. This has prompted unprecedented international efforts in mitigation of, and adaptation to, climate change. A critical component of these efforts is focused on improved quantification of how changes in global climate have been manifest regionally and how future changes may be manifest at the regional/local scale. This volume focuses on the Midwestern USA and provides detailed assessments of climate change, variability, and predictability using observational data and model simulations of the past one hundred years and the coming one hundred years. The contributions featured derive from a workshop held in October 2007 at Indiana University, which brought together climate scientists from across the Midwest to document the state of the art regarding the predictability, variability, and possible changes in the Midwestern climate and to identify remaining gaps in our knowledge and possible strategies to address those gaps.

Acknowledgments

The workshop from which this volume derives was funded by the Office of the Dean of the College of Arts and Science of Indiana University, Bloomington. The editor acknowledges additional financial support from the National Science Foundation.

Abbreviations and Acronyms

AGL	Above Ground Level
AMO	Atlantic Multi-decadal Oscillation
AO	Arctic Oscillation
AOGCMs	Atmosphere-Ocean General Circulation Models
AR4	Fourth Assessment Report (of the IPCC)
ASOS	Automated Surface Observing System
AWEA	American Wind Energy Association
BCC	Beijing Climate Center
BCCR	Bjerknes Centre for Climate Research
CA	Cluster Analysis
CAPE	Convective Available Potential Energy
CDD	Cooling Degree Days
CCCMA	Canadian Centre for Climate Modelling and Analysis
CCN	Cloud Condensation Nuclei
CCSR	Centre for Climate System Research
CLIVAR	CLimate VARiability and predictability
CMIP3	Coupled Model Intercomparison Project phase 3
CNA	Central North America
CNRM	Centre National de Recherches Météorologiques
COOP	CO-OPerative observer network
CPC	Climate Prediction Center
CRU	Climatic Research Unit (at the University of East Anglia)
CSIRO	Commonwealth Scientific and Industrial Research Organization
DoE	Department of Energy
DTR	Diurnal Temperature Range
ECMWF	European Center for Medium Range Weather Forecasts
EL	Environmental Level
ENSO	El Niño Southern Oscillation
EOF	Empirical Orthogonal Function
ERA-40	40-year reanalysis data set issued by ECMWF
GCM	General Circulation Model or Global Climate Model
GCIP	GEWEX Continental Scale International Project
GDP	Gross Domestic Product
GEV	Generalized Extreme Value
GEWEX	Global Energy and Water Experiment
GFDL	Geophysical Fluid Dynamics Laboratory
GHCN	Global Historical Climatology Network
GHG	GreenHouse Gases
GISS	Goddard Institute for Space Studies

HCU	Hydrologic Cataloguing Unit		NPP	Net Primary Productivity
HDD	Heating Degree Days		NREL	National Renewable Energy Laboratory
HRU	Hydrological Response Units		NWS	National Weather Service
HSG	Hydrologic Soil Groups		OLR	Outgoing Longwave Radiation
INM	Institute of Numerical Mathematics		OLS	Ordinary Least-Squares (regression)
IPCC	Intergovernmental Panel on Climate Change		PBL	Planetary Boundary Layer
IPSL	Institut Pierre Simon Laplace		PCA	Principal Component Analysis
LFC	Level of Free Convection		PCMDI	Program for Climate Model Diagnosis and Intercomparison
LI	Lifted Index		PCM	Parallel Climate Model
LLJ	Low-Level Jet		PDO	Pacific Decadal Oscillation
MAGIC	Midwest Assessment Group for Investigations of Climate		PNA	Pacific–North American
MCS	Mesoscale Convective System		RCM	Regional Climate Model
METROMEX	METROpolitan Meteorological EXperiment		SI	Seasonality Index
			SLP	Sea Level Pressure
MGA	Midwest Governors Association		SOM	Self-Organizing Maps
MIP	Model Intercomparison Projects		SPC	Storm Prediction Center
MIROC	Model for Interdisciplinary Research on Climate		SRES	Special Report on Emissions Scenarios (of the IPCC)
MIUB	Meteorologischen Instituts der Universität Bonn		SST	Sea Surface Temperature
			SWAT	Soil Water Assessment Tool
MPI	Max Plank Institute		TCMA	Twin Cities Metropolitan Area
MRI	Meteorological Research Institute of Japan		TF	Transfer Function
			Tmax	Maximum daily temperature
NAM	Northern Annular Mode		Tmin	Minimum daily temperature
NAO	North Atlantic Oscillation		TNH	Tropical Northern Hemisphere
NARCCAP	North American Climate Change Assessment Program		TPW	Total Precipitable Water
			UMRB	Upper Mississippi River Basin
NASA	National Aeronautics and Space Administration		UEA	University of East Anglia
NCAR	National Center for Atmospheric Research		UKMO	United Kingdom Meteorological Office
			USDA	US Department of Agriculture
NCDC	National Climatic Data Center		USGCRP	US Global Change Research Program
NCEP	National Center for Environmental Prediction		USGS	US Geological Survey
NOAA	National Oceanographic and Atmospheric Administration		WCRP	World Climate Research Programme
NNR	NCEP/NCAR Reanalysis		WGEN	Weather GENerator
NPO	North Pacific Oscillation		WH	"Warming Hole"

UNDERSTANDING CLIMATE CHANGE

1. Climate Variability, Predictability, and Change: *An Introduction*

S. C. PRYOR AND E. S. TAKLE

Introduction

CLIMATE SCIENCE

For about the last three decades scientists from a variety of related disciplines have applied three-dimensional, time-dependent models of fluid flow to study the atmosphere on time scales of minutes to a few days and spatial scales of meters to a few thousand kilometers. About two decades ago atmospheric modelers began routine simulations of the global circulation using three-dimensional grids on time scales beyond one year, and hence launched the science of climate modeling. Addition of this new tool to the list of methods used by those studying climate called for a new field of study now known as climate science. Evolution of this sub-discipline has resulted in unprecedented opportunities to simulate climates that evolve concurrently with changes in atmospheric greenhouse gases—a major driver of future climates—in a physically consistent way. The prospect of future anthropogenically induced changes to global and regional climate of magnitudes to have measurable impact has brought new urgency to the study of climate.

Concurrently, population growth and economic development have created new dependencies and risks relating to possible changes in means and extremes of climate from conditions observed during the instrumental record. Because of this, the future of climate science is now more than ever tied to the future of climate-sensitive societal activities. Herein we summarize the state of climate science research in the Midwestern USA.

GLOBAL CLIMATE CHANGE: AN OVERVIEW

The 2007 report from the Intergovernmental Panel on Climate Change contained the most unequivocal statements regarding global climate change of any such report issued to date (IPCC 2007). Their major findings include the following:

• We have "*very high confidence* that the global average net effect of human activities since 1750 has been one of warming, with a radiative forcing of +1.6 [+0.6 to +2.4] Wm^{-2}".
• "Warming of the climate system is unequivocal. Palaeoclimatic information supports the interpretation that the warmth of the last half

century is unusual in at least the previous 1,300 years."

- "At continental, regional and ocean basin scales, numerous long-term changes in climate have been observed." Although the committee also note, "Some aspects of climate have not been observed to change."

- "Most of the observed increase in global average temperatures since the mid-20th century is *very likely* due to the observed increase in anthropogenic greenhouse gas concentrations."

- "For the next two decades, a warming of about 0.2°C per decade is projected for a range of SRES emission scenarios. Even if the concentrations of all greenhouse gases and aerosols had been kept constant at year 2000 levels, a further warming of about 0.1°C per decade would be expected. . . . Continued greenhouse gas emissions at or above current rates would cause further warming and induce many changes in the global climate system during the 21st century that would *very likely* be larger than those observed during the 20th century."

- "There is now higher confidence in projected patterns of warming and other regional-scale features, including changes in wind patterns, precipitation and some aspects of extremes."

For the majority of scientists the conclusion is clear. The global climate system has evolved, and continues to evolve, principally (though not exclusively) due to human activities. We are at a nexus, where unequivocal evidence exists for anthropogenic forcing of climate change, and unprecedented international efforts are being engaged in mitigation of, and adaptation to, climate change. A critical component of these efforts

is focused on improved quantification of how changes in global climate have been manifest regionally and how future changes may be manifest at the regional/local scale. In this volume we present examples of these analyses.

PURPOSE OF THIS VOLUME

Herein we focus on climate variability, predictability, and change in the Midwestern USA (figure 1.1) and provide detailed assessments of past climate variability and possible future states. The contributions presented derive from a workshop held in October 2007 at Indiana University, which brought together climate scientists from across the Midwest to document the state of the art regarding the predictability, variability, and possible changes in the Midwestern climate, to identify remaining gaps in our knowledge, and to articulate possible strategies to address those gaps.

The climate system operates across a wide spectrum of time scales, from "short-term" climate variability that might reasonably be characterized using measures of inter-

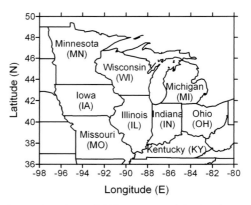

Figure 1.1. A map showing the Midwestern region and identifying the states that compose the Midwest.

annual to inter-decadal variability and that may or may not be predictable depending on the source, to longer-term evolution of the mean climate state and/or variability therein. We focus on a variety of parameters in the past and current climate over the Midwestern USA in order to quantify the historical climate and provide an assessment of variability, and assess the likelihood of change in the mean and variability of the climate over the past one hundred years and the coming one hundred years respectively. In doing so we are limiting our analysis and presentation to essentially the period for which observational records exist and for which simulations from multiple coupled Atmosphere-Ocean General Circulation Models (AOGCMs) are deemed to exhibit skill and are readily available (IPCC 2007).

"Climate change" refers to a shift in mean climate conditions in a specific region, whereas "climate variability" refers to temporal fluctuations around the mean. A further focus of this volume is on a key question in climate science: whether a change in the mean is always associated with a change in the variability (e.g., variance) of a given climate parameter, and indeed whether an increase in the mean de facto results in an increase in the variability around that mean and hence a change in climate extremes.

The Midwestern USA

Differing definitions of the Midwestern USA are present in the literature. We focus predominantly on the upper eight states shown in figure 1.1 and thus use a definition coincident with the definition of the Midwest used in the National Climate Assessment program (Easterling and Karl 2000). However, some chapters also include data from Kentucky and hence expand the definition to nine states (listed in alphabetical order: Illinois, Indiana, Iowa, Kentucky, Michigan, Minnesota, Missouri, Ohio, and Wisconsin). These nine Midwestern states comprise approximately 21% of the nation's population and economic activity (the cumulative state domestic product is approximately 21% of the gross domestic product for 2004) (table 1.1).

Farming, manufacturing, and forestry characterize the Midwest. Of these, agricultural activity dominates land use, with up to 89% of the land area of the states being used for agricultural purposes (table 1.1). As a result the Midwest produces most of the nation's corn and soybeans. Climate variability and extreme weather play a key role in dictating crop yields. For example, the region, like much of the Great Plains, experienced greatly depressed crop yields during the droughts of the 1930s. By some estimates yields of wheat and maize in the Great Plains were 50% lower during the 1930s than in the prior decade (Rosenzweig and Hillel 2008). Outbreaks of pests and disease are also critically linked to meteorology and climate. For example, the epidemic of southern leaf blight fungus of maize (*Helminthosporium maydis*) that affected the central United States during 1970–1971 led to a 15% reduction in national maize production and $1.09 billion in economic losses, and was spread from Mississippi into the Midwest by winds associated with tropical storm activity in the Gulf of Mexico (Rosenzweig and Hillel 2008). Equally, changes in land use and land cover associated with agricultural activities have led to changed surface energy and moisture budgets and thus changes in boundary layer properties and local to regional thermal and moisture regimes (Cotton and Pielke 2007).

The Great Lakes, which form the world's largest freshwater lake system, serve as recreational centers and a transportation link via the St. Lawrence Seaway to the Atlantic Ocean. Additionally, the Midwest is traversed by the Mississippi and Ohio rivers. Changes in hydrologic regimes may influence river discharge and lake levels and hence have a profound effect on natural and agro-ecosystems, tourism, and commerce.

Equally, over the entire nation, the United States of America uses more than 500 billion liters of freshwater per day, with over 40% going to cooling power plants and approximately the same amount being used for irrigation. Changes in hydrologic regimes (e.g., the amount and/or timing of precipitation; see chapter 9) may increase pressure on water supplies even in regions that historically have not exhibited water scarcity. The severity of this problem with respect to electricity production is exemplified by the drought in France during 2003 that caused loss of 15% of the nuclear electricity generation for five weeks due to reductions in the

Table 1.1. Descriptive statistics for each Midwestern state, along with estimates of their greenhouse gas (GHG) emissions

	Area of state (sq. miles)	Population in 2006	Gross state product (2004)	Land in farms in 2002	1990 Total GHG emissions (MMTCE)	Fraction of GHG from Energy/ Agriculture	2000*/2002** Total GHG emissions (MMTCE)
Illinois	55,583	12,831,970	478,966 (4.5)	27.3 (77)	66.1	86/3	
Indiana	35,867	6,313,520	208,834 (1.9)	15 (66)	61.3	93/3	
Iowa	55,869	2,982,085	100,853 (0.9)	31.7 (89)	17	72/18	32.8*
Kentucky	39,728	4,206,074	125,021 (1.2)	13.8 (54)	35.4	84/3	
Michigan	56,803	10,095,643	344,954 (3.2)	10.1 (28)	57.4	NA	62.59**
Minnesota	79,610	5,167,101	206,216 (1.9)	27.5 (54)	22.5	85/10	
Missouri	68,885	5,842,713	186,018 (1.7)	30.0 (68)	29.3	86/8	
Ohio	40,948	11,478,006	385,373 (3.6)	14.6 (58)	88.9	85/2	
Wisconsin	54,310	5,556,506	193,900 (1.8)	15.7 (45)	27.1	87/9	

Population estimates from the Census Bureau http://quickfacts.census.gov/qfd/states/.
GHG inventory data available from http://www.epa.gov/climatechange/emissions/state_ghginventories.html.
Gross state product is in millions of chained 2000 dollars (data from Werneke and DeBrandt [2006]). Number in parentheses is percent of U.S. gross domestic product.
Land in farms is given in millions of acres (data from Werneke and DeBrandt [2006]). Number in parentheses is percent of total state land that is agricultural.
The GHG inventory methodology changed between the 1990 inventories and the subsequent inventories shown in this table. Of the Midwestern states, only Iowa and Michigan have updated their inventories since 1990.

availability of water for cooling purposes (Hightower and Pierce 2008).

The Midwest is also home to a number of important metropolitan centers, including Chicago, Detroit, Indianapolis, and Columbus, which rank as the third, eleventh, twelfth, and fifteenth most populous urban areas in the United States. There is considerable evidence that these, and other, urban areas cause substantial changes in precipitation and severe weather regimes due to enhanced emission of cloud condensation nuclei (CCN), changes in surface roughness and low-level convergence, and addition of heat and moisture (see chapters 19, 20, and 22 herein and Cotton and Pielke 2007). Additionally, increasing urbanization in the region may increase the vulnerability of the populace to climate change via increased exposure to climate risk such as unhealthful temperatures (resulting from the urban heat-island effect; Schoof, Pryor, and Robeson 2007).

Climate Variability, Predictability, and Change in the Midwestern USA

GLOBAL AND CONTINENTAL CONTEXT

Climate change over the Midwestern USA is best interpreted in the spatial context of global and continental tendencies. Annual minimum and maximum temperatures for the continental United States are shown in figure 1.2. For the Midwest, the most notable feature is the north-south temperature gradient, which is due primarily to low temperatures to the north in winter (see chapter 2). Global projections of climate change imply larger temperature increases over the poles than over the equator, which would suggest that isotherms will migrate northward, with northern isotherms moving farther than southern isotherms over the Midwest. This seems to be generally true for the annual mean but not uniformly so for all seasons (see chapter 2) or for daily maximum and minimum temperatures (e.g., the warming hole discussed in chapter 3). Observations of mean U.S. temperature since 1975 show a notable warming trend (~1.1°C over 32 years) (NOAA 2008) that is consistent with temperature increases over other large land masses at mid to high latitudes in the Northern Hemisphere. Mid-continent areas are warming more than coastal areas, and winters have been warming more than summers. Globally, daily minimum temperatures increased more than daily maximum temperatures until about the mid 1980s. Since that time, global daily maximum and minimum temperatures have been rising at about the same rate (IPCC 2007). While the northern hemisphere land masses warmed by approximately 0.3°C per decade between 1979 and 2005, averaged over the entire Midwestern USA the warming rate was 0.5°C per decade, with most of this change being contributed from increases in fall and winter (see chapters 2 and 7). Nighttime temperatures in the Midwest are rising more than daytime temperatures, thereby leading to a reduced diurnal temperature range, unlike the global diurnal temperature range, which has been constant in recent years.

The historical climatological pattern for U.S. precipitation (figure 1.3) has high annual amounts in the eastern half of the United States and low values in the west (excepting coastal Washington and Oregon), with superposed highest regional

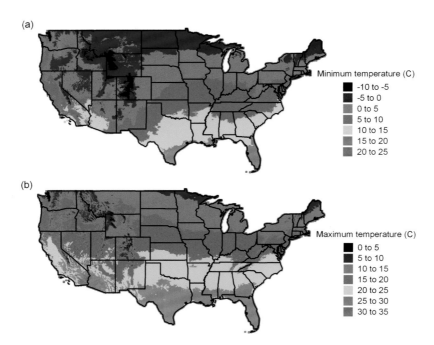

Figure 1.2. Climatological normals (1971–2000) for (a) annual minimum temperature (°C) and (b) annual maximum temperature (°C) over the contiguous U.S., computed using data from the PRISM Group, Oregon State University, http://www.prismclimate.org.

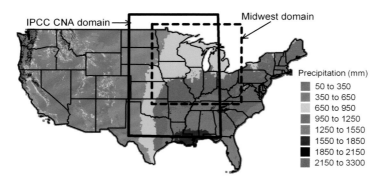

Figure 1.3. Annual average precipitation for 1971–2000 over the contiguous United States, computed using data from the PRISM Group, Oregon State University, http://www.prismclimate.org. Also shown is the Midwest domain used herein and the central North American region described in the Fourth Assessment Report of the IPCC.

precipitation in the southeast quarter and less but still high amounts in the northeast. This overall pattern puts the Midwest in the unique location of having a large climatological east-west gradient on its western boundary (separating the semi-arid or steppe climate to the west from the temperate continental climate over the Midwest) and north-south gradient throughout (separating the subtropical humid climate to the south from the temperate continental Midwest, see chapter 9). Future changes in precipitation, in contrast to temperature, are not likely to consist of a simple shift of current patterns. As discussed herein, changes in U.S. precipitation to date are more complex and show regionally specific behaviors and high variability. The United States has generally experienced a trend of increasing precipitation in the latter half of the twentieth century, but due to the high variability trend detection is very challenging (see chapters 9 and 12).

CLIMATE OF THE MIDWESTERN USA

As described in several of the chapters contained in this volume, the Midwestern USA has a strongly continental climate, though hot summers are mitigated in part by the presence of the Great Lakes (see chapter 21). Only the southwest portion of the region exhibits mean daytime maximum temperatures in excess of 32°C (approximately 90°F) (Burt 2004). The region experiences relatively cold winters (again mitigated in part by the Great Lakes) that result in mean daily minimum temperatures in January of below −18°C (approximately 0°F) in the northwest of the region (Minnesota and parts of Wisconsin, see chapter 7). Large diurnal variations in temperature are occasionally experienced. On 11 November 1911, Kansas City, Missouri, had a maximum temperature of 76°F and a minimum temperature of 11°F. The continental climate results in a relatively high energy demand, as evidenced in the climatological mean heating and cooling

Table 1.2. 1971–2000 Mean heating degree days (HDD) and cooling degree days (CDD) in each calendar month in the four most populous cities in the Midwest and the two largest cities in the U.S. (New York and Los Angeles)

		J	F	M	A	M	J	J	A	S	O	N	D	Ann
Chicago	HDD	1333	1075	858	513	232	49	6	9	112	401	759	1151	6498
	CDD	0	0	1	9	48	159	279	233	91	10	0	0	830
Detroit	HDD	1270	1074	886	527	219	41	5	12	121	426	742	1099	6422
	CDD	0	0	0	6	42	145	254	208	75	6	0	0	736
Indianapolis	HDD	1192	957	724	394	141	16	2	4	77	335	659	1020	5521
	CDD	0	0	2	10	69	221	331	272	122	14	1	0	1042
Columbus	HDD	1154	940	731	415	152	27	3	7	80	347	654	982	5492
	CDD	0	0	2	9	61	198	305	254	109	12	1	0	951
New York	HDD	1015	869	731	433	182	21	6	4	43	264	532	847	4947
	CDD	0	0	0	2	31	151	326	300	125	13	1	0	949
Los Angeles	HDD	252	205	200	141	78	19	1	0	2	21	121	234	1274
	CDD	4	6	6	15	19	58	135	175	154	81	22	4	679

Note: Heating degree day (HDD) and cooling degree day (CDD) are quantitative indices that reflect the demand for energy needed to heat or cool buildings. A base of 65°F is used in computing HDD and CDD.
Source: Data taken from http://cdo.ncdc.noaa.gov/cgi-bin/climatenormals/climatenormals.pl.

Table 1.3. Summary of extreme weather impacts on the Midwestern states

	Flood damage (1955–1999) average/year (million 1999 $US)	Tornado damage (1950–2006) (million 2006 $US)	Lightning fatalities (1959–1994)
Illinois	218.7	1754	85
Indiana	113.4	3951	74
Iowa	312.9	1709	65
Kentucky	118.8	1015	82
Michigan	35.56	1641	89
Minnesota	144.9	2285	53
Missouri	272.2	1975	79
Ohio	102.4	2149	115
Wisconsin	60.87	936	47

Source: Data taken from http://www.sip.ucar.edu/sourcebook/.

degree days for the four largest cities (table 1.2). In contrast to the huge seasonality in thermal regimes, much of the Midwest exhibits only moderate seasonality of precipitation receipt, with generally wetter spring and summer and low precipitation totals in fall and winter (see chapters 9 and 12). Precipitation climates are strongly influenced by the presence of the Great Lakes, and there is evidence for a large increase in lake-effect snowfall in the region of the Great Lakes since 1951, due in part to reduced ice cover since the early 1980s (Burnett et al. 2003; see also chapter 21).

As described in more detail in chapters 19–22, the region experiences a range of climate hazards (Burt 2004; Schmidlin and Schmidlin 1996; Visher 1944; Gallus, Snook, and Johnson 2008). Between 1959 and 1994, 689 people died as a result of lightning strikes, and economic losses due to tornado damage over the last fifty-six years exceed $17 billion (see table 1.3). The region's inhabitants are accustomed to major disruptions of transportation, agricultural, and industrial activities due to weather-related phenomena such as episodic flooding associated with prolonged and/or intense precipitation (see chapters 13 and 19). For example, one of the worst floods in United States history occurred in 1993 (June to August), with Illinois, Iowa, Missouri, and Wisconsin experiencing either record or near-record flooding (Bell and Janowiak 1995). Another devastating flood occurred in Ohio in March 1913, which resulted in 467 deaths and $147 million in damage (LaPenta et al. 1995). Equally, several weeks of heavy precipitation across the Midwest culminated in June 2008 in extensive flooding. As of June 20, 2008, twenty levees along the Mississippi River had been breached. Some parts of the Midwest also experience a high frequency of snowfall. For example, the western portion of the Michigan Peninsula experiences lake-effect snow and in individual years can have snow accumulations of over 2500 mm.

The Midwest has also experienced periods of intense and prolonged drought. The three-year drought of the late 1980s (1987–1989) covered 36% of the United States at its peak, but was focused on the north-central and Midwestern states. Particularly the summer of 1988 was characterized by both elevated temperatures and suppressed precipitation across much of the contiguous United States. On the basis of area-averaged temperature departures, the summer (June–August) of that year was the third warmest summer since 1931 for the country as a whole, and precipitation during June of that year was in the lowest 10th percentile of observed values across much of the Midwest (Ropelewski 1988). This drought was predicted, though the severity and extent were underestimated (Namais 1991). According to some estimates the 1980s' drought was the most expensive natural disaster of any kind to affect the United States. Combining losses in energy, water, ecosystems, and agriculture, the total cost was approximately $39 billion (Riebsame, Changnon, and Karl 1991).

The Midwestern USA is characterized by frequent passages of synoptic systems, and the seasonality of the climate is strongly influenced by the position and intensity of the polar jet stream (see chapters 17 and 18). Inter-annual and intra-annual variations in the intensity and tracking of synoptic scale systems and several surface parameters (e.g., temperature and precipitation) are explicable, at least in part, by a number of teleconnection indices—most notably the North Atlantic Oscillation (NAO), the Pacific–North American (PNA) index, and indices of the El Niño Southern Oscillation (ENSO) (Gershunov 1998; Leathers and Palecki 1992; Leathers, Yarnall, and Palecki 1991; Rosenzweig and Hillel 2008; Schoof 2004; Schoof and Pryor 2006). In accord with the dependence of thermal and hydrologic regimes on these indices, agricultural output is also linked to these hemispheric scale oscillations. Gross Domestic Product (GDP) during 1998 from agriculture in the United States was approximately $198 billion. The net effect of a strong El Niño was to suppress that by nearly $5 billion, while a strong La Niña depressed it by nearly $3 billion (Rosenzweig and Hillel 2008). Although the impacts on agro-ecosystems are not wholly repeatable, predictable, or consistently negative, during the strong El Niño of 1982–1983 the Midwest/Great Plains experienced hot and dry conditions resulting in soybean reductions associated with approximately

$3.4 billion in losses, and maize production losses of approximately $5.5 billion (Rosenzweig and Hillel 2008). Given the tremendous importance of these teleconnections to the agricultural and other industries, over the last decade the National Atmospheric and Oceanic Administration (NOAA) has invested resources in developing seasonal forecasts of both the teleconnection indices and the related regional climate impacts (Hartmann et al. 2002; Rosenzweig and Hillel 2008). The importance of these teleconnections to the climate of the Midwestern USA also means that, as described in chapters 3 and 17, accurate portrayal of these teleconnections is a critical component of developing accurate climate projections derived using AOGCMs.

CLIMATE MODEL PROJECTIONS OF FUTURE TRENDS OVER THE MIDWESTERN USA FROM THE IPCC AR4

The Midwest climate is dominated by the influence of mid-latitude cyclones. In future climate scenarios, AOGCMs generally indicate a slight poleward shift in storm tracks, an increase in the number of strong cyclones but a reduction in medium-strength cyclones. Atmospheric moisture transport and convergence are projected to increase, resulting in a widespread increase in annual precipitation over the northern half of the eastern United States, mostly due to increases of 5–20% in winter precipitation.

All global climate models reporting results for the IPCC AR4 produce global increases in surface air temperature for all future greenhouse gas emissions scenarios, and the projected increase in global mean temperature varies only modestly for different emissions scenarios for the first half of the twenty-first century. The mean of several global models for the A1B scenario (a mid-range emissions scenario) is about 0.28°C per decade increase in temperature (assuming a linear trend) over the twenty-first century. Summarizing from the AR4, all of North America is very likely to experience a continuation of the warming observed over the past twenty-five years, with magnitude that is likely to exceed the global average. In northern regions, warming is likely to be largest in winter. The lowest winter temperatures are likely to increase more than the average winter temperature in northern North America. Summer daily minimum temperatures are likely to increase more than the mean in the Midwest, although there is less certainty about daily maximum temperatures in summer. Annual precipitation is likely to increase in the north with most of the increase coming in winter and spring but decreases in summer. The length of the snow season and annual snow depth are very likely to decrease (although total

cold-season precipitation might not decrease since more might fall as rain).

The AR4 does not specifically look at the region we define as the Midwestern USA (dashed box in figure 1.3). However, the report does include a focus on a region labeled Central North America (CNA) (solid black box in figure 1.3), the results for which are shown in table 1.4. CNA encompasses a large portion of the southern United States, which will reduce confidence that table 1.4 is applicable to the Midwest. However, some general model results can be used to improve applicability of table 1.4 to the Midwest. For instance, the northeastern United States is likely to experience a larger increase in precipitation than the southeastern United States in this future climate. Also, the warming is likely to be higher in the northern half than the southern half of the CNA box, particularly in winter. This suggests that true numbers for the Midwest (except the number of years to attain significance) would be slightly larger than entries in table 1.4 for both temperature and precipitation. This would tend to weaken

Table 1.4. Distribution of model responses (50% being mean of all models) from a set of 21 AOGCMs for Central North America (CNA) for changes in temperature (°C) and precipitation (%) between the 1980–1999 period and the 2080–2099 period for the A1B emission scenario. T(yr) indicates the time in years for the response to reach the 95% significance level (assuming linear changes over the 100-year period). This region has precipitation increasing for all sub-categories of the middle half of the distribution for March-April-May. Frequencies (%) of extreme warm, wet, or dry seasons, averaged over all models, are shown only when 14 of 21 models agree on increase (bold and underline) or decrease. Table adapted from results presented in IPCC (2007).

Season	Temperature response						Precipitation response						Extreme season (%)		
	Min	25	50	75	Max	T(yr)	Min	25	50	75	Max	T(yr)	Warm	Wet	Dry
DJF	2.0	2.9	3.5	4.2	6.1	30	-18	0	5	8	14		71	**7**	
MAM	1.9	2.8	3.3	3.9	5.7	25	-17	2	7	12	17	>100	81	**19**	4
JJA	2.4	3.1	4.1	5.1	6.4	20	-31	-15	-3	4	20	>100	93		15
SON	2.4	3.0	3.5	4.6	5.8	20	-17	-4	4	11	24		91	**11**	
Annual	2.3	3.0	3.5	4.4	5.8	15	-16	-3	3	7	15		98		

the north-south precipitation gradient over the Midwest (figure 1.3). The median annual temperature increase for CNA is 3.5°C (table 1.4). By the reasoning in the previous paragraph we might expect the Midwest to experience temperature increases of more than 0.36°C per decade, with the increase over this value coming from the fall and winter months. Similarly, the increases in winter and spring precipitation likely to occur in the northern regions of CNA would likely increase the annual increase in the Midwest above the CNA-mean of 3%.

The western United States, particularly the southwestern United States, is projected by AR4 models to have notable decreases in annual precipitation with multi-year or even multi-decadal periods of continuous drought. This, coupled with precipitation increases to the east, would suggest a larger east-west gradient along the western border of the Midwestern USA. It also is plausible that the pattern of dryness to the west periodically could encroach on the Midwest from the west and create higher inter-annual variability of precipitation in this region.

Risk-management decisions and other applications involving climate information for the Midwest would benefit substantially from having future climate projections that were developed for a region smaller than the CNA. Climate scenarios for regions smaller than the CNA are now being developed as described in multiple chapters within this volume. The projections of changes in climate of the Midwestern USA for the next few decades being produced under activities such as the North American Regional Climate Change Assessment Program (NARCCAP) will provide the complement

to observed trends of the past thirty years to deliver the best available guidance on future trends of climate for use in studying the impacts of climate change and for developing climate risk assessment tools.

REGIONAL CONTRIBUTION TO, AND RESPONSE TO, ANTHROPOGENIC CLIMATE CHANGE

As described above, the Midwestern USA has relatively high energy demands. and the nine Midwestern states shown in table 1.1 produced nearly one-third of the national greenhouse gas emissions (as measured in units of metric ton of carbon equivalents) in 1990, the last year for which data are uniformly available for all states.

Concerned with rising demand for energy and energy prices and increasing dependence on imported energy, in combination with increased recognition of the need to address climate change while sustaining and enhancing economic growth and job creation, in late 2007 ten Midwestern leaders (Governor Jim Doyle of Wisconsin, Governor Tim Pawlenty of Minnesota, Governor Rod Blagojevich of Illinois, Governor Mitch Daniels of Indiana, Governor Chester J. Culver of Iowa, Governor Jennifer Granholm of Michigan, Governor Kathleen Sebelius of Kansas, Governor Ted Strickland of Ohio, Governor M. Michael Rounds of South Dakota, and Premier Gary Doer of Manitoba) signed the Midwestern Regional Greenhouse Gas Reduction Accord. Indiana, Ohio, and South Dakota signed the agreement as observers to participate in the formation of the regional cap-and-trade system.

The stated goal for the Midwestern

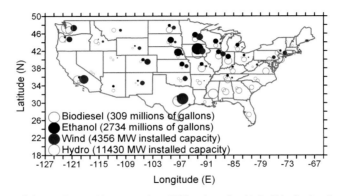

Figure 1.4. A map showing installed capacity for renewable energy supplies in 2007 for each state from: biodiesel (data from http://www.biodiesel.org/), ethanol (data from http://www.ethanol.org/), wind power installations (data from http://www.awea.org/), and hydroelectricity (data http://www.hydro.org/). The state data are plotted around the geographic center of population but are slightly offset to improve legibility. The colors used for each source are given in the legend along with the highest value for any state. The area of the circles scales linearly with the installed capacity.

Regional Greenhouse Gas Reduction Accord is to develop "a regional strategy to achieve energy security and reduce greenhouse gas emissions that cause global warming." The accord will:

• Establish greenhouse gas reduction targets and timeframes consistent with MGA (Midwest Governors Association) member states' targets;
• Develop a market-based and multi-sector cap-and-trade mechanism to help achieve those reduction targets;
• Establish a system to enable tracking, management, and crediting for entities that reduce greenhouse gas emissions; and
• Develop and implement additional steps as needed to achieve the reduction targets, such as low-carbon fuel standards and regional incentives and funding mechanisms. http://www.midwesterngovernors.org/

The Midwestern USA depends heavily on electricity from traditional coal-fired plants and on largely imported petroleum for agricultural, transportation, and industrial sectors. However, as identified by the governors, the Midwest also exhibits characteristics that may be useful in measures to mitigate climate change by reducing carbon dioxide emissions. These include:

• Existing and growing wind energy developments, due to the moderately good wind resource (see chapters 15 and 16).
• Substantial existing, and potential for expanded, biofuel production (figure 1.4).

However, these activities, while reasonable components of mitigation strategies, are themselves vulnerable to a changing climate. Hence, questions can be raised regarding whether the climate change commitment we have already undertaken, and that we have yet to make, altered the feasibility of both traditional and renewable energy sources?

REGIONAL SENSITIVITY TO CLIMATE VARIABILITY AND CHANGE

In the context of climate variability, predictability, and change, the Midwest is unique with respect to:

(1) Potential importance in current policy measures to reduce climate change. For example, U.S. ethanol production climbed to almost 5 billion gallons in 2006, up nearly 1 billion gallons from 2005, and production is slated to exceed 10 billion gallons by 2009 (Westcott 2007). According to U.S. Department of Agriculture Statistics, in 2005/2006 14% of corn grown in the United States went into ethanol production, and that fraction is projected to increase (Westcott 2007). As described above (figure 1.4), the Midwestern states contain a very large fraction of current ethanol production capacity. With the bio-economy gaining momentum, additional stress will be placed on agricultural yields in the region (McNew and Griffith 2005; Wang, Saricks, and Wu 1999), which in turn exhibit climate sensitivity (see, for example, discussions in chapters 4 and 7). The demand for increased and inter-annually consistent agricultural production in the Midwest may change the resilience of the Midwest to changes in the mean climate state and variability therein. It may also increase the need for accurate long-range forecasts, the pay-offs for accurate forecasts, and the costs associated with inaccurate forecasts (Changnon 2002).

(2) The sign/magnitude of recent climate change/evolution. For example, parts of the Midwest have experienced cooling temperatures over the last century in contrast to the global mean, and there is a well-documented warming hole over the southwestern portions of the Midwest (Pan et al. 2004) (see also chapter 3 of this volume). Understanding of the mechanisms,

and consequently improved simulations, of the warming hole will improve confidence in climate projections.

(3) The ability of coupled AOGCMs to simulate current and possible future climate states. For example; AOGCM simulations over the central/Midwestern USA exhibit much greater divergence than over many parts of the world (see, for example, chapters 6 and 8). This divergence includes teleconnections with sea-surface temperatures in the central and south Pacific (Joseph and Nigam 2006). Resolving reasons for the model divergence will greatly enhance the capabilities of AOGCMs.

(4) The frequency and consequences of extreme events. Insurance losses from weather-related risk (Kunreuther and Michel-Kerjan 2007) and many of the hazardous weather events occur within the Midwest (Kunkel, Pielke, and Changnon 1999) (table 1.3). For example, Hurricane Katrina excepted, the major extreme weather-related economic losses within the United States over the last decades have been focused on the Midwest (see chapter 20). Such events are extremely important to the regional economy, but changes in the frequency of inherently rare events are difficult to quantify in the historical records and very challenging to simulate in the context of climate change.

SYNTHESIS OF MIDWESTERN USA
ANALYSIS IN THE 2000 CLIMATE
CHANGE IMPACTS ASSESSMENT

In the Climate Change Impacts on the United States assessment published in 2000 (Easterling and Karl 2000) the following were identified as "key issues" in the Midwestern USA:

(1) Reduction in lake and river levels. During the twentieth century, rapid lake-level fluctuations of 1 to 1.5 m were observed in Lake Michigan (Polderman and Pryor 2004). Significant drops in the level of the Great Lakes during the late 1990s affected tourism and shipping industries and forced local communities to undertake expensive dredging operations, but also permitted reestablishment of diverse wetlands and expanded fish spawning grounds. Conversely, abnormally high lake levels pose a risk to sensitive shoreline environments by increasing their susceptibility to erosion from higher wave energy during strong storms (Meadows et al. 1997). Recent research reports that Lake Superior, the world's largest lake, experienced record low water levels in 2007 (Holden 2007) and that Lakes Huron and Michigan were also at near-record lows in that year (Sellinger et al. 2008). While these recent declines do not prove a climate link, changing precipitation regimes and/or evapo-transpiration may influence both lake and river levels (see chapter 10).

(2) Health and quality of life in urban areas. Major regional heat waves in the Midwest during both 1980 and 1995 (Changnon, Kunkel, and Reinke 1996) have focused concern on heat-related mortality and morbidity, particularly in light of high regional relative humidity, which tends to exacerbate heat stress (see chapter 5). There is also concern that climate change may cause enhancement of air pollutant concentrations (Leung and Gustafson 2005;

Mickley et al. 2004) in a region where several major urban areas already exceed the National Ambient Air Quality Standards (Pryor and Spaulding 2009), or assist the spread of infectious diseases into the region (Ebi et al. 2006).

(3) Agricultural shifts. Inter-annual variability in agricultural production over the past decades in the United States is closely linked to climate variability. For example, the drought of 1988 was accompanied by estimated economic losses of $56 billion. The floods of 1993 in the Mississippi River Basin caused over $23 billion in agricultural losses (Rosenzweig et al. 2000). Changes in the timing/amount of precipitation may alter the need for irrigation, or, under scenarios of intensification of extreme precipitation events, prompt increased soil erosion. Equally, changes in climate parameters may alter the introduction and spread of plant pathogens (Pan et al. 2006). Primary weather-related impacts on Midwest agriculture are dependent on precipitation and/or temperatures during key phenological stages during the spring and summer months for crops such as maize and soybeans and, to a lesser extent, during winter and fall months for crops such as winter wheat. The primary limitation on yields appears to be water stress caused by anomalously low or high precipitation (Wu, Hubbard, and Wilhite 2004). These yield limitations can be enhanced by anomalously warm or cool temperatures, especially when drought conditions lead to low precipitation combined with high temperatures (Hubbard and Wu 2005).

(4) Changes in semi-natural (uncultivated) and natural ecosystems.

Phenological changes in native and perennial plant species have been observed in response to changes in thermal regimes (Walther et al. 2002). In at least one study terrestrial equilibrium net primary production (NPP) over parts of the Midwestern USA exhibited a high degree of climate sensitivity (Moldenhauer and Ludeke 2002). As discussed in chapter 4, significant changes in frost-free season length have been observed across the Midwest, and may continue in the current century. Changes in thermal regimes have already been observed to be associated with changes in plant hardiness across the Midwest (chapter 7).

While the state of our knowledge regarding climate variability, predictability, and change has evolved since 2000, these four key vulnerabilities remain. Although regional sensitivity to climate change is often framed in terms of vulnerability, at least in terms of the latter two points raised above, climate evolution may also be associated with new opportunities. Global climate change and variability will result in both "winners" and "losers" (O'Brien and Leichenko 2003). It may be that the Midwestern USA will experience a mixture of these effects, depending on the socioeconomic or environmental sector under consideration.

Structure and Content of This Book

While the impacts of climate change act as motivation for much of the research presented in this volume, we focus solely upon the physical climate and structure the chapters along four thematic lines:

- Thermal regimes
- Hydrologic regimes
- Atmospheric circulation and flow regimes
- Climate hazards.

Each section is preceded by an overview or synthesis chapter designed to familiarize the reader with the topic under study and highlight results of both prior research and the presentations contained in this volume. The individual chapters present unique research focused on a variety of aspects of the physical climate of the Midwest, designed to quantify the current climate in which we live, past climate evolution, and possible future climate states. We conclude the volume with a chapter (number 23) that describes some ongoing efforts to improve our understanding of climate variability, change, and predictability along with recommendations for possible additional research avenues that could or should be pursued.

ACKNOWLEDGMENTS

The workshop from which this volume derives was funded by the Office of the Dean of the College of Arts and Science of Indiana University. SP acknowledges additional financial support from the NSF Geography and Regional Science program (grants # 0618364 and 0647868).

REFERENCES

Bell, G. D., and J. E. Janowiak. 1995. "Atmospheric Circulation associated with the Midwest Floods of 1993." *Bulletin of the American Meteorological Society* 76: 681–695.

Burnett, A. W., et al. 2003. "Increasing Great Lake-Effect Snowfall during the Twentieth Century: A Regional Response to Global Warming?" *Journal of Climate* 16: 3535–3542.

Burt, C. C. 2004. *Extreme Weather: A Guide and Record Book*. Hong Kong: Twin Age, 304.

Changnon, S. A. 2002. "Impacts of the

Midwestern Drought Forecasts of 2000." *Journal of Applied Meteorology* 41: 1042–1052.

Changnon, S. A., K. Kunkel, and B. Reinke. 1996. "Impacts and Responses to the 1995 Heat Wave: A Call to Action." *Bulletin of the American Meteorological Society* 77: 1497–1506.

Cotton, W. R., and R. A. Pielke. 2007. *Human Impacts on Weather and Climate.* 2nd ed. Cambridge: Cambridge University Press, 330.

De Sherbinin, A., A. Schiller, and A. Pulsipher. 2007. "The Vulnerability of Global Cities to Climate Hazards." *Urbanization and Environment* 19: 39–64.

Easterling, D. R., and T. Karl. 2000. "Potential Consequences of Climate Variability and Change for the Midwestern United States." In *Climate Change Impacts on the U.S.: The Potential Consequences of Climate Variability and Change,* ed. N. A. S. Team, 167–188.

Ebi, K. L., et al. 2006. "Climate Change and Human Health Impacts in the United States: An Update on the Results of the U.S. National Assessment." *Environmental Health Perspectives* 114: 1318–1324.

Gallus, W. A., N. A. Snook, and E. V. Johnson. 2008. "Spring and Summer Severe Weather Reports over the Midwest as a Function of Convective Mode: A Preliminary Study." *Weather and Forecasting* 23: 101–113.

Gershunov, A. 1998. "ENSO Influence on Intraseasonal Extreme Rainfall and Temperature Frequencies in the Contiguous United States: Implications for Long-Range Predicability." *Journal of Climate* 12: 3192–3203.

Hartmann, H. C., et al. 2002. "Confidence Builders—Evaluating Seasonal Climate Forecasts from User Perspectives." *Bulletin of the American Meteorological Society* 83: 683–698.

Hightower, M., and S. A. Pierce. 2008. "The Energy Challenge." *Nature* 452: 285–286.

Holden, C. 2007. "Lean Times for Lake Superior." *Science* 318: 893.

Hubbard, K. G., and H. Wu. 2005. "Modification of a Crop-Specific Drought Index for Simulating Corn Yield in Wet Years." *Agronomy Journal* 97: 1478–1484.

IPCC. 2007. *Climate Change 2007: The Physical Science Basis.* Contribution of Working Group I to the Fourth Assessment Report of the Intergovernmental Panel on Climate Change. [Ed. S. Solomon, et al.] New York: Cambridge University Press.

Joseph, R., and S. Nigam. 2006. "ENSO Evolution and Teleconnections in IPCC's Twentieth-Century Climate Simulations: Realistic Representation?" *Journal of Climate* 19: 4360–4377.

Kunkel, K. E., R. A. Pielke, and S. A. Changnon. 1999. "Temporal Fluctuations in Weather and Climate Extremes that Cause Economic and Human Health Impacts: A Review." *Bulletin of the American Meteorological Society* 80: 1077–1098.

Kunreuther, H. C., and E. O. Michel-Kerjan. 2007. "Climate Change, Insurability of Large-Scale Disasters, and the Emerging Liability Challenge." *University of Pennsylvania Law Review* 155: 1795–1842.

LaPenta, K. D., et al. 1995. "The Challenge of Forecasting Heavy Rain and Flooding throughout the Eastern Region of the National Weather Service. Part I: Characteristics and Events." *Weather and Forecasting* 10: 78–90.

Leathers, D. J., and M. A. Palecki. 1992. "The Pacific/North American Teleconnection Pattern and United States Climate. Part II: Temporal Characteristics and Index Specification." *Journal of Climate* 5: 707–716.

Leathers, D. J., B. Yarnal, and M. A. Palecki. 1991. "The Pacific/North American Teleconnection Pattern and the United State Climate. Part I: Regional Temperature and Precipitation Associations." *Journal of Climate* 4: 517–528.

Leung, L. R., and W. I. Gustafson, Jr. 2005. "Potential Regional Climate Change and Implications to U.S. Air Quality." *Geophysical Research Letters* 32, doi:10.1029/2005GL022911.

McNew, K., and D. Griffith. 2005. "Measuring the Impact of Ethanol Plants on Local Grain Prices." *Review of Agricultural Economics* 27: 164–180.

Meadows, G., et al. 1997. "The Relationship between Great Lakes Water Levels, Wave Energies, and Shoreline Damage." *Bulletin of the American Meteorological Society* 78: 675–683.

Mickley, L. J., et al. 2004. "Effects of Future Climate Change on Regional Air Pollution Episodes in the United States." *Geophysical Research Letters* 31, doi:10.1029/2004GL021216.

Moldenhauer, O., and M. K. B. Ludeke. 2002. "Climate Sensitivity of Global Terrestrial Net Primary Production (NPP) Calculated Using the Reduced-Form Model NNN." *Climate Research* 21: 43–57.

Namias, J. 1991. "Spring and Summer 1988 Drought over the Contiguous United States—Causes and Prediction." *Journal of Climate* 4: 54–65.

NOAA. 2008. "2007 Annual Climate Review: U.S. Summary." Available online at: http://www.ncdc.noaa.gov/oa/climate/research/2007/ann/us-summary.html.

O'Brien, K. L., and R. M. Leichenko. 2003. "Winners and Losers in the Context of Global Change." *Annals of the Association of American Geographers* 93: 89–103.

Pan, Z., et al. 2004. "Altered Hydrologic Feedback in a Warming Climate Introduces a Warming Hole." *Geophysical Research Letters* 31, doi:10.1029/2004GL020528.

Pan, Z., et al. 2006. "Long-Term Prediction of Soybean Rust Entry into the Continental United States." *Plant Disease* 90: 840–846.

Polderman, N. J., and S. C. Pryor. 2004. "Linking Synoptic-Scale Climate Phenomena to Lake-Level Variability in the Lake Michigan-Huron Basin." *Journal of Great Lakes Research* 30: 419–434.

Pryor, S. C., and A. M. Spaulding. 2009. "Air Quality in Indiana." In J. A. Oliver, *Indiana's Weather and Climate*. Bloomington: Indiana University Press.

Riebsame, W. E., S. A. Changnon, and T. Karl. 1991. *Drought and Natural Resources Management in the United States: Impacts and Implications of the 1987–89 Drought*. Boulder, Colo.: Westview, 174.

Ropelewski, C. F. 1988. "The Global Climate for June–August 1988: A Swing to the Positive Phase of the Southern Oscillation, Drought in the United States, and Abundant Rain in Monsoon Areas." *Journal of Climate* 1: 1153–1174.

Rosenzweig, C., et al. 2000. "Climate Change and U.S. Agriculture: The Impacts of Warming and Extreme Weather Events on Productivity, Plant Diseases and Pests." Center for Health and the Environment, Harvard Medical School. Available online at: http://www.med.harvard.edu/chge/, p. 47.

Rosenzweig, C., and D. Hillel. 2008. *Climate Variability and the Global Harvest: Impacts of El Niño and Other Oscillations on Agroecosystems*. New York: Oxford University Press, 259.

Schmidlin, T. W., and J. A. Schmidlin. 1996. *Thunder in the Heartland*. Kent, Ohio: Kent State University Press, 362.

Schoof, J. T. 2004. "Generation of Regional Climate Change Scenarios Using General Circulation Models and Empirical Downscaling." Ph.D. diss., Indiana University, Bloomington, 288.

Schoof, J. T., and S. C. Pryor. 2006. "An Evaluation of Two GCMs: Simulation of North American Teleconnection Indices and Synoptic Phenomena." *International Journal of Climatology* 26: 267–282.

Schoof, J. T., S. C. Pryor, and S. M. Robeson. 2007. "Downscaling Daily Maximum and Minimum Temperature in the Midwestern USA: A Hybrid Empirical Approach." *International Journal of Climatology* 27: 439–454.

Sellinger, C. E., et al. 2008. "Recent Water Level Declines in the Lake Michigan–Huron System." *Environmental Science and Technology* 42: 367–373.

Visher, S. S. 1944. *Climate of Indiana*. Bloomington: Indiana University Press, 512.

Walther, G.-R., et al. 2002. "Ecological Responses to Recent Climate Change." *Nature* 416: 389–395.

Wang, M., C. Saricks, and M. Wu. 1999. "Fuel Ethanol Produced from Midwest U.S. Corn: Help or Hindrance to the Vision of Kyoto?"

Journal of the Air and Waste Management Association 49: 756–772.

Werneke, D., and K. A. DeBrandt, eds. 2006. *State Profiles: The Population and Economy of Each U.S. State.* Lanham, Md.: Bernan, 562.

Westcott, P. 2007. "U.S. Ethanol Expansion Driving Changes throughout the Agricultural Sector." *Amber Waves* 5: 10–16.

Wu, H., K. G. Hubbard, and D. A. Wilhite. 2004. "An Agricultural Drought Risk-Assessment Model for Corn and Soybeans." *International Journal of Climatology* 24: 723–741.

2. Overview: *Thermal Regimes*

J. T. SCHOOF

Introduction

The Intergovernmental Panel on Climate Change (IPCC) estimates that global mean surface air temperatures have risen by 0.74°C±0.18°C over the last one hundred years (1906–2005), with an accelerated warming rate of approximately 0.13°C±0.03°C per decade during the last fifty years (Trenberth et al. 2007). Closer examination reveals that land areas have warmed more than oceans, particularly during recent decades, and that the greatest warming has occurred in the Northern Hemisphere during winter and spring. In order to adapt to and mitigate regional impacts of climate change, it is important to understand thermal variations and changes at the regional scale. The focus of this chapter is to synthesize the historical and potential future evolution of thermal regimes over the Midwestern USA.

The thermal climate of the Midwestern USA is characterized by large diurnal and seasonal variations. The latter are due to inter- and intra-seasonal migration of the polar jet stream, which often acts as a focal point for the development of baroclinic cyclones over the region. Winter cold spells are typically driven by advection of cold air from Arctic regions in the wake of these cyclones. The subtropical Bermuda high, which is most intense during the summer months, is also an important control on thermal and moisture variations in the Midwest, and expansion of this semi-permanent feature is often associated with prolonged or extreme warm spells. As shown in figure 2.1, the seasonal mean temperature variations over the Midwest follow the expected north-south gradient, with mean temperatures ranging from less than −10°C during winter at northern locations to more than 25°C at southern locations during summer.

As described in chapter 1, human activities in the Midwestern USA are highly susceptible to temperature variations and changes. Temperature variations and changes can readily impact the region via changes in heat-related stress and mortality, shifts in agricultural productivity, and reductions in water levels on lakes and rivers. The 1995 heat wave and recent reductions in lake levels serve as prime examples of regional susceptibility.

The 1995 heat wave was responsible for hundreds of fatalities throughout the central United States, but 87% of the fatalities were in the Midwest and 65% were in the city of Chicago (Kunkel et al. 1996). The event resulted from a combination of high air temperatures and high dew point temperatures and was likely exacerbated by urban heat island influences in Chicago (Kunkel et al. 1996). Climate modeling experiments (e.g., Meehl and Tebaldi 2004) have indicated that future heat waves in North America will become more intense, more frequent, and longer lasting in the second half of the twenty-first century, raising concerns about preparation for such events within a climate change context.

During the decade 1997–2006, water levels in the coupled Lake Michigan–Lake Huron basin decreased by approximately 1.1 m (Sellinger et al. 2008). The changes appear to be linked to longer scale periodicity (for example the ~30 year periodicity discussed by Polderman and Pryor 2004), but may also be linked to some combination of changes in precipitation and warming-induced evaporation that may or not be linked to anthropogenic climate change. It is, however, noteworthy that the observed decreases are consistent with modeling studies based on projections of greenhouse-gas-induced climate change (e.g., Croley 1990; Smith 1991). These lake-level declines adversely affect shipping on regional waters, which leads to required dredging and subsequent release of toxins.

These examples indicate that it is critically important to understand temperature variability in the Midwestern region of the United States. Specifically, greater attention needs to be paid to understanding (1) historical regional temperature variations with explicit consideration of the responsible forcing mechanisms and (2) potential future

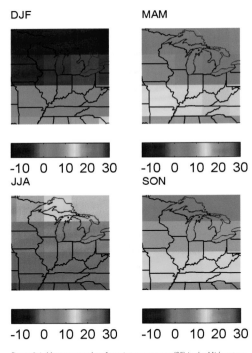

Figure 2.1. Mean seasonal surface air temperatures (°C) in the Midwestern USA (1888–2005) based on the CRUTEM3 5°x5° gridded dataset (Brohan et al. 2006). Results are shown for winter (DJF), spring (MAM), summer (JJA), and autumn (SON).

temperature changes as projected by coupled Atmosphere-Ocean General Circulation Models (AOGCMs). In this chapter, an overview of these foci is provided. Specific aspects of thermal variability and change follow in subsequent chapters.

Historical Variations in Regional Air Temperature

The Midwestern USA has an extensive record of station-based temperature data beginning in the late 1880s, allowing analysis of more than a century of data. For consistency with the future projections presented in the following section, the historical variability is described here using the CRU/Hadley Centre gridded land-surface air temperature version 3; CRUTEM3 (Brohan et al. 2006).

CRUTEM3 data is available globally on a 5°×5° grid and is subjected to extensive quality control as described by Jones and Moberg (2003), Brohan et al. (2006), and references therein. Here the focus is on region-wide variability, and the grid boxes covering 35–50°N latitude and 95–80°W longitude are averaged to provide a Midwestern USA temperature series (figure 2.2). Higher resolution spatial variations in Midwestern temperature variations are considered in several of the subsequent chapters.

The Midwestern USA surface air temperature series (figure 2.2) exhibits broad similarity to the Northern Hemisphere surface air temperature trend for land areas presented by Trenberth et al. (2007). For the gridded data product used here, Brohan et al. (2006) report a linear trend of 0.09°C per decade for Northern Hemisphere land areas over the period 1901–2005. The upward temperature trend is not linear with respect to time and exhibits accelerated warming since 1979, with a linear trend of 0.33°C per decade. For the time series presented in figure 2.2, the 1901–2005 linear trend is 0.07°C per decade, while the 1979–2005 trend is larger than the hemispheric land average of 0.55°C

per decade. Both are significant at the 95% confidence level, and are shown in table 2.1 along with seasonal and hemispheric trends.

It has become increasingly clear that at the hemispheric scale cold season temperatures are increasing faster than warm season temperatures. Seasonal anomaly time series over the Midwest for winter (DJF), spring (MAM), summer (JJA), and autumn (SON) exhibit agreement with larger-scale analyses indicating the largest changes in autumn and winter since 1979 (figure 2.3). The linear trends are positive in each season over both 1901–2005 and 1979–2005, with accelerated warming in the latter period. As shown in table 2.1, the long-term (1901–2005) trend is statistically significant (with α=0.05) for winter and spring, while the shorter-term (1979–2005) trend is statistically significant only for autumn. It is noteworthy that although the winter trend for the 1979–2005 period is larger than the trend in other seasons (0.82°C per decade), it is accompanied by large variability around the trend line. This results in a 95% confidence interval that narrowly includes zero. The 1979–2005 linear trends presented here and their statistical significance are also consistent with the data

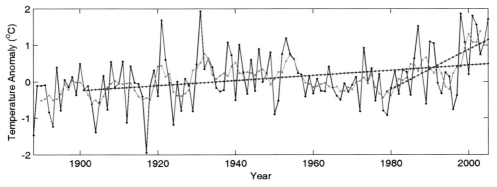

Figure 2.2. CRUTEM3 annual temperature anomaly time series for the Midwestern USA (35–50°N, 80–95°W) for 1888–2005. Anomalies are computed relative to the 1961–1990 data. Also shown are a five-year moving average (red), and 1901–2005 and 1979–2005 linear trends (blue).

Table 2.1. Linear trends (°C per decade, with 95% confidence intervals) applied to surface air temperature time series for the Northern Hemisphere land areas (after Brohan et al. 2006) and for the Midwestern USA. Also shown are the seasonal surface air temperature trends for the Midwest. Linear trends that are significantly different from zero (with α=0.05) are in italics.

	1901–2005	1979–2005
NH Land (Brohan et al. 2006)	*0.09±0.03*	*0.33±0.09*
Midwestern USA (figure 2.2)	*0.07±0.04*	*0.55±0.36*
Midwestern USA DJF (figure 2.3)	*0.11±0.11*	0.82±0.85
Midwestern USA MAM (figure 2.3)	*0.01±0.07*	0.33±0.59
Midwestern USA JJA (figure 2.3)	0.04±0.06	0.26±0.49
Midwestern USA SON (figure 2.3)	0.02±0.06	*0.57±0.46*

of Smith and Reynolds (2005) presented by Trenberth et al. (2007).

Several recent papers have focused on specific aspects of seasonal time series. For example, there has been increasing focus on variations and changes related to the frost-free season length in the United States (Easterling 2002; Kunkel et al. 2004) and in the Midwest (e.g., Robeson 2002) and sub-

sequent changes in biological response (e.g., Zhao and Schwartz, 2003). The long-term trends depicted in figure 2.3 are consistent with the reported increases in growing season length due to greater warming during spring than autumn, resulting in a longer period of time between the last spring freeze and first autumn freeze. There has also been extensive research conducted on the nature

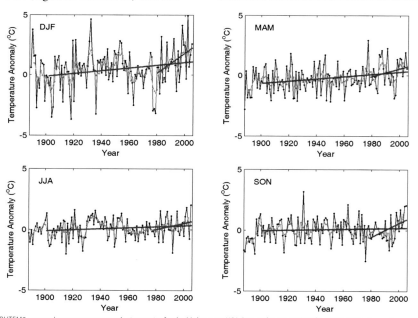

Figure 2.3. CRUTEM3 seasonal temperature anomaly time series for the Midwestern USA (35–50°N, 80–95°W) for 1888–2005. Anomalies are computed relative to the 1961–1990 data. Also shown are a five-year moving average (red), and 1901–2005 and 1979–2005 linear trends (blue).

of the weak summer temperature trend depicted in figure 2.3, which is negative at some western locations and thought to be linked to agricultural land-use changes, such as the change from natural savannah to cropland (Bonan 2001; Kalnay and Cai 2003). This cooling trend, referred to as a "warming hole," has not been widely reproduced by AOGCMs (Kunkel et al. 2006), but has been reproduced in regional climate modeling experiments by Pan et al. (2004). Both of these phenomena (frost-free season length changes and warming holes) are discussed in further detail in subsequent chapters.

Regional Surface Air Temperature Projections

Given the regional sensitivity to surface air temperature described above, there is large demand for information related to future thermal regimes. The primary tools used to address this demand are AOGCMs, which are based on the known physical principles that govern the climate system. Due to coarse and varying spatial resolution and differences in sub-grid scale parameterization and other model characteristics, AOGCMs exhibit a wide range of variability, even when driven with identical forcings. It is therefore important to consider output from multiple models and consider inter-model differences. An additional large uncertainty associated with future projections from AOGCMs is the magnitude of anthropogenic greenhouse gas forcing. To account for a range of scenarios, the IPCC has developed a portfolio of standard emissions scenarios, known as SRES (Nakićenović and Swart 2000), for modeling groups to utilize.

The scenarios used here correspond roughly to low (B1), medium (A1B), and high (A2) emissions of greenhouse gases (Meehl et al. 2007). An additional scenario is considered in which greenhouse gas emissions are held constant at levels observed in the year 2000. This scenario is referred to as the committed climate change scenario since it refers to change that is likely to occur regardless of future policy changes regarding anthropogenic emissions.

In preparation for the Fourth Assessment Report (AR4) of the Intergovernmental Panel on Climate Change (IPCC 2007), an unprecedented array of model simulations has been assembled. The IPCC AR4 models are used here to assess potential pathways for temperature change in the Midwestern USA. Specifically, results were analyzed for the ten models listed in table 2.2. For each of these models, results were available for the late twentieth century with known forcings, for the committed climate change scenario, and for the SRES B1, A1B, and A2 scenarios. Model projections of mean surface air temperature were examined for two future periods corresponding to the middle (2046–2065) and end (2081–2100) of the twenty-first century. A more complete description of these models, including references, is included in Meehl et al. (2007). The resolutions of these models range from 4°×5° (GISS-MODEL E-R) to 1.9°×1.9° (CSIRO-MK3.x and MPI-ECHAM5). For each model, anomalies were computed using 1961–1990 mean values. To compare these models with the observations presented above, all grid points within the box from 80–95°W and 35–50°N were averaged to yield a single value for the study region. This approach results in

Table 2.2. List of IPCC AR4 coupled AOGCM projections used in this analysis. The first column provides the model ID. The second column specifies the modeling center and country of origin.

Model ID	Organization (Country)
BCCR-BCM2.0	Bjerknes Centre for Climate Research (Norway)
CCCMa-CGCM3.1	Canadian Centre for Climate Modelling and Analysis (Canada)
CNRM-CM3	Météo-France / Centre National de Recherches Météorologiques (France)
CSIRO-MK3.0	Commonwealth Scientific and Industrial Research Organisation (Australia)
CSIRO-MK3.5	Commonwealth Scientific and Industrial Research Organisation (Australia)
GISS-MODEL E R	NASA Goddard Institute for Space Studies (USA)
IPSL-CM4	Institut Pierre Simon Laplace (France)
MIROC-3.2 (medium resolution)	Center for Climate System Research, National Institute for Environmental Studies, and Frontier Research Center for Global Change (Japan)
MIUB-ECHO-G	Meteorological Institute of the University of Bonn, Meteorological Research Institute of the Korea Meteorological Administration (KMA), and Model and Data Group (Germany / Korea)
MPI-ECHAM5	Max Planck Institute for Meteorology (Germany)

slightly different areas considered for each model, but the spatial differences are generally small, with the domains varying by approximately 1° of latitude and longitude at each border. Within the individual emissions scenarios, there is a large amount of variability among models for reasons previously discussed. Therefore, in order to provide the most reliable estimates of regional temperature response to anthropogenic greenhouse gas emissions, the results from all models have been averaged for each of the emission scenarios. The results are presented in terms of a multi-model ensemble with additional information provided regarding the range of mean values for individual models during each period.

During the overlapping period (1961–2000), there is general agreement between the observations and AOGCM ensemble, although through the averaging process, some of the simulated variability has been compromised (figure 2.4). The observations have a mean anomaly of 0.08°C for the period 1961–2000 relative to 1961–1990, reflecting the warmth of the 1990s relative to

the rest of the 1961–2000 period. The mean anomaly value from the AOGCM ensemble is 0.13°C (table 2.3). The AOGCM ensemble also reproduces the slight upward trend in temperature during this period.

As expected, the results from the transient climate projections (figure 2.4, table 2.3) indicate that the largest projected temperature changes correspond to scenarios with the largest greenhouse gas emissions. The committed climate change scenario reflects a 1.1°C warming by mid-century with little additional warming post-2065. Since the committed climate change scenario represents change induced by previously emitted greenhouse gases, the SRES scenarios present plausible pathways that may result from implementation of different policy decisions related to greenhouse gas emissions. In this context, "medium" and "high" emission scenarios result in almost twice as much additional (above our commitment) warming as the "low" emissions scenario. It is noteworthy that although the ensemble means for the 2046–2065 period differ by less than 1°C among the SRES sce-

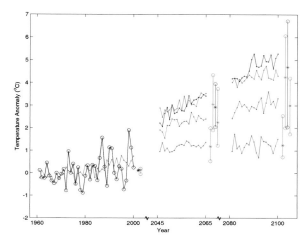

Figure 2.4. Midwestern regional temperature projections based on a multi-model ensemble from the IPCC AR4 models. All values are based on anomalies from 1961 to 1990 means. The black line depicts observations from the CRUTEM3 gridded data (Brohan et al. 2006). The corresponding green line represents the multi-model ensemble from twentieth-century climate simulations. For the climate projections, four series are presented: the green series represents the committed climate change scenario, the magenta series is for SRES B1, the red series is for SRESA1B, and the blue series is for SRESA2. The vertical bar next to each time period indicates the range of mean values for that period among the AOGCM simulations.

narios, the range among the means of all models is greater than 3.5°C. Additionally, the means of the SRES scenarios represent a considerably warmer climate than the mean of the committed climate change scenario, with only a small degree of overlap between individual model means from the committed climate change scenario and projections from SRES B1. For the "medium" and "high" emissions scenarios (SRES A1B and SRES A2), the coldest model is still warmer than the warmest model projection under committed greenhouse gas concentrations.

The results for the last twenty years of this century indicate that most of the warming we have already committed to through emissions of greenhouse gases will occur by mid-century, with little additional warming under the committed greenhouse gas scenario. Each of the SRES scenarios shows considerable additional warming relative to mid-century, but for each scenario there is a large range of responses among the models (figure 2.4). The magnitude of the inter-model differences is directly and linearly related to the magnitude of the change being con-

Table 2.3. Average and range of temperature changes (°C) in the Midwestern USA for observed data (CRUTEM3) for 1961–2000, and IPCC AR4 models for 1961–000, 2046–2065, and 2081–2100. For the future periods, results are shown for four different emissions scenarios. Numbers in parentheses for the model simulations are the lowest and highest individual model mean anomalies as depicted in figure 2.4.

	CRUTEM3 1961–2000	AOGCM 1961–2000	AOGCM 2046–2065	AOGCM 2081–2100
	0.1	0.1 (−0.1, 0.2)		
COMMIT			1.1 (0.6, 1.6)	1.1 (0.7, 1.6)
SRES B1			2.1 (1.2, 3.0)	3.0 (1.8, 4.1)
SRES A1B			2.9 (1.9, 4.3)	4.3 (2.9, 6.1)
SRES A2			2.9 (2.1, 3.9)	4.7 (2.9, 6.7)

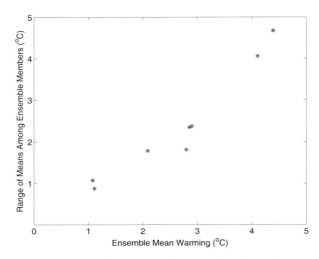

Figure 2.5. The relationship between the ensemble mean Midwestern USA regional temperature change from ten AOGCMs and the range of the mean values among the individual ensemble members. The AOGCMs used are listed in Table 2.2.

sidered (figure 2.5). The largest changes are associated with SRES A2 scenario, for which the 2081–2100 ensemble mean is 4.7°C above the 1961–1990 value. The lowest temperature change for any model in the "medium" (SRES A1B) or "high" (SRES A2) scenario is approximately 2.9°C. The "low" emissions scenario (SRES B1) is associated with an ensemble mean increase of more than 3°C.

Concluding Remarks and Summary of the Individual Chapters in the Thermal Theme

In this overview of the thermal climate of the Midwestern USA the focus has been on historical evolution of regional temperatures and projections based on coupled climate (AOGCM) model experiments driven by different scenarios of greenhouse gas emissions. Temperatures in the region have increased over the last century with a linear trend of approximately 0.07°C

per decade. Since the late 1970s the region has experienced accelerated warming at a rate of 0.55°C per decade, in accord with hemispheric temperature trends during the same time period. Seasonal trend analyses show that the largest warming since the early twentieth century has occurred during winter and spring with smaller, and statistically insignificant, trends during summer and autumn. During the recent warming (1979–2005), the largest linear trends occur during the winter (0.82°C per decade) and autumn (0.57°C per decade), although only the autumn trend is statistically significant at $\alpha=0.05$.

Projections from the IPCC AR4 suite of models universally exhibit twenty-first-century warming for the average conditions over the Midwestern USA regardless of the emissions scenario or AOGCM employed. For the mid-century period (2046–2065), mean temperature changes range from 0.56°C to 4.25°C depending on the model and emissions scenario employed. Late-century simulations (2081–2100) show larger warming, ranging from 0.71°C to 6.72°C. If the committed climate change scenarios are not included, these ranges narrow to 1.22°C to 4.25°C and 1.82°C to 6.72°C for the 2046–2065 period and 2081–2100 period, respectively.

The results presented here are based on a wide range of AOGCM simulations encompassing ten different models and four emissions scenarios. The findings indicate that further anthropogenic greenhouse gas emissions, even on the lower end of the scenario spectrum, will likely result in considerable warming in the Midwestern region during this century. Additional research is needed to assess the spatial variability and impact of these changes. The following chapters explore the details of temperature variability and change in the region. Specifically, chapter 3 provides an investigation into the lack of strong summer warming in the regional temperature record from the west of the region. Chapter 4 focuses on historical and projected regional changes in the length of the frost-free season. Chapter 5 describes the regional variations of historical equivalent temperature records, which are critically important for understanding how recent warm temperature extremes are related to their historical counterparts. Chapter 6 focuses on the important problem of temperature variability, using multiple variability measures, climate models, and emissions scenarios, and this section concludes with chapter 7, which provides a case study of historical and projected temperature regimes over Wisconsin in the broader context of regional change.

REFERENCES

Bonan, G. B. 2001. "Observational Evidence for Reduction of Daily Maximum Temperature by Croplands in the Midwest United States." *Journal of Climate* 14: 2430–2442.

Brohan, P., et al. 2006. "Uncertainty Estimates in Regional and Global Observed Temperature Changes: A New Dataset from 1850." *Journal of Geophysical Research* 111, D12106, doi:10.1029/2005JD006548.

Croley, T. E. 1990. "Laurentian Great-Lakes Double-CO_2 Climate Change Hydrological Impacts." *Climatic Change* 17: 27–47.

Easterling, D. R. 2002. "Recent Changes in Frost Days and the Frost-Free Season in the United States." *Bulletin of the American Meteorological Society* 83: 1327–1332.

IPCC. 2007. *Climate Change 2007: The Physical Science Basis.* Contribution of Working Group I to the Fourth Assessment Report of the Intergovernmental Panel on Climate Change. [Ed. S. Solomon, et al.] New York: Cambridge University Press.

Jones, P. D., and A. Moberg. 2003. "Hemispheric and Large-Scale Surface Air Temperature Variations: An Extensive Revision and an Update to 2001." *Journal of Climate* 16: 206–223.

Kalnay, E., and M. Cai. 2003. "Impact of Urbanization and Land-Use Change on Climate." *Nature* 423: 528–531.

Kunkel, K. E., et al. 1996. "The July 1995 Heat Wave in the Midwest: A Climatic Perspective and Critical Weather Factors." *Bulletin of the American Meteorological Society* 77: 1507–1518.

Kunkel, K. E., et al. 2004. "Temporal Variations in Frost-Free Season in the United States: 1895–2000." *Geophysical Research Letters* 31, doi:10.1029/2003GL018624.

Kunkel, K. E., et al. 2006. "Can CGCMs Simulate the Twentieth-Century 'Warming Hole' in the Central United States?" *Journal of Climate* 19: 4137–4153.

Meehl, G. A., and C. Tebaldi. 2004. "More Intense, More Frequent, and Longer Lasting Heat Waves in the 21st Century." *Science* 13: 994–997.

Meehl, G. A., et al. 2007. "Global Climate Projections." In *Climate Change 2007: The Physical Scientific Basis.* Contribution of Working Group I to the Fourth Assessment Report of the Intergovernmental Panel on Climate Change, [ed. S. Solomon, et al.]. New York: Cambridge University Press.

Nakićenović, N., and R. Swart, eds. 2000. "Special Report on Emissions Scenarios." A Special Report of Working Group III of the Intergovernmental Panel on Climate Change. New York: Cambridge University Press.

Pan, Z., et al. 2004. "Altered Hydrologic Feedback in a Warming Climate Introduces a 'Warming Hole.'" *Geophysical Research Letters* 31, doi:10.1029/2004GL02528.

Polderman, N. J., and S. C. Pryor. 2004. "Linking Synoptic-Scale Climate Phenomena to Lake-Level Variability in the Lake Michigan-Huron Basin." *Journal of Great Lakes Research* 30: 419–434.

Robeson, S. M. 2002. "Increasing Growing-Season Length in Illinois during the 20th Century." *Climatic Change* 52: 219–238.

Sellinger, C. E., et al. 2008. "Recent Water Level Declines in the Lake Michigan-Huron System." *Environmental Science and Technology* 42: 367–373.

Smith, J. B. 1991. "The Potential Impacts of Climate Change on the Great-Lakes." *Bulletin of the American Meteorological Society* 72: 21–28.

Smith, T. M., and R. W. Reynolds. 2005. "A Global Merged Land and Sea Surface Temperature Reconstruction Based on Historical Observations (1880–1997)." *Journal of Climate* 18: 2021–2036.

Trenberth, K. E., et al. 2007. "Observations: Surface and Atmospheric Climate Change." In *Climate Change 2007: The Physical Scientific Basis.* Contribution of Working Group I to the Fourth Assessment Report of the Intergovernmental Panel on Climate Change. [Ed. S. Solomon, et al.] New York: Cambridge University Press.

Zhao, T., and M. D. Schwartz. 2003. "Examining the Onset of Spring in Wisconsin." *Climate Research* 24: 59–70.

3. Global Climate Change Impact on the Midwestern USA—A Summer Cooling Trend

Z. PAN, M. SEGAL, X. LI, AND B. ZIB

Introduction

Global warming and associated climate change are the consensus now among scientists and policymakers. Instrumental records from land stations and ships indicate that the global annual mean surface air temperature warmed by 0.6°C during the twentieth century. Spatially, the warming was greater in high latitudes than in the tropics, more significant over land than over the ocean, and stronger in the Northern Hemisphere than the Southern Hemisphere (Folland et al. 2001). Temporally, the warming occurred more during winter than summer and more in night than day. The latter is reflected in the well-documented reduction in daily temperature range (DTR) that decreased almost everywhere for the past few decades because of relatively greater warming during night than daytime (Easterling et al. 1997; Karl et al. 1993).

Warming at the global scale does not imply uniform or universal warming. Several studies have documented cooling in the eastern United States (e.g., Kalnay and Cai 2003; Easterling et al. 1997; Robinson, Reudy, and Hansen 2002). Most of these studies analyzed either annual or daily mean temperatures over eastern and/or central United States as a whole. However, underlying causes for cooling are different during day and night, in winter and summer. For example, evapo-transpiration reduces the daytime temperature, but has little effect on night temperatures (Dai, Trenberth, and Karl 1999). Similarly, the cold season temperature is more highly correlated with low-frequency ocean-atmosphere oscillations than summer temperature. Here we focus on summer daytime temperatures.

Observed Datasets Analyzed Herein

The main source of daily observed station data used in this study is the Global Historical Climatology Network (GHCN) as compiled into monthly means and interpolated onto regular latitude/longitude grids by the Climate Research Unit of the University of East Anglia. The dataset includes monthly mean surface daily maximum/minimum temperatures, precipitation, and cloud cover on a 0.5°×0.5° latitude/longitude grid for the period 1901–2000 (New, Hulme, and Jones 2000; Mitchell and Jones 2005, Vose, Easterling, and Gleason 2005). Since data before the 1950s were somewhat sparse

(New, Hulme, and Jones 2000 for details), our analyses mainly focus on the temperature changes after the 1950s, although prior data are also used to determine longer-term trends. This surface dataset, which has already gone through rigid quality control, is supplemented by raw original station temperature on a daily basis within the cooling regions. To examine the formation and maintenance mechanisms of the surface cooling, some upper-air data are also analyzed using the NCEP/NCAR reanalysis data (NNR, Kalnay et al. 1996).

North-Central and South-Central U.S. Cooling Patterns

Since temperature variation is not monotonic, but fluctuates, temporal trend magnitudes depend on the evaluation period. While longer periods give larger sample sizes, they may obscure underlying physical processes during different periods. For example, the second half of the twentieth century, an often-used period of recent studies, includes a period of global slight cooling before 1975 and strong warming after that. To reduce arbitrary choice of lengths of periods, we evaluated three durations: one hundred years (1901–2000), fifty years (1951–2000), and two periods of twenty-five years (1951–1975 and 1976–2000). The one-hundred-year data duration represents the longest available dataset and the fifty-year period corresponds to the data-rich period. The separation of the second half-century into two equal twenty-five-year periods is not chosen for simplicity, but is based on the following considerations:

1. The year 1976 is the turning point of two climate epochs and is the time when the Pacific Decadal Oscillation (PDO) shifted from a negative to a positive phase (Hu and Feng 2001; Miller et al. 1994).

2. The global temperature trend changed from a slight decrease to a strong increase around 1975 (Folland et al. 2001).

3. The year 1979 is the beginning of the satellite era when the NNR data used in this study exhibit a discontinuity (Kalnay and Cai 2003).

Figure 3.1 shows the linear trends of summer mean daily maximum surface temperature (hereafter Tmax) during the different periods. During the whole twentieth century the south-central United States cooled by 0.5–2.0°C while most of the United States slightly warmed. During the period 1951–2000 the south-central U.S. cooling was more extensive, with most parts being cooled by 1.0–1.5°C (figure 3.1a). The most extensive and strong cooling occurred in the 1951–1975 period when the cooling spread over all the southeastern states including Texas (Figure 3.1b). A strip of 3°C cooling ran from South Carolina to eastern Texas. During the last twenty-five years (1976–2000), the peak global warming period, the cooling was shifted to the central section of the United States, with the largest cooling of over 2.0°C located at the Iowa-Nebraska-South Dakota border (figure 3.1c). This well-defined region of cooling oriented in the NW-SE direction, in contrast with NE-SW direction in 1951–2000 (figure 3.1b). While central-eastern U.S. cooling is evident for most parts of the twentieth century, the cooling magnitude and location differ for different periods. The overall cooling within the fifty-year periods consists of two distinct

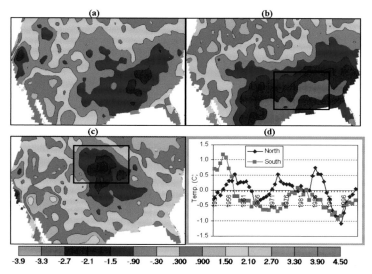

Figure 3.1. Summer mean daily maximum surface temperature (Tmax) trends (°C) during various periods of the twentieth century. (a) 1951–2000; (b) 1951–1975; and (c) 1976–2000. (d) Time series of Tmax anomaly from the fifty-year mean (1951–2000) in the north-central (40–50°N by 105–90°W) and south-central U.S. (30–40°N by 100–85°W), as delineated by the rectangles in (b) and (c).

periods and locations: the north-central U.S. cooling in 1976–2000, hereafter referred to as north-central cooling center (NCC), and south-central cooling in 1951–1975, hereafter referred to as south-central cooling center (SCC). The time series of Tmax in NCC and SCC as indicated by the dark rectangles in figures 3.1b and 3.1c show that cooling in NCC was mainly concentrated in the 1986–1996 period, while the southern cooling in SCC was persistent during 1954–1967 (figure 3.1d).

To further characterize the patterns of Tmax trends, we performed empirical orthogonal function (EOF) analysis of the whole fifty-year period when observed temperature data were more dense and reliable. Figure 3.2 displays four leading eigenvectors of the Tmax covariance. The first EOF shows a clear pattern of negative center, resembling the NCC during 1976–2000. The second EOF shows a well-organized positive area, matching the SCC during 1951–1975.

The EOF analysis reveals that the first four modes explain over 90% of Tmax correlation variance, with the first EOF contributing 36% of the variance. The amplitude of the first EOF shows that the cooling with this eigenvector is rather sporadic, with individual peaks corresponding to abnormal conditions such as the 1988 drought and the 1993 flood. The amplitude of the second EOF shows a steady decrease from positive to negative during the first half of the fifty-year period as shown in Figure 3.2b, suggesting that the southern cooling is largely caused by the positive-to-negative shift of the second EOF. The fourth EOF has a negative center, coinciding with the NCC. Its amplitude was persistently positive during the early 1990s, which may also contribute to the sharp northern cooling.

The cooling in NCC occurred in the last quarter-century in spite of pronounced warming at the global scale. In fact, the 0.6°C global warming was mostly contributed by the last quarter of the century. Hence, the 1976–2000 period may be most representative of Midwestern USA response to global warming of the twentieth century, implying the need to study the north-central U.S. cooling.

Ridging supported by the warming induces northerly winds that advect colder air (figure 3.3a). During the fifty-year period, the Tmax gradient between the western ridge and the NCC cooling is about 1°C. From the hydrostatic equation, the resultant geopotential gradient of about 5 m over the same distance is very close to the observed value of 4 m (figure 3.3a). Following the geostrophic approximation, the geopotential gradient would produce northerly winds of approximately 1 ms^{-1} (Pan et al. 1999), which is similar to the difference in 850 hPa winds between the two periods (figure 3.3b). The 0.8–1.0 ms^{-1} northerly winds coupled with the north-south horizontal temperature gradients result in strong cold-air advection from the north to the north-central United States. The largest wind increase coincides with the cooling center. To the south, the southerly component is increased, bringing more moisture into the central United States. The overall wind change pattern in the fifty-year period created a net convergence and enhanced precipitation and cooling in the central United States.

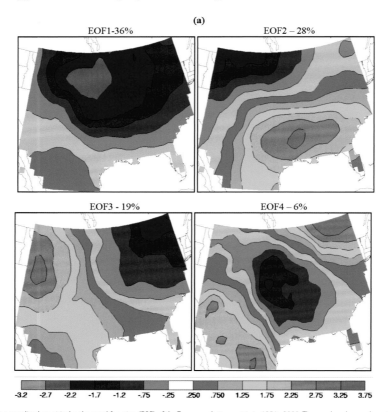

Figure 3.2. (a) The normalized empirical orthogonal function (EOF) of the Tmax correlation matrix in 1951–2000. The number shown above the frame is the variance explained by the EOF.

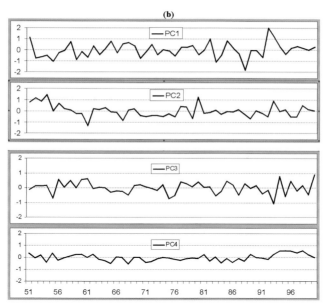

Figure 3.2. (continued)(b) Time series of the annual EOF coefficients.

Global Pattern of "Warming Holes"

It seems natural to ask whether there are any similar major cooling regions (referred to as "warming holes" [WH]) in other parts of the world. To answer this question, we expand our analysis to the global scale. Figure 3.4 shows the global daily Tmax change averaged over summer (JJA) and winter (DJF) in 1991–2000 versus 1951–1960. There are well-defined, extensive areas of cooling, that is, WHs, two in each summer hemisphere. The two Northern Hemisphere WHs are located in the central United States and central China during boreal summer, whereas the two Southern Hemisphere WHs are located in central South America and northern Australia in austral summer. These four WHs hereafter are referred to as USWH, ChWH, SAWH, and AuWH, respectively. A coherent cooling area at the very northern edge of South America exists during both seasons. This cooling area differs, however, from the four WHs because of its lack of seasonal variations, and thus it is not considered a WH. There are also some scattered cooling regions over the continents, but they are not as extensive or contiguous.

The well-defined ChWH is centered in the Wuhan province in central China, corroborating the results of Liu et al. (2004). It has a similar cooling magnitude to the USWH (figure 3.4a) but with less areal coverage. The SAWH is the smallest in size and is located in Paraguay and northern Argentina, with a typical cooling of 0.6–1.0°C over a small area (figure 3.4b). The AuWH is an elongated area, with a peak cooling of 0.6°C. In the winter hemisphere, the "hole" regions experience either slight cooling or warming, meaning that the WHs occur only in the corresponding summers.

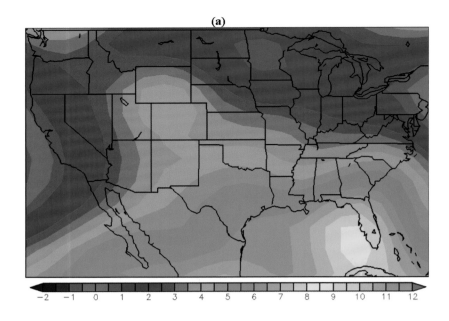

(a)

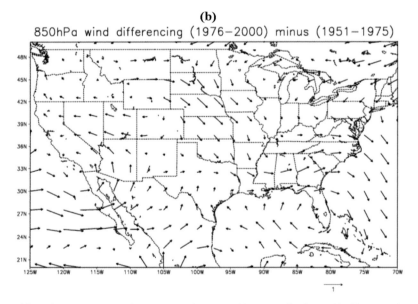

(b)

Figure 3.3. Mean difference between 1976–2000 and 1951–1975 periods in summer. (a) geopotential height at 700 hPa; (b) vector wind at 850 hPa. Data are taken from the NCEP-NCAR Reanalysis.

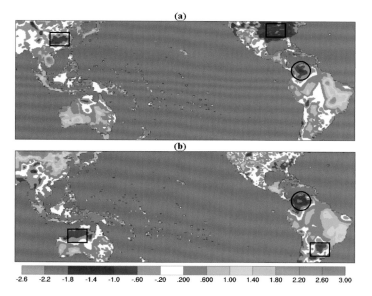

Figure 3.4. Difference in daily maximum surface temperature (Tmax) in °C between 1991–2000 and 1951–1960 averaged in boreal summer (a) and boreal winter (b), showing "global warming holes" (WHs) in the central United States, central China, southern South America, and northern Australia. The rectangles surrounding the WHs indicate approximate regions where the mean WH quantities were averaged. The USWH: 100–85°W, 35–45°N. ChWH: 105–120°E, 25–35°N. SAWH: 65–50°W, 25–35°S. AuWH: 122–137°E, 18–28°S.

Figure 3.4 shows only a "snapshot" of Tmax difference between the two decades (1950s and 1990s). To put the differences in the perspective of a longer-term trend, we present the time series of the mean summer daily Tmax change averaged over each WH during the second half of the twentieth century (figure 3.5). The extent of WHs is defined as a 15° longitude by 10° latitude region, roughly covering the cooling areas indicated in figure 3.4. In all four WHs, Tmax decreased persistently for the fifty-year period. The USWH Tmax shows a clear cooling trend (0.90°C/50yr). The ChWH experienced a similar cooling trend (0.75°C/50yr) with smaller inter-annual variability than the USWH, while the Tmax in SAWH decreased by 0.45°C in the fifty years. The AuWH showed little change (−0.05°C/50yr), mainly because it is considerably smaller than the 15°×10° averaging area. The actual cooling at some individual stations was much greater than the WH averages. For example, the cooling at certain sites in China reached a rate of 4–7°C/100yr during 1958–2000 (Liu et al. 2004), and Tmax at some stations in northern Argentina declined at 4–8°C/100yr between 1959 and 1998 (Rusticucci and Barrucand 2004).

According to Pan et al. (2004), increased cloudiness and precipitation are the main local contributors to WH formation in the central United States. A quantification of their interrelation is provided below. By blocking solar irradiance, cloud cover affects surface energy budgets and thus temperature in the daytime (Dai, Trenberth, and Karl 1999). The cloud cover over all WHs has increased during the 1951–2000 period (figure 3.6a), except for the ChWH, where aerosols played key roles (Qian and Giorgi 1999). Cloud cover increased by 4.2% in

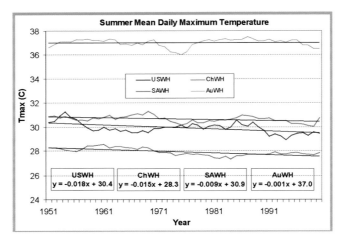

Figure 3.5. Time series of the mean summer surface maximum temperature averaged over WHs as defined in figure 3.4. The series were smoothed using five-point moving average. The slope of linear regression has units of °C/yr.

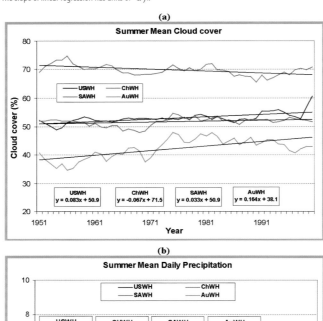

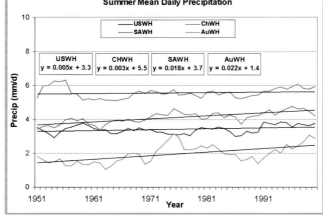

Figure 3.6. As figure 3.5, but for (a) cloud cover (trend in %/yr) and (b) precipitation (trend in mm/yr).

USWH, 1.7% in SAWH, and 8.2% in AuWH. Independently, the change in outgoing longwave radiation (OLR) based on NNR data over the second half of the twentieth century showed a well-defined local minimum over the USWH as an example, where it decreased by 6 W m^{-2} (~5%) during the same fifty-year period, consistent with the cloudiness trend (not shown). The increased cloud cover would translate into more precipitation. All WHs showed precipitation increase during the fifty-year period, with the USWH increasing by 0.25 mm d^{-1}, ChWH by 0.15 mm d^{-1}, SAWH by 0.9 mm d^{-1}, and ChWH by 1.1 mm d^{-1} for the century (figure 3.6b). Increased precipitation resulted in enhanced surface latent heat flux, which further contributes to the cooling trend. The Tmax also correlates well with cloud cover and precipitation.

Possible Mechanisms Responsible for the Global WHs Formation

Various mechanisms could be responsible for the WH formation, and they may contribute differently to each WH. However, their combined effects should result in the cooling trend. Three possible important mechanisms are suggested below.

LOW-LEVEL JETS

Globally there are four major low-level jet (LLJ) areas overland (Stensrud 1996; Laing and Fritsch 1997). They are the USLLJ over the central Great Plains of the United States, ChLLJ in southern China, SALLJ in central South America, and AuLLJ in northern Australia, as reflected by the strong wind areas in figure 3.7. The USLLJ and ChLLJ have a sharp peak occurrence in early summer, while the SALLJ and AuLLJ have a weaker seasonal cycle with a peak occurrence in the austral summer. Mesoscale convective systems (MCSs) that are critical to warm-season precipitation receipt predominantly occur downstream of LLJs (Laing and Fritsch 1997). Thus warm-season precipitation processes over the central regions in North and South America are strongly modulated by LLJs. More than half of warm-season rainfall in central United States is associated with MCSs (Fritsch, Kane, and Chelius 1986). A similar situation occurs in South America (Liebmann et al. 2004). In south-central China, the warm-season precipitation is associated with Mei-yu front (Chen et al. 1998), frequently along with the ChLLJ. Figure 3.7 depicts the 850 hPa wind components along the LLJ orientated at the four WH locations. (We use the 850 hPa wind to approximate LLJs. The winds in the Northern Hemisphere were averaged over the May–July period, which is one month earlier than the Tmax averaging period to partly account for the earlier occurrence of LLJs and likely time lag in the subsequent surface hydrology.) For clarity, figure 3.7 shows only the meridional or zonal components in approximately the same direction as the LLJs—southerly wind component for the USLLJ and ChLLJ, northerly for the SALLJ, and easterly for the AuLLJ. The USLLJ and ChLLJ are shown as red areas in the southern United States and China, respectively, while the SALLJ and AuLLJ are represented by blue areas in central South America and northern Australia, respectively. The SALLJ appears to locate somewhat to the north of SAWH, partly due to the approximation of the actual northwesterly SALLJ by the northerly wind only in this study (Veraa et al. 2006). Interestingly, the southerly winds are also strong in the region corresponding to the well-defined cooling region at the northern tip of South America. In fact, LLJ occurrence was also sometimes observed in this region (Stensrud 1996). This further supports a relationship between WHs and LLJs even though the LLJ in this case is not as prominent as the other four LLJs. The change in LLJ-associated moisture convergence can be produced by changes in either flow convergence or troposphere vapor content. For example, the LLJ winds strengthened (weakened) to the south (north) of the USWH from the 1950s to 1990s, resulting in a net flow convergence over the WH (figure 3.3). Likewise, the atmospheric precipitable water showed a clear increase to the south of the ChWH (not shown).

CHANGES IN AGRICULTURAL PRACTICES

Changes in agricultural practices in the second half of the twentieth century modified plant transpiration and therefore latent heat flux (Xue, Fennessy, and Sellers 1996). For example, in the Midwestern USA, where corn and soybean are the currently dominant summer crops, several changes have been introduced during this period: (1) a change from forage crops to higher transpiration corn and soybean that was completed by the 1970s, (2) increased use of chemical fertilizers up to the 1980s, and (3) increased plant density (stabilized in 1970s). Changes (2) and (3) led to an increase in biomass and presumably also transpiration. Thus, the direct reduction in sensible heat flux and the indirect effects of increasing clouds and pre-

cipitation suppress Tmax (Trenberth et al. 2003). Expansion of the agricultural irrigation area in the central United States during the second half of the twentieth century (Segal et al. 1998) is likely to have contributed locally somewhat to the trend of decreasing Tmax. Kalnay and Cai (2003), by comparing observed and NNR computed trends, indicated the cooling effect from agricultural changes. As in the United States, changes in agricultural practices in China and South America have also likely contributed to the corresponding WH cooling.

CHANGES IN EL NIÑO FREQUENCY

In the recent thirty years, there has been a trend of increasing frequency of El Niño events, and summer precipitation in the central United States during El Niño years tends to increase (Trenberth et al. 2003). Likewise in AOGCM sensitivity analyses, Robinson, Reudy, and Hansen (2002) found that the temperature in the central United States decreased in response to the observed warming of the tropical Pacific in the second half of the twentieth century. Their sensitivity study indicated that warming of the tropical Pacific promotes moisture advection from the Gulf of Mexico to the central United States, thus enhancing cloudiness and reducing solar radiation, and thereby reducing daytime surface temperatures. Grimm, Barros, and Ooyle (2000) and studies cited therein found increased rainfall in the SAWH region in the austral summer during a warm phase of El Niño events. In contrast, Wu, Hu, and Kirtman (2003) found that precipitation in the ChWH area is affected only weakly by the El Niño cycles.

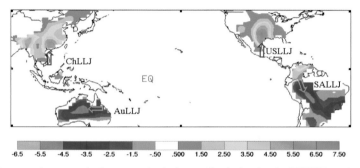

Figure 3.7. The 850 hPa mean wind components (ms⁻¹) approximately in the direction of low-level jets (based on NCEP/NCAR reanalysis data). Southerly averaged in May–July for USLLJ and ChLLJ; northerly in December–February for SALLJ; and easterly in December–February for AuLLJ. All data are averaged over 1961–1990.

Concluding Remarks

This study focuses mainly on the daily maximum temperatures in the summer (JJA) during the second half of the twentieth century. The robust central-eastern U.S. cooling trends, termed U.S. "warming hole" (USWH), existed throughout most of the twentieth century, but more pronounced cooling occurred after 1950. The cooling consists of the north-central United States in the last quarter of the twentieth century and widespread south-central cooling mainly during the third quarter of the century. The southern cooling occurred during a period of slight global cooling, and is thus partly explained by the global cooling component. The northern cooling occurred during the peak of the global warming period, and thus was more reflective of the current global warming effects. Therefore, the cooling mechanisms are likely different for the two locations. The south-central U.S. cooling has been linked to the tropical sea surface temperature (Robinson, Reudy, and Hansen 2002) and possibly to more frequent El Niño events in the past decades. The central U.S. cooling has also been associated with intense agriculture cultivation (Kalnay and Cai 2003).

The climate change pattern shows strong warming in the western mountain region of North America, which produces a sharp pressure gradient and thus northerly wind. In summer, winds are more meridional than in the winter, which may partly explain why the cooling occurs only in the summer. The moderate increase in northerly flow supported by the western ridging would cool the north-central United States, as observed in the wind trend in the lower troposphere during the fifty-year period. The Pacific Decadal Oscillation (PDO) signal is strongest during winter (October–March), when flow patterns are more zonal, which may advect warmer surface air from Northwest Pacific. This may explain why the cooling is absent during winter. More meridional flow in summer with a weaker PDO signal may balance out the zonal flow with a stronger PDO signal in winter. These qualitative arguments need quantitative analyses and numerical experimentation that are beyond the scope of this study.

If the scenario developed herein is accurate, the north-central U.S. cooling may continue (Pan et al. 2004). The north-central cooling in the last quarter of the twentieth century is in broad agreement with the model-simulated future summer warming minimum over the central United States in a scenario climate (Pan et al. 2004; Liang et al. 2006; Kunkel et al. 2006). In these studies, it is found that the warming by mid-twenty-first century in summer is 2°C cooler in the central United States than surroundings. If indeed the observed cooling in this study and the less warming (warming hole) are the same system, it would suggest that this cooling would likely continue into the future as global warming intensifies.

Three other WHs are also identified: in south-central China, south-central South America, and northern Australia, respectively. All the WHs (except for the Australian one) occur in the eastern slopes of major mountain ranges, where low-level jets are prominent. A tentative global WH formation mechanism follows:

1. Increased moisture convergence due to increased atmospheric water vapor and/or increased LLJ strength or frequency enhances cloudiness and precipitation.

2. Increased cloudiness attenuates solar radiation and surface heating during daytime (Tmax).

3. Deep soil planted with intensive crops (except AuWH) provides soil water storage and transpiration, and thus suppresses afternoon heating in summer.

Other factors contributing to the cooling in the WHs likely include changes in agricultural practices that lead to increased plant transpiration, reduced daytime warming, and the increased frequency of El Niño events that increased precipitation over the USWH and SAWH. Atmospheric aerosols are believed to partly contribute to the cooling associated with the ChWH (Qian and Giorgi 1999).

It should be pointed out that the cooling described in this chapter occurred only during June–August and was evident *only in* daytime maximum temperature. The night minimum temperature in winter has generally increased in the WH regions over the past decades as widely reported. Furthermore, the summer daytime cooling should not affect the lengthening of growing season attributed to the warming in the early spring and late fall, as shown in the next chapter.

ACKNOWLEDGMENTS

We are thankful to William Gutowski, Gene Takle, and Raymond Arritt for their input during the early stage of the study. This research was partly supported by the U.S. Department of Energy's Office of Science (BER) through the Midwestern Regional Center of the National Institute for Climatic Change Research at Michigan Technological University.

REFERENCES

Chen, C., et al. 1998. "The Intensification of the Low-Level Jet during the Development of Mesoscale Convective Systems on a Mei-Yu Front." *Monthly Weather Review* 126: 349–371.

Dai, A., K. E. Trenberth, and T. R. Karl. 1999. "Effects of Clouds, Soil Moisture, Precipitation, and Water Vapor on Diurnal Temperature Range." *Journal of Climate* 12: 2451–2473.

Easterling, D. R., B. et al. 1997. "Maximum and Minimum Temperature Trends for the Globe." *Science* 277: 364–367.

Folland, C. K., et al. 2001. "Observed Climate Variability and Change." In *Climate Change 2001: The Scientific Basis,* [ed. J. H. Houghton et al.], 99–182. Cambridge: Cambridge University Press.

Fritsch, J. M., R. J. Kane, and C. R. Chelius. 1986. "The Contribution of Mesoscale Convective Weather Systems to the Warm-Season Precipitation in the United States." *Journal of Climate and Applied Meteorology* 25: 1333–1345.

Grimm, A. M., V. R. Barros, and M. E. Ooyle. 2000. "Climate Variability in Southern South

America Associated with El Niño and La Niña Events." *J. Climate* 13: 35–58.

Hu, Q., and S. Feng. 2001. "Variations of Teleconnection of ENSO and Interannual Variation in Summer Rainfall in the Central United States," *Journal of Climate* 14: 2469–2480.

Kalnay, E., and M. Cai. 2003. "Impact of Urbanization and Land-Use Change on Climate." *Nature* 423: 528–531.

Kalnay, E., et al. 1996. "The NCEP/NCAR 40–Year Reanalysis Project." *Bulletin of the American Meteorological Society* 77: 437–471.

Karl, T. R., et al. 1993. "Asymmetric Trends of Daily Maximum and Minimum Temperature." *Bulletin of the American Meteorological Society* 74: 1009–1022.

Kunkel, K. E., et al. 2006. "Can CGCMs Simulate the Twentieth-Century 'Warming Hole' in the Central United States?" *Journal of Climate* 19: 4137–4153.

Laing, A. G., and J. M. Fritsch. 1997. "The Global Population of Mesoscale Convective Complexes." *Quarterly Journal of the Royal Meteorological Society* 123: 389–405.

Liang, X.-Z., et al. 2006. "Regional Climate Model Downscaling of the U.S. Summer Climate and Future Change." *Journal of Geophysical Research* 111, doi:10.1029/2005JD006685.

Liebmann, B., et al. 2004. "Subseasonal Variations of Rainfall in South America in the Vicinity of the Low-Level Jet East of the Andes and Comparison to Those in the South Atlantic Convergence Zone." *Journal of Climate* 17: 3829–3842.

Liu, B., et al. 2004. "Taking China's Temperature: Daily Range, Warming Trends, and Regional Variations." *Journal of Climate* 17: 4453–4462.

Miller, A. J., et al. 1994. "The 1976–77 Climate Shift of the Pacific Ocean." *Oceanography* 7: 21–26.

Mitchell, T. D., and P. D. Jones. 2005. "An Improved Method of Constructing a Database of Monthly Climate Observations and Associated High-Resolution Grids." *International Journal of Climatology* 25: 693–712.

New, M. G., M. Hulme, and P. D. Jones. 2000. "Representing 20th Century Space-Time Climate Variability. II: Development of 1901–1996 Monthly Terrestrial Climate Fields." *Journal of Climate* 13: 2217–2238.

Pan, Z., et al. 2004. "Altered Hydrologic Feedback in a Warming Climate Introduces a 'Warming Hole.'" *Geophysical Research Letters* 31, L17109, doi:10.1029/2004GL02528.

Pan, Z., et al. 1999. "A Method for Simulating Effects of Quasi-Stationary Wave Anomalies on Regional Climate." *Journal of Climate* 12: 1336–1343.

Qian, Y., and F. Giorgi. 1999. "Interactive Coupling of Regional Climate and Sulfate Aerosol Model over EASTERN ASIA." *Journal of Geophysical Research* 104: 6477–6499.

Robinson, W. A., R. Reudy, and J. E. Hansen. 2002. "General Circulation Model Simulations of Recent Cooling in the East-Central United States." *Journal of Geophysical Research* 107, doi:10.1029/2001JD001577.

Rusticucci, M., and M. Barrucand. 2004. "Observed Trends and Changes in Temperature Extreme over Argentina." *Journal of Climate* 17: 4099–4107.

Segal, M., et al. 1998. "On the Potential Impact of Irrigated Areas in North America on Summer Rainfall Caused by Large Scale Systems." *Journal of Applied Meteorology* 37: 325–331.

Stensrud, D. J. 1996. "Importance of Low-Level Jets to Climate: A Review." *Journal of Climate* 9: 1698–1711.

Trenberth, K. E., et al. 2003. "The Changing Character of Precipitation." *Bulletin of the American Meteorological Society* 84: 1205–1217.

Veraa, C. J., et al. 2006. "The South American Low Level Jet Experiment." *Bulletin of the American Meteorological Society* 87: 63–77.

Vose, R. S., D. R. Easterling, and B. Gleason. 2005. "Maximum and Minimum Temperature Trends for the Globe: An Update through 2004." *Geophysical Research Letters* 33, doi:10.1029/2005GL024379.

Wu, R., Z.-Z. Hu, and B. P. Kirtman. 2003. "Evolution of ENSO-Related Rainfall Anomalies in East Asia." *Journal of Climate* 16: 3742–3758.

Xue, Y., M. J. Fennessy, and P. J. Sellers. 1996. "Impact of Vegetation Properties on U.S. Summer Weather Prediction." *Journal of Geophysical Research* 101: 7419–7430.

4. Historical and Projected Changes in the Length of the Frost-Free Season

J. T. SCHOOF

Introduction

During the last century, the length of the frost-free season (time between the last spring freeze and first autumn freeze) has increased at most mid- and high-latitude locations in both hemispheres (Easterling 2002; Kunkel et al. 2004; Trenberth et al. 2007). Recent studies (Christidis et al. 2007; Meehl, Arblaster, and Tebaldi 2007) have attributed the observed changes in frost-free season length to anthropogenic greenhouse gas emissions, raising interest in impacts on physical and biological systems, including agriculture (Chmielewski, Muller, and Bruns 2004), plant phenology (Schwartz and Reiter 2000; Cleland et al. 2007), animal behavior (Parmesan and Yohe 2003), marine ecosystems (Winder and Schindler 2004), and hydrology (Beniston 2003; Groisman et al. 2004).

The Midwestern USA is a major agricultural center, and is therefore potentially sensitive to changes in the length of the frost-free season. In a study focused on a single Midwestern state, Illinois, Robeson (2002) found that, averaged over multiple stations, the length of the frost-free season increased by approximately one week during the twentieth century. However, Robeson (2002) also found that time series of frost-free season lengths are highly variable, with inconsistent trends in the date of the first autumn freeze. Here, the work of Robeson (2002) is expanded to (1) include analysis of historical data from other locations in the Midwestern USA and (2) examine the potential evolution of frost-free season statistics in the study region in the future using output from AOGCMs.

The primary tools available for examining evolution of the climate system, including frost-free season length, are coupled Atmosphere-Ocean General Circulation Models (AOGCMs). The archive of AOGCMs assembled for the Fourth Assessment Report of the Intergovernmental Panel on Climate Change (IPCC AR4) and Coupled Model Intercomparison Project (CMIP3; Meehl et al. 2007) provides an unprecedented opportunity to investigate potential evolution of conditions in simulations of twenty-first-century climate. However, since coupled model simulations are generally considered to be most realistic at large scales and away from the surface, model output must be downscaled to investigate evolution of surface climate parameters.

The objectives of this investigation are thus (1) to analyze observed trends in frost-free season length for multiple locations in the Midwestern USA, (2) to evaluate the ability of AOGCMs to reproduce observed frost-free season length statistics, (3) to downscale AOGCM output using statistical methods and investigate the improvement provided relative to direct AOGCM output, and (4) to summarize the information that coupled models can provide about future changes in the length of the frost-free season in the study region.

Data Description

To meet the stated research objectives, several types of data were required. For the investigation of observed changes in frost-free season characteristics, minimum daily temperature (T_{min}) data were needed for multiple locations in the Midwestern USA. In addition, data from multiple AOGCMs were required to investigate the characteristics of the frost-free season and their future trajectories from a range of models. Downscaling of the AOGCM data requires knowledge of the relationship between the surface climate parameter of interest (T_{min}) and upper-level predictors that are well simulated by the suite of AOGCMs used. To meet this need, observed (reanalyzed) upper air data were used.

Daily minimum surface air temperature records were extracted from the Global Historical Climatology Network (GHCN; Peterson and Vose 1997). Although historical temperature data are susceptible to a number of biases, these data are generally considered to be of adequate quality for detection of climatic trends. The GHCN data do not include stations likely to be influenced by urbanization. For this study, data were examined for the period 1961–2000, corresponding to the availability of coupled climate model output. Within the study area and time period, fifty-three stations were available for analysis (figure 4.1).

Gridded upper air data 2.5°×2.5° resolution were extracted from the European Center for Medium-Range Weather Forecasting (ECMWF) reanalysis (ERA-40; Uppala et al. 2005). The ECMWF ERA-40 data have a dual purpose for this analysis. First, the data are used to construct transfer functions that relate upper air variables to daily minimum surface air temperature at each station. Second, the ERA-40 data are used to evaluate the fidelity of the AOGCM simulations. While good agreement between the ERA-40 and AOGCM data does not guarantee that the future AOGCM simulations are valid, it enhances the confidence with which projections can be regarded. For the analyses presented here, the ERA-40 data were bilinearly interpolated to the station locations shown in figure 4.1.

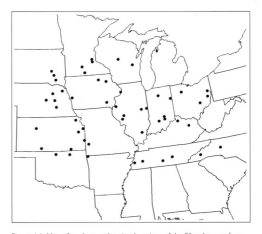

Figure 4.1. Map of study area showing locations of the fifty-three surface climate stations used for the frost-free season length analysis.

Daily output from multiple modeling centers were available from the IPCC AR4 / WCRP CMIP3 data portal for three time periods: 1961–2000, 2046–2065, and 2081–2100, corresponding to the recent historical period, the mid-twenty-first century, and the late twenty-first century, respectively. This resource is representative of current-generation AOGCMs (table 4.1). The results presented here use output from multiple AOGCMs from this archive and focus on results using a single emissions scenario—the IPCC SRES A2 scenario (IPCC 2000). Output from these models was also interpolated to the station locations shown in figure 4.1.

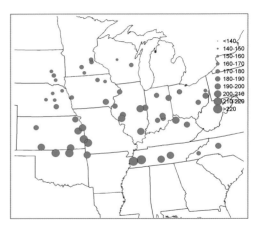

Figure 4.2. Observed mean frost-free season length (days) at fifty-three stations in the Midwestern USA, 1961–2000.

Observed Frost-Free Season Length Characteristics

Observed mean frost-free season lengths (1961–2000) show the expected north-south gradient with values less than 150 days for northern stations to more than 200 days for southern stations (Figure 4.2). Figure 4.3 shows the contribution to the mean frost-free season from the last spring freeze and the first autumn freeze. A common feature of these maps is that frost-free season de-

scriptors exhibit substantial variability, even for nearby stations.

The beginning and end (and therefore the length) of the frost-free season are each governed by a single freeze event and therefore exhibit considerable variability from year to year (Figure 4.4). This large variability makes time series trends of frost-free season characteristics difficult to assess using standard techniques (e.g., ordinary least squares, OLS). Nevertheless, analysis of the 1961–2000 frost-free season data indicates an OLS trend toward longer frost-free seasons at most locations

Table 4.1. Description of IPCC AR4 / WCRP CMIP3 AOGCM simulations used to investigate the evolution of the frost-free season length in the Midwestern USA

Institution	Model Name
Bjerknes Centre for Climate Research (Norway)	BCCR BCM2
Canadian Centre for Climate Modelling and Analysis (Canada)	CCCMA CGCM3
Geophysical Fluid Dynamics Laboratory (USA)	GFDL CM2.0
Goddard Institute for Space Studies (GISS)	GISS MODEL E (Russell)
Institut Pierre Simon Laplace (France)	IPSL CM4
Max Planck Institute for Meteorology (Germany)	MPI ECHAM5
Meteorological Institute of the University of Bonn (Germany)	MIUB ECHO
Meteorological Research Institute (Japan)	MRI CGCM2

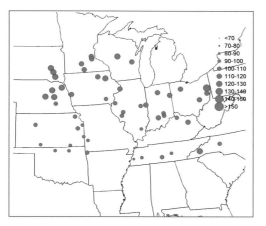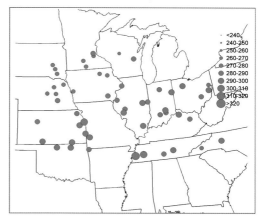

Figure 4.3. Spring and autumn contributions to the mean frost-free season length as indicated by the mean day of year of the last spring freeze (left) and the first autumn freeze (right).

in accord with previous studies. However, a smaller number of stations exhibit negative trends indicating a decreasing frost-free season length through time. Averaged over all stations in the study region, the mean trend in the length of the frost-free season is 0.16 days/year and the standard deviation of the trend is 0.32 days/year. Previous studies (Robeson 2002; Christidis et al. 2007) found that most of the change in frost-free season length was attributable to changes in the timing of spring, perhaps as a result of feedback from reductions in snow cover (see Groisman et al. 2001 and Cayan et al. 2001). The results of the analysis presented here also reflect a stronger spring contribution to frost-free season length changes. Averaged over all stations, the trend in the day of year that the last spring freeze occurs is –0.26 days/yr (standard deviation is 0.19 days/yr), while that for the first autumn freeze is –0.09 days/yr (standard deviation is 0.21 days/yr). Time series (and trends) for two stations, representing large positive and negative frost-free season length trends, are shown in Figure 4.4. The station depicted in Figure 4.4a exhibits a large negative frost-free sea-

son length trend that is controlled primarily by earlier autumn freeze events, with little change in the timing of the last spring freeze from 1961 to 2000. The station depicted in Figure 4.4b exhibits a large positive frost-free season length trend with contributions from both an earlier last spring freeze and a later first autumn freeze. These stations are representative of nearly all of the stations analyzed in that (i) trends in the date of the last spring freeze are nearly uniformly negative (excepting two stations) and (ii) overall frost-free season length trends vary by station, governed primarily by variable trends in the date of the first autumn freeze.

AOGCM Simulated Frost-Free Season Length Characteristics

The first step toward using AOGCMs to investigate potential future evolution of frost-free season length statistics was to consider AOGCM performance with respect to these statistics during the same period as the observed data (1961–2000). As stated previously, AOGCM performance is typically not optimal near the surface where poor resolution

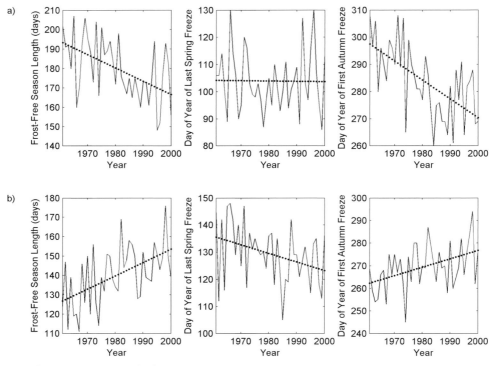

Figure 4.4. Time series and linear trends in frost-free season length and its spring and autumn contributions as derived from observed data collected at two example stations: a) Jacksonville, Illinois, and b) Marshfield, Wisconsin. The left column shows the frost-free season length while the center and right columns show the spring and autumn contributions, respectively.

results in inadequate representation of surface characteristics and surface-atmosphere exchange in model simulations. Due to these shortcomings, each of the AOGCMs considered exhibits some bias with respect to daily surface minimum air temperature and the resulting frost-free season length statistics. The observations indicate a mean frost-free season length of 176.1 days and an interannual variability (as represented by the standard deviation) of 15.5 days, while the models indicate a range of 117.5 to 217.6 days for the mean and 13.6 to 21.6 days for the variability. Further, no individual model reproduces the observed mean frost-free season length (Figure 4.5 top) or standard deviation (Figure 4.5 bottom), even when averaged over all stations, although the latter is better reproduced than the former. With respect to the mean, a few

of the models exhibit particularly large bias. For example, the GFDL CM2.0 AOGCM has a cold bias over the study region resulting in both shorter and more variable frost-free season lengths. These results do not preclude use of these AOGCMs for investigation of frost-free season length evolution. However, they do justify application of downscaling techniques prior to analyzing future frost-free season length characteristics.

Downscaled Frost-Free Season Length Characteristics

AOGCM downscaling is based on the concept that AOGCM biases are generally smaller away from the surface. Predictors chosen for downscaling should have four characteristics: i) they should be strongly related to the surface variable of interest (in this case, T_{min}), ii) they should be widely available from a suite of AOGCMs, iii) they should be well simulated by the AOGCM, and iv) they should carry the climate change signal. While it not possible to determine a priori which predictors will are likely to change as the climate system evolves under warming, some simple considerations can provide insight into proper predictor choices.

Since minimum daily surface air temperature (T_{min}) is the variable of interest in this study, it is reasonable to consider upper level temperature as a potential predictor. Upper air temperatures have the additional advantage that they may act as a surrogate for changes in the radiative properties of the atmosphere (Schubert 1988). Similarly, mid-troposphere moisture content is strongly related to surface T_{min} primarily through control of radiative loss from the surface. By including both upper air temperature and humidity in the downscaling process, it is possible to account indirectly for other pro-

cesses that are also likely to affect T_{min}. For the reasons outlined above, the predictors chosen for the downscaling conducted here were temperature and specific humidity at the 700 mb level.

To evaluate the ability of the AOGCMs to simulate the two predictor variables, Taylor diagrams (Taylor 2001) were constructed (Figure 4.6). The Taylor diagram is a widely used model diagnostic tool that allows simultaneous evaluation of three statistics. The x and y-axes show the ratio of simulated to observed standard deviations, a measure of the spatial variability present in the mean field. The radial axis is the correlation coefficient which signifies the strength of the relationship between of the observed and AOGCM simulated map patterns. When plotted in this fashion, the distance between any point and the origin (the point where the ratio of standard deviations and the correlation coefficient are both equal to 1) is proportional to the root mean square error. The utility of Taylor diagrams for model diagnostic studies

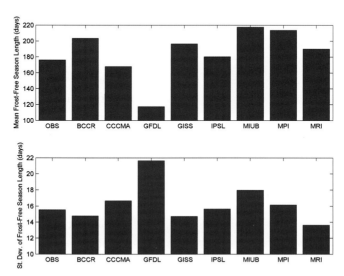

Figure 4.5. Mean (top) and standard deviation (bottom) of frost-free season length derived from observations and direct AOGCM output of daily minimum surface air temperature. Results are averaged over all stations and shown for each model described in table 4.1.

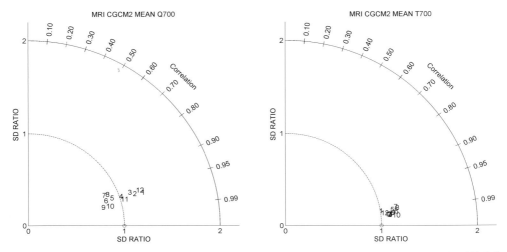

Figure 4.6. Taylor diagrams for MRI CGCM2 700 mb specific humidity (left) and 700 mb air temperature (right) relative to the ERA-40 reanalysis fields. Each number represents a calendar month.

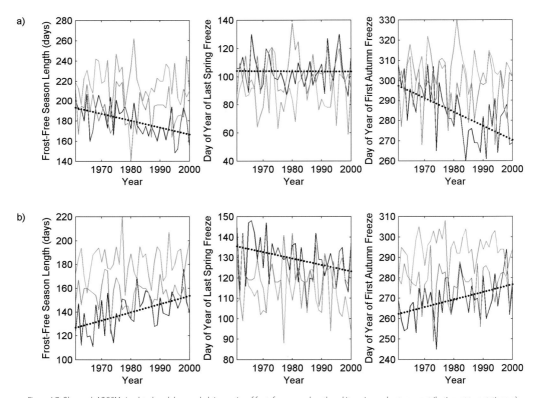

Figure 4.7. Observed, AOGCM simulated, and downscaled time series of frost-free season length and its spring and autumn contributions at two stations: a) Jacksonville, Illinois, and b) Marshfield, Wisconsin. The black series and linear trend are as in figure 4.4. The red line depicts the direct AOGCM output (MPI ECHAM5) while the green line shows the value downscaled from MPI ECHAM5.

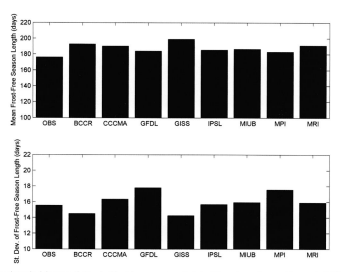

Figure 4.8. Mean (top) and standard deviation (bottom) of frost-free season length derived from observations and downscaled AOGCM output. Results are averaged over all stations and shown for each model described in table 4.1.

has led to their use by both the IPCC and the climate downscaling community (e.g., Pryor, Schoof, and Barthelmie 2005). The Taylor diagrams comparing AOGCM simulations for a single model with ECMWF reanalysis data are shown in Figure 4.6 and indicate that the model simulations from MRI CGCM2 are in good agreement with ERA-40 reanalysis data. In the diagrams, each number represents a month of the year. For 700 mb specific humidity, monthly mean fields are correlated at greater than 0.9 throughout the year with slight overestimation of spatial variability during the cold half of the year and slight underestimation of spatial variability during the warm half of the year. The Taylor diagram for 700 mb air temperature depicts even stronger correlation (>0.95 during all months) and slightly overestimated spatial variability throughout the year. These results are representative of the models listed in Table 4.1 and indicate that both 700 mb specific humidity and 700 mb air temperature are adequately simulated for use as downscaling predictors.

The downscaling was conducted by deriv-

ing transfer functions, statistical equations that relate the predictors to the daily T_{min} at each station. The transfer functions were constructed separately for each station and for each calendar month using the observed station data and ECMWF reanalysis data interpolated to the station locations. The AOGCM simulations were then substituted into the equations resulting in new surface T_{min} values consistent with the large-scale climate simulated by the AOGCMs. This approach is therefore analogous to the Perfect-Prog approach to weather forecasting in which it is assumed that the AOGCMs accurately represent the ERA-40 variables used as predictors in the downscaling transfer functions. Downscaling results for two representative stations and a single AOGCM (MPI ECHAM5) are shown in Figure 4.7. Use of the downscaling approach substantially reduces the bias in the mean frost-free season length. However, neither the direct AOGCM output nor the downscaled results reproduce the underlying trends evident in observed frost-free season length statistics. While trends in predictor variables are not investigated extensively in this preliminary analysis of AOGCM-simulated frost-free season length statistics, some initial testing of MPI ECHAM5 output indicates trends in the predictor variables, and particularly in 700 mb specific humidity, are of the same sign, but generally smaller than those in the ERA-40 reanalysis data. This characteristic of the AOGCM data is likely responsible for the lack of trends in the downscaled frost-free season length data.

Averaged over all stations, the frost-free season characteristics from the downscaled data exhibit much better agreement with observations than the direct AOGCM output discussed above (cf. Figure 4.5 and Figure 4.8). However, as shown in Figure 4.8, the mean frost-free season length of the down-scaled AOGCM simulations is slightly over-estimated by each model. To achieve a greater understanding of the overall model behavior, a multi-model ensemble was constructed by averaging the results from all of the AOGCM simulations. The ensemble results differ from those shown in Figure 4.2 by an average of 12.8 days. Examination of the spring and autumn contributions to this difference suggests that errors in the AOGCM ensemble frost-free season length are approximately equally divided between the last spring freeze (occurs 6.1 days too early on average) and the first autumn freeze (occurs 6.7 days too late on average). The spatial pattern of frost-free season length, including the last spring freeze and first autumn freeze, is well produced by the multi-model ensemble.

Projections of Frost-Free Season Length Characteristics

Given the large range of biological and physical impacts associated with the ob-served changes in frost-free season length, there is considerable interest in the potential future evolution of frost-free season length characteristics. The analysis presented above suggests that there are substantial differences between individual AOGCM simulations and observed data with respect to frost-free season length statistics. The future frost-free season length statistics presented here are therefore based on the multi-model ensemble as determined from downscaled AOGCM output. Differences between future and current climates are quantified only as differences in the ensemble through time rather than as differences between observations and the future ensemble behavior. Future model simulations are considered for two periods: 2046–2065 and 2081–2100. These are periods for which output from multiple models are available from the IPCC AR4 / WCRP CMIP3 archive. Downscaled projections for these periods are constructed by substituting the AOGCM projections of 700 mb specific humidity and air temperature from each model into the transfer functions derived from observations. These projections were then averaged over models to produce the ensemble average. While there were some stations in the study region with negative frost-free season length tendencies in this historical record (see Figure 4.4a, for example), the multi-model ensemble projection produces only positive changes in frost-free season length for 2046–2065 relative to 1961–2000. Furthermore, the ensemble average results, averaged over all stations, indicate a substantive increase of approximately 2 weeks (15.8 days), indicating a sustained trend toward longer frost-free seasons in the study region (Figure 4.9a).

As shown in Figure 4.9a, the spatial pattern of projected change shows only minor

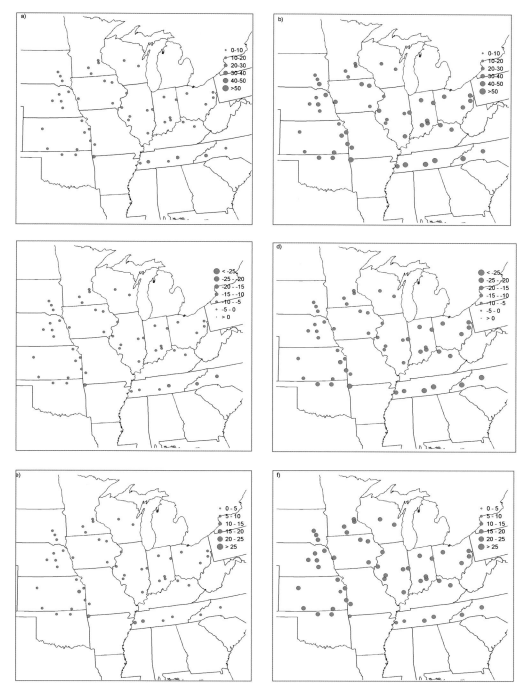

Figure 4.9. Projected changes in frost-free season length (days) based on a multi-model ensemble of downscaled minimum surface air temperatures. The maps depict differences between future and current ensemble values for 2046–2065 (a, c, e) and 2081–2100 (b, d, f) for frost-free season length (a, b), last spring freeze (c, d) and first autumn freeze (e, f).

variations. Indeed, the standard deviation of the station values is only 2.6 days suggesting a somewhat spatially coherent change over the study region both in terms of sign and magnitude. As noted previously here and in other studies, the majority of frost-free season length change in the historical record is attributed to earlier arrival of spring with the timing of autumn freezes either adding to growing season length, or in some cases, reversing the spring trend completely (see Figure 4.4). Interestingly, in the ensemble projections presented here, the autumn contribution to frost-free season length increases is of similar magnitude to the spring contribution, corresponding to widening, rather than translation, of the seasonal cycle. As shown in Figure 4.9c and 4.9e, the magnitude of changes in the timing of the last spring freeze and the first autumn freeze are nearly the same. Averaged over all stations, these are equal to –7.8 days and 7.9 days, respectively.

The results for the late twenty-first-century simulation period (2081–2100) indicate further increases in frost-free season length. Differences between the multi-model ensemble average in 2081–2100 and 1961–2000 are shown in Figure 4.9b, indicating an approximate doubling of the increase discussed above and shown in Figure 4.9a for the 2046–2065 simulations. Averaged over all stations, the difference in mean frost-free season length for 2081–2100 minus 1961–2000 is approximately 1 month (32.1 days). As with the 2046–2065 analysis, there are only small spatial variations in this change (the spatial standard deviation for the 2081–2100 case is 4.5 days). Additionally, the increase in frost-free season length is approximately evenly divided between the spring and autumn with multi-model ensemble means of

–15.4 days and 16.7 days for change in timing of the last spring freeze and first autumn freeze, respectively (Figures 4.9d and 4.9f). These results indicate that the frost-free season length increases that have been reported in the observed record may continue or even accelerate slightly in the remaining part of the twenty-first century with commensurate impacts on physical and biological systems.

Concluding Remarks

In this chapter, historical and projected changes in the length of the frost-free season have been examined for 53 surface locations in the Midwestern region of the United States. The results of the historical analysis showed that many, but not all, stations exhibited trends toward longer frost-free seasons during the latter part of the twentieth century. Nearly all stations had negative trends in the date of the last spring freeze, in accord with previous studies that have reported earlier arrival of spring in many mid- and high-latitude locations in both hemispheres. For some locations, the date of the first autumn freeze had positive trends (later autumn freezes) resulting in large increases in frost-free season length. However, many stations exhibited negative trends (earlier autumn freezes) that often offset, and occasionally reversed, the trends resulting from spring-time changes. It is noteworthy that the results presented here do not contradict the results presented in the previous chapter. Although summer cooling has occurred in parts of the Midwest, summer temperatures remain well above freezing and therefore have no effect on the frost-free season length statistics discussed here. Understanding the evolution of such changes under further anthropogenic global warming represents an important step toward understanding climate change impacts in the Midwestern United States.

To investigate the evolution of frost-free season length under changed climate conditions, simulations from eight AOGCMs were obtained from the IPCC AR4 / GCRP CMIP3 data portal. All of the simulations were conducted under the A2 emissions scenario, corresponding to large increases in atmospheric carbon dioxide concentrations. The resulting simulations therefore represent a high-end estimate of likely changes given current knowledge about the evolution of atmospheric carbon burdens based on projections of socio-economic factors. Due to the low-resolution of AOGCMs, and the resulting shortcomings associated with their surface climate characteristics, the model simulations of surface minimum daily air temperature were downscaled from mid-tropospheric specific humidity and air temperature by deriving regression equations for each month based on reanalysis data and surface observations from the historical record. The downscaled values were then used to compute the frost-free season length statistics for each AOGCM. Although the downscaling technique resulted in major improvement over the native AOGCM output, the individual models still exhibited significant biases. This prompted use of a multi-model ensemble which exhibited good overall agreement with observations (1961–2000), including the spatial pattern of mean frost-free season length and its spring and autumn contributions. However, the observed trends in frost-free season length statistics were not reproduced by the direct AOGCM output or the downscaled AOGCM results.

Results of transient twenty-first-century AOGCM simulations indicate a further increase in frost-free season length in the Midwest consistent with the increases in greenhouse gases that the AOGCM simulations are based on. The magnitude of the increase, relative to 1961–2000, is roughly two weeks by mid-century (2046–2065) and more than four weeks by 2081–2100. In both periods the projected increases exhibit little spatial variability and, in contrast to the observed frost-free season trends, have nearly equal contributions from earlier last spring freezes and later first autumn freezes. The model behavior is therefore consistent with a broadening of the seasonal cycle that has not been uniformly observed in the study region historically, but is consistent with other studies that have focused on AOGCM-based projections of frost-free season length (e.g., Christidis et al. 2007).

There are two major caveats to the research results presented here. First, the results presented here are based on the results of a single emissions scenario consistent with large increases in greenhouse gas emissions and atmospheric concentrations. True estimates of the uncertainty can only be attained through consideration of multiple models under multiple emissions scenarios. Second, the work here focused on the freezing (0°C) threshold and therefore may provide only a surrogate for related impacts on agriculture since individual crops are sensitive to different thresholds. Despite these caveats, and other uncertainties in the future evolution of frost-free season characteristics, the results suggest substantial increases in the length of the frost-free season during the next century. Additional research is needed to investigate and quantify the likely impacts of these changes on physical and biological systems.

REFERENCES

Beniston, M. 2003. "Climatic Change in Mountain Regions: A Review of Possible Impacts." *Climatic Change* 59: 5–31.

Cayan, D. R., et al. 2001. "Changes in the Onset of Spring in the Western United States." *Bulletin of the American Meteorological Society* 82: 399–415.

Chmielewski, F.-M., A. Muller, and E. Bruns. 2004. "Climate Changes and Trends in Phenology of Fruit Trees and Field Crops in Germany, 1961–2000." *Agricultural and Forest Meteorology* 121: 69–78.

Christidis, N., et al. 2007. "Human Contribution to the Lengthening of the Growing Season during 1950–99." *Journal of Climate* 20: 5441–5454.

Cleland, E. E., et al. 2007. "Shifting Plant Phenology in Response to Global Change." *Trends in Ecology and Evolution* 22: 357–365.

Easterling, D. R. 2002. "Recent Changes in Frost Days and the Frost-Free Season in the United States." *Bulletin of the American Meteorological Society* 83: 1327–1332.

Groisman, P. Ya., et al. 1994. "Changes of Snow Cover, Temperature, and Radiative Heat Balance over the Northern Hemisphere." *Journal of Climate* 7: 1633–1656.

Groisman, P. Ya., et al. 2004. "Contemporary Changes of the Hydrological Cycle over the Contiguous United States: Trends Derived from in situ Observations." *Journal of Hydrometeorology* 5: 64–85.

IPCC. 2000. *Special Report on Emissions Scenarios.* Cambridge: Cambridge University Press.

Kunkel, K., et al. 2004. "Temporal Variations in Frost-Free Season in the United States: 1895–2000." *Geophysical Research Letters* 31, doi:10.1029/2003GL018624.

Meehl, G. A., J. M. Arblaster, and C. Tebaldi. 2007. "Contributions of Natural and Anthropogenic Forcing to Changes in Temperature Extremes over the United States." *Geophysical Research Letters* 34, doi:10.1029/2007GL030948.

Meehl, G. A., et al. 2007. "The WCRP CMIP3 Multimodel Dataset: A New Era in Climate Change Research." *Bulletin of the American Meteorological Society* 88: 1383–1394.

Parmesan, C., and G. Yohe. 2003. "A Globally Coherent Fingerprint of Climate Change Impacts across Natural Systems." *Nature* 421: 37–42.

Peterson, T. C., and R. S. Vose. 1997. "An Overview of the Global Historical Climatology Network Temperature Database." *Bulletin of the American Meteorological Society* 78: 2837–2849.

Pryor, S. C., J. T. Schoof, and R. J. Barthelmie. 2005. "Climate Change Impacts on Wind Speeds and Wind Energy Density in Northern Europe: Empirical Downscaling of Multiple AOGCMs." *Climate Research* 29: 183–198.

Robeson, S. M. 2002. "Increasing Growing-Season Length in Illinois during the 20th century." *Climatic Change* 52: 219–238.

Schubert, S. 1998. "Downscaling Local Extreme Temperature Changes in Southern-Eastern Australia from the CSIRO Mark2 GCM." *International Journal of Climatology* 18: 1419–1438.

Schwartz, M. D., and B. E. Reiter. 2000. "Changes in North American Spring." *International Journal of Climatology* 20: 929–932.

Taylor, K. E. 2001. "Summarizing Multiple Aspects of Model Performance in a Single Diagram." *Journal of Geophysical Research* 106: 7183–7192.

Trenberth, K. E., et al. 2007. "Observations: Surface and Atmospheric Climate Change." In *Climate Change 2007: The Physical Scientific Basis.* Contribution of Working Group I to the Fourth Assessment Report of the Intergovernmental Panel on Climate Change, [ed. S. Solomon, et al.], 235–336. New York: Cambridge University Press.

Uppala, S. M., et al. 2005. "The ERA-40 Re-analysis." *Quarterly Journal of the Royal Meteorological Society* 131: 2961–3012.

Winder, M., and D. E. Schindler. 2004. "Climatic Effects on the Phenology of Lake Processes." *Global Change Biology* 10: 1844–1856.

5. Long-Term Midwestern USA Summer Equivalent Temperature Variability

J. C. ROGERS, S.-H. WANG, AND J. S. M. COLEMAN

Introduction

Evaluations of long-term regional and global change typically involve analysis of air temperature and atmospheric moisture. Surface air temperature variability is often assessed separately from humidity (e.g., Mann and Jones 2003; Jones and Moberg 2003) and is the focus of widely publicized statements regarding the relative warmth of each passing year, used as a quantifier of the status and extent of global change. Although different measures of atmospheric moisture exist, dew point temperature is widely used in research studies, and climatologies of its spatial variability have been developed (Robinson 2000). Trends in summertime tropospheric water vapor content (Ross and Elliott 1996; 2001) and surface dew point temperatures (Sandstrom, Lauritsen, and Changnon 2004) have been evaluated, as have their economic and human impacts (Gaffen and Ross 1998; Sparks, Changnon, and Starke 2002). The eastern United States is typically characterized by high specific humidity and dew points during summer months. Its variability can range, however, from summers that are hot and dry, such as those of the 1930s, to those that are hot and humid, producing

high human heat stress (Kunkel et al. 1996). This has created considerable interest in impacts of summer humidity and has led to development by the National Weather Service of temperature-humidity indices that help gauge the heat stress impact of daily weather on humans.

Air temperature and humidity are also evaluated together when considering the energy content of the atmosphere. For example, the total energy content of the air (h), per unit mass, can be written:

$$h = c_p T + Lq + gz + \tfrac{1}{2}V^2 \qquad (5.1)$$

where T is the observed air temperature, c_p is the specific heat of air at constant pressure (1005 J kg^{-1} K^{-1}), L is the latent heat of vaporization (2.5×10^6 J kg^{-1}), q is the specific humidity (g kg^{-1}), z is the elevation of the air, and V its velocity.

The kinetic energy term is generally an order of magnitude smaller than the others, and near the earth's surface (where z≅0) we are left with h=c_pT + Lq whose right-hand-side terms account for variations due to both air temperature and moisture content of the air. This total surface energy content is potentially a more comprehensive variable for analysis of global change as it represents the total surface warming and energy content (Pielke, Davey, and Morgan 2004).

Haltiner and Martin (1957) and Davey, Pielke, and Gallo (2005) define the isobaric equivalent air temperature (T_E) at the earth's surface as:

$$T_E = h / c_p \qquad (5.2)$$

leading to:

$$T_E = T + Lq/c_p \qquad (5.3)$$

The units of T_E are °C, and the Lq/c_p term will hereafter be abbreviated "Lq" in the text. T_E defines the total surface energy content, and Pielke, Davey, and Morgan (2004) were among the first to study it using two years of observational data from Fort Collins, Colorado. They showed the annual cycles of T_E and T, illustrating that T_E exceeds T by a large amount during the summer growing season when the annual cycle in specific humidity (Lq) peaks over the United States. Davey, Pielke, and Gallo (2005) used over 150 stations in the eastern United States from 1982–1997 to evaluate monthly T and T_E trends in light of station land surface conditions. They found that T and T_E trends generally have the same sign, while the relative magnitudes of the trends vary from station to station with type of land cover. Sites dominated by grass and shrub had T_E trends exceeding those of T, while the converse occurred in dominantly forested and agricultural settings.

Rogers, Wang, and Coleman (2007) examined summertime moist static energy parameters at Columbus, Ohio, evaluating decadal trends in mean summer T, T_E, and Lq and showing how local inter-annual variations in the parameters are linked to surface dryness and total summer rainfall.

The record at Columbus is somewhat unique in having complete dew point temperatures and station sea level pressures back to 1882, permitting long time series of Lq. Mean summer T and Lq were found to have opposite correlations to summer rainfall at Columbus, producing a net weak link between T_E and rainfall. These interrelationships varied through time, with Lq strongly linked to rainfall before 1936 but weakly thereafter. While nearly 47% of summers 1882–1936 were warm and dry, 77% have been warm and humid since 1937.

The purpose of this chapter is to provide further description of concepts associated with, and climatological analysis of, total surface energy content. Six relatively long-term Midwestern USA summer climate records are evaluated and compared in terms of their total surface energy content variability and with regard to the role that basic data parameters such as maximum, minimum, and dew point temperatures play in creating that variability. One of the six station records is that of Columbus, Ohio, which was previously analyzed to 1882 but which here is evaluated in the same context as the other five stations where records begin in the 1930s and 1940s. The chapter expands upon the analyses of Rogers, Wang, and Coleman (2007; hereafter RWC07), describing the basic moist static energy concepts, but also delving into comparative statistical analyses among the stations, as well as examining atmospheric circulation variability and changes in atmospheric moisture fluxes.

Data and Methods

Climate data for five non-Ohio Midwestern stations including maximum, minimum, and mean summer air temperatures (T_{max}, T_{min}, T), as well as mean dew point temperature (T_d), were obtained from the TD3200 Surface Land Daily Cooperative Summary for the Day available from the National Climatic Data Center (NCDC). Monthly and summer seasonal averages were then computed from the daily values. The stations, shown in figure 5.1, include Belleville Scott Air Force base (IL; data period of record 1938–1999), Fort Wayne airport (IN; 1942–2006 with missing data in 1946–1947 and 1965–1972), Mount Clemens Selfridge Field Air Force base (MI; 1937–1999), Milwaukee Mitchell Field airport (WI; 1948–2006, missing 1965–1972), and Minneapolis–St. Paul airport (MN; 1945–2006, missing 1965–1972). Stations were chosen for their comparative long length of records (especially T_d) and will be subsequently identified solely by their two-letter state identifier. Data for the non-Ohio Midwestern stations were used as they appear in the TD3200 dataset without further evaluation for inhomogeneities since the purpose of this chapter is an exploratory overview of similarities among the stations in trends and variability among energy content parameters. Columbus (OH; 1937–2005) data were obtained from various sources as described in RWC07 and were corrected for a major station move from the city downtown area to a more rural airport location in 1949 (see RWC07).

In calculating the mean Lq term, the vapor pressure of the air (e) is obtained from observed monthly mean surface T_d using Bolton's (1980) empirical relationship, accurate to within 0.3% between $T_d = -35°C$ to $+35°C$:

$$e = 6.112 exp\left(\frac{17.67T_d}{T_d + 243.5}\right) \qquad (5.4)$$

from which q is obtained:

$$q = \frac{.622e}{p - .378e} \qquad (5.5)$$

where p is the monthly mean sea level pressure in hPa.

A unit change in specific humidity ($\Delta q = 1$ g/kg, or 0.001) represents about 2.5 times (i.e., 2,500,000 J kg^{-1}×0.001÷1005 J kg^{-1} K^{-1}) the contribution to the total energy (T_E) that a unit change in air temperature ($\Delta T = 1°C$) does. The relationship between incremental changes in T_d and T is not a fixed relationship, however, a fact that warrants further comment as T_d is the data used here. Table 5.1 shows the relative change in Lq/c_p as T_d changes in 1° and 2°C increments over a range of typically occurring Midwestern USA summertime values between 10°C and 24°C. If mean T_d increases from 10°C to 12°C the moist energy increases by 2.7°C (table 5.1), but if T_d goes from 22°C to 24°C the Lq term changes by 5.3°C. For simplification, if one assumes summer mean T is fixed at

Figure 5.1. Locations of climatic stations used in this study.

30°C for table 5.1, then the ratio Lq:T_E obtained via $T_E=T+Lq/c_p$ will vary. Lq is only 39% of T_E for $T_d=10$°C (and constant T) but it becomes 61% of T_E at a 24°C dew point. The idea behind these concepts, particularly the Lq:T_E ratio, is applied in this study through evaluation of the inter-annual variability and trends in the summer T and T_d.

Elements of the atmospheric moisture budget are compared during humid and drier summers using NCEP/NCAR reanalysis data (Kalnay et al. 1996) that has a resolution of 2.5°×2.5°. The atmospheric moisture budget is represented as follows:

$$P - E = -\frac{\partial w}{\partial t} - \nabla \bullet \frac{1}{g} \int_{P_{top}}^{P_s} \overline{qV} dp \qquad (5.6)$$

where E is the column evaporation rate, P is the precipitation rate, w is the precipitable water, g is the earth's gravitational constant, p is the pressure integrated from the surface (p_s) to the upper boundary of the humidity

Table 5.1. Changes in Lq/c_p occurring at selected 1°C and 2°C intervals of T_d values between 10°C and 24°C (column 2) and changes in the ratio of Lq/c_p:T_E (column 3) assuming a constant T = 30°C. Lq/c_p is calculated as described in the text using a constant p = 1012 hPa.

T_d(°C)	Lq/c_p (°C)	Lq/T_E ratio
10	18.8	39%
12	21.5	42%
14	24.6	45%
15	26.2	47%
16	28.0	48%
17	29.8	50%
18	31.8	51%
19	33.9	53%
20	36.0	55%
22	40.8	58%
24	46.1	61%

data ("top"; 300 hPa), V is the horizontal wind components, and q is the specific humidity. The precipitable water, w, is given as:

$$w = \frac{1}{g} \int_{P_{top}}^{P_s} q dp \qquad (5.7)$$

When P–E is positive (negative), precipitation (evaporation) exceeds evaporation (precipitation) and surface conditions will be wet (dry). Results are obtained for the 850 hPa moisture flux components $\overline{qu}$ and $\overline{qv}$ and for moisture flux divergence, the full second right-hand-side term of the equation for the atmospheric moisture budget.

Results

The full 1882–2005 time series of OH T, Lq, and T_E, reproduced here from RWC07, indicate OH air temperature (T; figure 5.2) peaks between 1930 and 1944, during fifteen persistently warm summers between 1930 and 1944, but then trend downward to the mid-1960s, after which the trend reverses to the 1990s. The same characteristics appear in Lq and T_E variability (figure 5.2), although the distinctive peak in the 1930s is largely missing from these. The drought of 1930–1936 is one of the main features of figure 5.2. While air temperatures were consistently high during this drought, the humidity (Lq) was normal to below normal and T_E largely responded to the relatively low humidities. While 1934 and 1936 are the hottest OH summers (high T values), they are not particularly humid. In contrast, the summer of 1995 was relatively hot (about seventh warmest in RWC07), but extremely humid, having the highest Lq and T_E on the

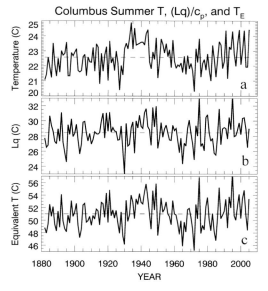

Figure 5.2. Time series of Columbus, Ohio, mean summer (a) air temperature (T), (b) moist enthalpy (Lq/c_p), and (c) equivalent temperature (T_E) in °C. Figure reproduced from Rogers et al. (2007) is provided courtesy of the American Meteorological Society.

124-year record, and it is associated with one of the worst heat waves on record in the Midwestern USA (Kunkel et al. 1996).

While the total energy content parameters are not shown for the other stations, all of their time series were significantly correlated to that of OH in figure 5.2. For example, correlations to OH summer Lq range from r=0.46 with MN Lq (1945–2005), the

location farthest from OH, to r=0.72 with nearby IN Lq (1942–2005). For both OH T and OH T_E, correlations with all other stations are equal to or exceed r=0.57. Another broad comparison (figure 5.3) reveals the relation between OH mean summer air temperatures (T) and those averaged over the contiguous United States, with data available from 1895 to 2005 from the NCDC. The two time series are correlated r=0.65 and are both characterized by the downward temperature trends from the 1930s to the 1960s and the four-decade upward trend since the mid-1960s. Relative to the single OH time series, the national average air temperature time series has substantially reduced variability.

Table 5.2 shows the correlations occurring for each individual Midwestern station between energy and climate parameters. Among the surface total energy parameters, the correlation between Lq and T_E (written Lq/T_E in table 5.2) exceeds 0.90 at each location, while the T/T_E correlations are lower and vary somewhat among sites. The T/Lq correlation is always lowest among the three parameter comparisons. MI and WI have the strongest T/Lq relationships but still have a coefficient

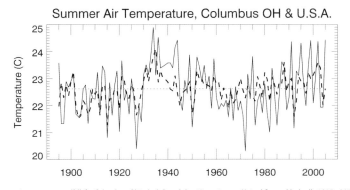

Figure 5.3. Mean summer air temperatures (°C) for Columbus, Ohio (solid) and the 48 contiguous United States (dashed), 1895–2005. The latter are from http://lwf.ncdc.noaa.gov/oa/climate/research/2005/cmb-prod-us-2005.html.

of variation of only around 50%, leaving much unexplained variation in these parameters. RWC07 used running twenty-year correlations to show that both mean summer T and T_E, and Lq and T_E, are consistently statistically significantly correlated through time. The same was not true, however, of the relation between T and Lq, which were shown to be poorly correlated prior to 1937 but significantly positively correlated after the Dust Bowl era, a situation confirmed here to some extent at all stations, but particularly at MI and WI. Among other basic climate data, mean summer T_{min} and T_d have relatively high correlations (table 5.2) everywhere but at IN, correlations that are higher than those obtained linking mean summer T_d with either T (not shown) or T_{max} data (column 4 in table 5.2). Mean summer T_{max} and T_{min} also are significantly correlated at all sites (table 5.2) but there is considerable unexplained variance in the relation between these parameters (highest r^2, at MI, is only 59%).

Table 5.3 shows the least squares linear regression slopes for several parameters for the six Midwestern climate sites. For 1965–2006 (note the time period varies depending on data availability at each site), upward decadal trends in Lq and T_E occur at all stations, although the level of statistical significance of the trends varies. The decadal slopes of T and Lq are additive in producing the decadal T_E trends, and all are based on non-normalized data. MI is the only site with significant upward trends in T, Lq, and T_E, while trends in the latter two parameters are significant at IL despite a weak T trend. The T_E slopes at IN and MN are not significant due to near-zero T trends and despite significant Lq trends. All sites except MN exhibit downward trends in T, Lq, and T_E from years starting in the 1940s to 1964. These downward trends are consistent with those at OH, the only site where regression slopes reach statistical significance.

Some sites have significant trends from 1965 to 2006 (table 5.3) in the variables that are the basis for T and Lq (i.e., T_{max}, T_{min}, T_d). T_{min} increases significantly at MI, MN, WI, and OH (table 5.3). More broadly, the upward trends in T_d (with which T_{min} is well correlated in table 5.2) are more positive than those in T (dry static energy) at IL, IN, MI, and MN, while the reverse is true at WI and OH. This leads to small trends in the Lq: T_E ratio (far right column in table 5.3) at the latter two sites, while the ratios at IL and IN are large and significant (95% confidence) and those at MN reach 90% significance.

Time series variability of T_{max}, T_{min}, T

Table 5.2. Correlations between the summer mean values of total energy parameters T, Lq, and T_E and between data parameters T_{max}, T_{min}, and T_d occurring at individual Midwestern climate sites over their periods of record. Correlations that are significant with 95% (99%) confidence are in italics (bold italics).

	T/T_E	Lq/T_E	T/Lq	T_{max}/T_d	T_{min}/T_d	T_{max}/T_{min}	Period of Record
IL	*0.65*	*0.94*	0.35	0.17	*0.64*	*0.72*	1938–1999
IN	*0.69*	*0.93*	0.36	0.26	0.36	*0.42*	1942–2006
MI	*0.87*	*0.97*	*0.72*	*0.55*	*0.8*	*0.77*	1937–1999
MN	*0.8*	*0.91*	*0.48*	0.25	*0.68*	*0.56*	1945–2006
WI	*0.9*	*0.95*	*0.72*	*0.54*	*0.76*	*0.7*	1948–2006
OH	*0.84*	*0.96*	*0.66*	*0.46*	*0.78*	*0.67*	1937–1999

Table 5.3. Least square linear regression trends (°C per decade) for time periods 1940–1964 and 1965–2006 for weather stations in six states (two-letter identifiers used). The far-right column indicates the years used in obtaining the trend slopes. The Lq:T_E ratio trend has units of percent per decade. Trends that are significant with 95% (99%) confidence are in italics (bold italics).

1940–1964	T_{max}	T_{min}	T	T_d	Lq	T_E	Lq:T_E	Summers
IL	−0.42	−0.31	−0.35	−0.37	−0.76	−1.11	−0.22	1940–1964
IN	0.00	−0.63	−0.31	−0.36	−0.67	−0.98	−0.23	1942–1964*
MI	−0.12	−0.34	−0.22	−0.48	−0.84	−1.06	−0.52	1940–1964
MN	0.08	−0.28	−0.02	0.20	0.33	0.32	0.34	1945–1964
WI	−0.35	*−1.11*	−0.63	−0.37	−0.64	−1.27	0.16	1948–1964
OH	−0.26	**−0.77**	−0.51	−0.61	−1.15	**−1.66**	−0.31	1940–1964
1965–2006	T_{max}	T_{min}	T	T_d	Lq	T_E	Lq:T_E	Summers
IL	0.22	0.26	0.22	**0.47**	**0.97**	**1.19**	*0.52*	1965–1999
IN	0.24	−0.50	−0.06	*0.39*	*0.70*	0.64	*0.70*	1973–2006
MI	0.35	**0.44**	*0.33*	*0.40*	*0.68*	*1.01*	0.27	1965–1999
MN	−0.30	*0.52*	0.05	*0.41*	*0.67*	*0.72*	*0.60*	1973–2006
WI	−0.07	**0.64**	0.27	0.20	0.35	0.62	0.01	1973–2006
OH	0.23	**0.52**	**0.38**	0.27	0.48	0.86	0.02	1965–2005

*Excludes 1946 and 1947 due to missing data.

and T_d (figure 5.4) is illustrated, by way of example, for MI and IL. The sizable (non-significant), 1940–1964 downward IL trends in these parameters (table 5.3, figure 5.4c and 5.4d) appears to be largely caused by the sharp decline to colder and drier (lower T_d) summers starting around 1960. T_{min} in particular (figure 5.4d) is normal to above normal throughout the 1950s and then sharply declines. The weak IL upward trend, 1965–1999, is apparent in T but is much more evident in T_d. The MI time series (figure 5.4a and 5.4b) of the four temperature and dew point parameters also become decidedly negative around 1960, and T_{min} and T_d in particular fall to a sharp minimum in 1965. That event helps define the sharp upward significant MI trends in T_{min} and T_d from 1965–1999 (table 5.3).

If we extend the post-1964 trends for an "even" forty-year period at IL (table 5.3), T_d would increase by 1.9°C (4 decades × 0.47°C), while T increases by 0.9°C. The mean IL summer T_d (1965–1999) is about 18°C (figure 5.4d), and we assume, for the purpose of discussion involving table 5.1, that the IL Td changes from 17°C to 19°C over the forty years, a rounded 2°C change instead of 1.9°C. With these assumptions, the 2°C rise in T_d has the energy equivalence of a 4.1°C rise in the air temperature T (table 5.1), whereas only a 0.9°C change took place. The resulting Lq:T_E ratio increase at IL, about 2.1% (0.52% × 4), is roughly in line with the 3% rise between 17°C and 19°C implied by table 5.1, which employed an assumption that T remains constant. These comparisons help show the details behind the large increase in energy held by moisture at some locations (IL, IN, MN), relative to the dry static energy content. Note also that as time progresses the mean summer IL T_d (figure

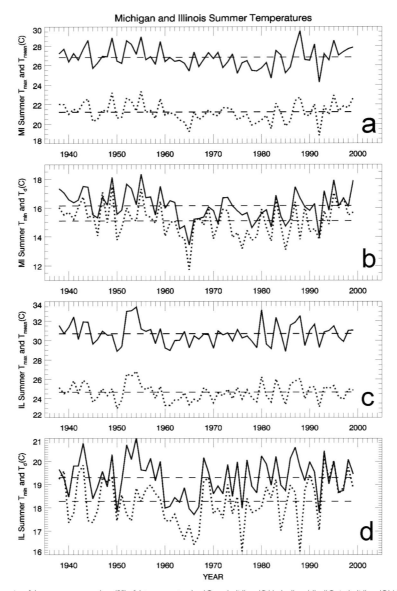

Figure 5.4. Time series of the summer mean values (°C) of data parameters (a, c) Tmax (solid) and T (dashed) and (b, d) Tmin (solid) and Td (dashed) occurring at Midwestern USA climate sites in (a, b) MI and (c, d) IL.

5.4d) approaches the numerical value of mean summer T_{min}, whereas T_d was typically much lower than T_{min} closer to 1965.

Variations in atmospheric moisture flux were examined based upon the covariability occurring between T and Lq, the energy parameters comprising T_E. Atmospheric moisture flux was determined for two sets of OH years characterized by unusually high summer Lq and T values (Lq+/T+) and those with negative departures of both Lq and T (Lq−/T−), using NCEP/NCAR reanalysis data since 1949. Mean summer temperature (figure 5.5a) and specific humidity (figure 5.5b) differences (positive minus negative) are positive and statistically significant (figure 5.5a and 5.5b) over much of the eastern United States and east-central Canada when stratified by the OH data. The largest mean differences are broadly centered around the Great Lakes. These figures illustrate the broad regional or even continental-scale associations that could likely be attributed to composites made at any of the Midwestern stations. The positive net mean 850 hPa height differences (Lq+/T+ versus Lq–/T–) are small (figure 5.5d) but significant from the Great Lakes to northern Quebec and indicative of increased ridging in the warmer, more humid summers, compared to the long-term summer mean 850 hPa field (figure 5.5c).

The net 850 hPa total moisture flux differences also have an anticyclonic flux maximum centered on the Great Lakes (figure 5.5e), and the vectors represent the direction of the fluxes in Lq+/T+ summers, while the reverse direction of the vectors would occur in Lq–/T– summers. Figure 5.5e effectively also helps summarize key results associated with individual maps of $\overline{qu}$ and $\overline{qv}$ (not shown). The vectors, for example, indicate Lq+/T+ is associated with a maximum strong southerly meridional ($\overline{qv}$) moisture flux in the Plains and Midwest (figure 5.5e) and a northerly flux over New England and the coastal western Atlantic, both of which represented statistically significant mean differences in $\overline{qv}$ data. Also significant was the easterly zonal flux (dominated by $\overline{qu}$) differences occurring over Tennessee and the Carolinas (figure 5.5e). Overall, however, the net flux differences are small over the Great Lakes and Ohio, where T and q differences are maximized (figure 5.5a and b). The

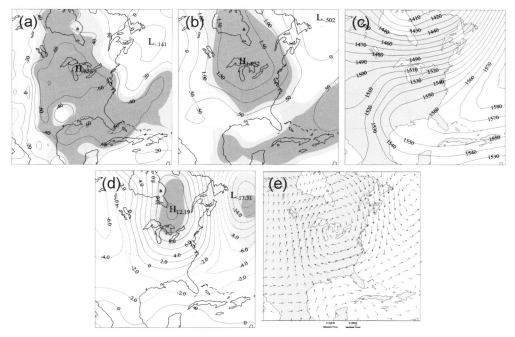

Figure 5.5. Mean differences in (a) air temperature (°C) and (b) specific humidity (g kg⁻¹) obtained by subtracting the averages during summers when Lq and T are both below normal from those when Lq and T are both above normal at OH. (c) Summer mean 850 hPa geopotential heights 1949–2004 for Lq+/T+ cases. (d) Mean 850 hPa height differences (m) between Lq+/T+ versus Lq–/T–, and (e) similarly obtained net mean total moisture flux. Mean differences that are statistically significant with 95% and 99% confidence are lightly and heavily shaded, respectively.

analysis is based on composites drawn from OH data, and it is little surprise that results exhibit a maximum in the mean differences (figure 5.5a–e) near OH. However, results also suggest that the largest moisture fluxes occur along the edges of a broader air mass regime having similar T and q anomalies over a large area (figure 5.5a and b).

Concluding Remarks

This chapter describes the concepts behind the total surface energy content of the air and examines how it varies during summer season in the Midwestern USA. The total energy content consists of the dry (T) and moist static energy (Lq), effectively representing the additional contribution of humidity to the energy content of air as measured by its temperature. Summer total surface energy content is evaluated at six Midwestern locations by examining relatively long time series (starting in the 1930s and 1940s) of equivalent temperature (T_E) and its static energy components T and Lq. The total energy content is evaluated both in terms of its decadal-scale trends and also the contributions by T_{max} and T_{min}, which comprise the dry static energy T, as well as of T_d, the widely used parameter that is synonymous with Lq. We also investigate variations associated with atmospheric circulation and moisture fluxes.

The analysis shows a Midwestern upward trend in Lq occurring since the mid-1960s, which is closely linked to rising summer mean dew point temperatures. However, trends in mean summer T are weak everywhere except MI and OH, while trends in T_E are significant only at MI and IL, although they are notably positive at all

six stations (table 5.3). Analyses of the data comprising the energy parameters reveal that the magnitude of T trends are largely due to those in T_{min} rather than T_{max} and that mean summer T_{min} is best linked to mean summer T_d and Lq. T_{min} trends at WI and OH create T trends that exceed those of T_d. However, T trends at IL, IN, MI, and MN are small, despite (or regardless of) the trend of T_{min}, and they are exceeded by T_d. This results in a small but statistically significant percentage increase in Lq relative to T_E (Lq:T_E) at IL and IN, and nearly so at MN. The impact of this increase in moisture content of the air has been assessed in other studies and can include increasing reliance on air conditioning and increasing human heat stress (Gaffen and Ross 1998; Sparks, Changnon, and Starke 2002).

The composites of figure 5.5 are drawn from anomalies in T and Lq at one station but reveal that a large portion of eastern North America seemingly shares in similar summer mean air mass T and q anomalies (figure 5.5a and b). The composite station in OH lies near the core of highest variation, but the fluxes of moisture are greatest on the edges of the air mass, over the Plains, and off the eastern seaboard. Kalkstein, Sheridan, and Graybeal (1998) note that the magnitude of dew point temperature increases in recent decades is largest in their moist tropical air mass (similar to the standard "maritime tropical" air mass). Sandstrom, Lauritsen, and Changnon (2004) find the decadal increases in summer dew points are not necessarily linked to advection increases from the Gulf of Mexico, but rather to changes in regional moisture sources, potentially due to vegetation changes. Davey, Pielke, and Gallo (2005) find that T and T_E trends generally have the same sign, while the relative magnitudes of the trends vary with type of land cover. Total energy content parameters are shown here to provide a quantitative measure of moist static energy increases relative to the total energy content ($Lq:T_E$ ratio). Research into the relation between temporal changes in land use, atmospheric circulation, and variations in the $Lq:T_E$ ratio may prove very useful in further evaluating human impacts of United States moisture variability. The use of numerical model sensitivity studies of the effects of increasing greenhouse emissions on twenty-first-century climate may also provide additional insight into potential future moist static energy changes.

REFERENCES

Bolton, D. "The Computation of Equivalent Potential Temperature." *Monthly Weather Review* 108: 1046–1053.

Davey, C. A., R. A. Pielke Sr., and K. P. Gallo. 2005. "Differences between Near-Surface Equivalent Temperature and Temperature Trends for the Eastern United States." *Global and Planetary Change* 54: 19–32.

Gaffen, D. J., and R. J. Ross. 1998. "Increased Summertime Heat Stress in the U.S." *Nature* 396: 529–530.

Haltiner, G. J., and F. L. Martin. 1957. *Dynamical and Physical Meteorology.* New York: McGraw Hill.

Jones, P. D., and A. Moberg. 2003. "Hemispheric and Large-Scale Surface Air Temperature Variations: An Extensive Revision and Update to 2001." *Journal of Climate* 16: 206–223.

Kalnay, E., et al. 1996. "The NCEP/NCAR 40-year Reanalysis Project." *Bulletin of the American Meteorological Society* 77: 437–471.

Kalstein, L. S., S. C. Sheridan, and D. Y. Graybeal. 1998. "A Determination of the Character and Frequency Changes in Air Masses using a Spatial Synoptic Classification." *International Journal of Climatology* 18: 1223–1236.

Kunkel, K. E., et al. 1996. "The July 1995 Heat Wave in the Midwest: A Climatic Perspective and Critical Weather Factors." *Bulletin of the American Meteorological Society* 77: 1507–1518.

Mann, M. E., and P. D. Jones. 2003. "Global Surface Temperatures over the Past Two Millennia." *Geophysical Research Letters* 30, doi: 10.1029/2003GL017814.

Pielke, R. A., Sr., C. Davey, and J. Morgan. 2004. "Assessing 'Global Warming' with Surface Heat Content." *EOS* 85(21): 214–215.

Robinson, P. J. 2000. "Temporal Trends in United States Dew Point Temperatures." *International Journal of Climatology* 20: 985–1002.

Rogers, J. C., S.-H. Wang, and J. S. M. Coleman. 2007. "Evaluation of a Long-Term (1882–2005) Equivalent Temperature Time Series." *Journal of Climate* 20: 4476–4485.

Ross, R. J., and W. P. Elliott. 1996. "Tropospheric Water Vapor Climatology and Trends over North America: 1973–1993." *Journal of Climate* 9: 3561–3574.

———. 2001. "Radiosonde-Based Northern Hemisphere Tropospheric Water Vapor Trends." *Journal of Climate* 14: 1602–1612.

Sandstrom, M. A., R. G. Lauritsen, and D. Changnon. 2004. "A Central-U.S. Summer Extreme Dew-Point Climatology (1949–2000)." *Physical Geography* 25: 191–207.

Sparks, J., D. Changnon, and J. Starke. 2002. "Changes in Frequency of Extreme Warm-Season Surface Dewpoints in Northeastern Illinois: Implications for Cooling-System Design and Operation." *Journal of Applied Meteorology* 41: 890–898.

6. Estimating Changes in Temperature Variability in a Future Climate

G. S. GUENTCHEV, K. PIROMSOPA, AND J. A. WINKLER

Introduction

Changes in the frequency of climate extremes can arise from (1) a shift in the mean of the distribution of the climate variable, also referred to as the "location" parameter of a distribution, (2) an increase or decrease in the variability of the climate distribution, or what is referred to as the "shape" parameter, and (3) changes in both the mean and variability (Wigley 1988). Katz and Brown (1992) demonstrated that the frequency of climate extremes is likely to be more dependent on changes in the variability of a distribution than changes in the mean. They further argued that "experiments using climate models need to be designed to detect changes in climate variability." In spite of this recommendation, until recently temporal changes in the shape parameter of climate distributions (referred to below as "variability") were rarely explicitly analyzed for simulations from climate models. Most previous research focused instead on changes in the frequency of extreme events over certain thresholds (e.g., Meehl and Tebaldi 2004; Weisheimer and Palmer 2005) rather than directly calculating traditional variability measures, such as variance and standard deviation. As a result, the relative contribution of changes in the mean as opposed to changes in variability to projected changes of climate extremes was often unclear.

In the last few years, however, several authors have investigated projected changes in temperature variability for Europe from simulations by global or regional climate models. For example, Rowell (2005) compared inter-annual temperature variability between a control and a future (2070–2100) period, as simulated by the HadAM3P model assuming the A2 greenhouse gas emissions scenario. Based on this single model simulation and the one future time slice, he concluded that by the end of the twenty-first century the inter-annual variability of maximum and minimum temperature in Europe will increase during summer but not during winter. Using a larger set of Intergovernmental Panel on Climate Change (IPCC) A2 and B2 simulations, Scherrer et al. (2005) found that inter-annual temperature variability for Central Europe is projected to increase in summer and decrease in winter, which generally concurs with Rowell's (2005) analyses. The findings of Raible et al. (2006), based on one regional and two global simulations, also suggest that changes in inter-annual temperature variability are likely to be largest in summer, at least for the Alpine region of Europe. Beniston and Goyette (2007) analyzed regional climate model simulations from the PRUDENCE Project for the end of the twenty-first century (2070–2100) and found that for Switzerland the inter-annual variability of maximum temperature, but not necessarily minimum temperature, is expected to increase in the future. Other recent studies have investigated the relative impact of changes in the location and shape parameters on the frequency of temperature extremes for Europe. Santos and Corte-Real (2006), using A2 and B2 simulations from the HadCM3 global model, concluded that the projected increase in warm extremes and decrease in cold extremes in Europe for 2070–2100 are primarily due to changes in mean temperature rather than to changes in variance. Goubanova and Li (2007) found that simulated changes in temperature extremes for the Mediterranean basin are also associated with a shift in the location parameter, but additionally concluded that changes in inter-annual variability must also be considered to explain simulated changes in the frequency of cold extremes. On a broader spatial scale, Räisänen (2002) investigated global inter-annual variability of temperature using nineteen experiments from the Coupled Model Intercomparison Project and found that temperature variability is expected to increase in winter over the extratropical Northern Hemisphere and high-latitude Southern Ocean.

As seen above, the majority of previous analyses of future changes in temperature variability focused on inter-annual variability. Also, previous studies have gener-

ally been limited to only a small number of climate simulations, usually for a single time slice in the future. Furthermore, potential changes in temperature variability have not been studied in as much detail for locations outside Europe.

This study expands on existing research in a number of ways:

• First, projected changes in variability are investigated for the Great Lakes region of North America, an area where substantial changes in the frequency of temperature extremes are expected. Earlier analyses completed as part of the U.S. National Assessment suggest a 10–40% reduction by 2099 in the frequency of very cold days (minimum temperature $\leq$ –20°C), an increase on the order of one to two weeks by 2034 and one to two months by 2099 in growing season length, and 1 to 20 additional hot days (maximum temperature $\geq$ 35°C) per year by 2099, based on simulations from two General Circulation Models (GCMs) (Winkler et al. 2000; Winkler et al. 2002).

• Second, alternative measures of climate variability are included in the analysis. In addition to inter-annual temperature variability, we also evaluate (1) "daily" variability, which is simply the variability of the daily temperature values (also often referred to as "inter-diurnal" variability; Gregory and Mitchell 1995) and (2) "synoptic" variability, which is the variation of daily deviations after removing the annual cycle from the time series and reflects the day-to-day fluctuations in weather patterns.

• Third, variability changes are assessed for a large ensemble of local climate scenarios developed using simulations from four GCMs driven by two different SRES greenhouse gas emissions scenarios (Nakicenovic et al. 2000).

• Fourth, potential changes of temperature variability are analyzed across the entire twenty-first century, in contrast to most previous studies that focused on a single future time slice.

• Fifth, the consistency of the variability of the temperature series with related free atmosphere variables is evaluated.

Methods

A large number of local climate change scenarios of maximum and minimum temperature were developed for Michigan and the surrounding Great Lakes area as part of the Pileus Project. The goal of the Pileus Project was to create web-based risk management tools for the specialized agriculture and tourism industries that incorporate current and potential future climate. Climate change scenarios were constructed for fifteen locations (figure 6.1) to serve as input to the decision-making tools (which can be found at www.pileus.msu.edu).

Simulations from four GCMs from the IPCC Third Assessment were used in the scenario development—CGCM2 (Flato

and Boer 2001), ECHAM4 (Roeckner et al. 1996), HadCM3 (Gordon et al. 2000), and NCAR CSM1.x (Meehl and Washington 1995). These are coarse scale models with grid resolutions of approximately 3.75° latitude by 3.75° longitude, 2.8° latitude by 2.8° longitude, 2.5° latitude by 3.75° longitude, and 2.8° latitude by 2.8° longitude for the CGCM2, ECHAM4, HadCM3, and NCAR CSM1.x models, respectively.

For each GCM, two simulations driven by different SRES greenhouse gas emissions scenarios (A2 and B2) were available. For one set of local climate scenarios, referred to as the "gridpoint" scenarios, the model-simulated maximum and minimum temperatures were simply interpolated to station locations using a sixteen-point Bessel scheme that weights the gridpoint values by the distance from the station location and the strength of the north-south and east-west temperature gradients.

A second, larger set of local climate scenarios was developed using empirical downscaling functions that relate, using multiple regression, free atmosphere variables (500

Figure 6.1. "Front page" of the Future Scenarios Tool of the Pileus Project. The map shows the locations for which temperature scenarios are available. Users must first choose a location and then the temperature parameter of interest. The tool can be accessed at http://pileus.msu.edu/tools/t_future.htm.

mb geopotential height and sea-level pressure) to local maximum and minimum temperature. These "transfer function" scenarios reflect the variability of free atmosphere variables rather than the GCM-simulated temperature series, and provide a means for evaluating the consistency of the simulated variability between the free atmosphere and temperature series. The empirical downscaling functions were developed following the recommendations of Winkler et al. (1997), who showed that quasi-subjective decisions made by the climate analyst when designing empirical transfer functions can have an influence on the resulting scenarios. Three "user decisions" were considered: (1) adjustment for deviations of model predictor variable simulations during a control period from observations, (2) definition of the seasons for which separate specification equations are derived, and (3) removal of the annual cycle from predictor variables and predictands. A total of eight different transfer function variants were constructed separately for maximum and minimum temperature. The transfer functions were initially developed using observations for 1970–1989 and were validated for the 1960–1969 and 1990–1999 periods.

Observations of maximum and minimum temperature were obtained from the Michigan State Climatologist's Office. These observations had earlier undergone (1) inspection for changes in location, time of observation, and instrumentation, (2) filtering for obvious errors, and (3) checks for discontinuities. The free-atmosphere variables used in the transfer function development were obtained from the NCAR/NCEP reanalysis fields (Kalnay et al. 1996). A detailed description of the Pileus Project temperature

scenario development is available in Winkler (2009).

Variability measures were calculated for thirty-year overlapping periods (1990–2019, 1991–2020, . . . , 2070–2099) to the end of the twenty-first century. Daily variability is defined as the standard deviation of the daily values of either maximum or minimum temperature for each thirty-year period. Synoptic variability was calculated by first pre-whitening the temperature series by subtracting the mean maximum or minimum temperature of each thirty-year period for every Julian day from the respective daily temperature value. The standard deviation of the daily deviations was then calculated by overlapping period. Inter-annual variability is the standard deviation of the thirty yearly temperature means for each period. All three variability measures were calculated annually and seasonally. Variability values for the overlapping future periods were expressed as "deltas," or differences between the respective future period and 1990–2019, which is the first thirty-year period of the scenario time series.

For each overlapping period, the number of scenarios with a projected increase in temperature variability was summed across all locations and GCMs to obtain a measure of model agreement on the sign of the projected change. The magnitude of the projected change in variability was not considered. For the gridpoint scenarios, there are a total of 60 cases (scenarios) included in each analysis (4 GCMs×15 stations), and there are 480 cases (4 GCMs×15 stations×8 transfer function variants) for the transfer function scenarios. A thirty-year period is expected to experience increased temperature variability if more than 50% of the cases are in agreement that the sign of the projected change is positive. When interpreting the results, it is important to keep in mind that the cases are not independent because of spatial autocorrelation among the stations and because each set of transfer function variants is developed using the same four GCMs. Nonetheless, this approach is useful for summarizing the variability changes across the large ensemble of scenarios.

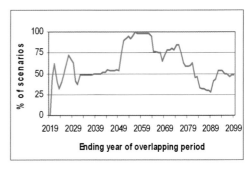

Figure 6.2. Percent of local minimum temperature scenarios suggesting an increase in inter-annual variability at the annual scale compared to a 1990–2019 control period for thirty-year overlapping periods. The scenarios were constructed by interpolating GCM-simulated temperature series (A2 emissions scenario) to station locations. Results for a given period are shown for the last year of each thirty-year period—e.g., results for 1990–2019 are presented in the graph at year 2019.

Results

In this section we present results to address the question: does the interpretation of simulated future changes in temperature variability depend on the choice of the (1) future time slice, (2) emissions scenario, (3) temperature parameter, (4) temporal aggregation period (annual and seasonal), (5) variability measure, and/or (6) downscaling method?

An important implication of this analysis is that the projected change in temperature variability varies with time across the twenty-first century. This finding is illustrated by the plot of the inter-annual variability of the gridpoint scenarios for minimum temperature developed using the A2 greenhouse gas emissions scenario (figure 6.2). For the overlapping periods ending from 2030 to 2049, there is little agreement among estimates of the inter-annual variability of daily minimum temperature derived from spatial interpolation of output from the four GCMs under the A2 emission scenario, with half of the scenarios exhibiting evidence of declines, and half increases. On the other hand, for the periods ending from approximately 2049 to 2080 almost all the scenarios project an increase in inter-annual variability. Later

in the century, for periods ending between 2080 and 2095, the majority of scenarios suggest a decrease in inter-annual variability. For the last period (2070–2099) of the century, which was used for previous studies in Europe, there is no agreement among the scenarios. Hence, interpretation of changes in inter-annual temperature variability is very dependent on the choice of future time slice.

The choice of greenhouse gas emissions scenario can also influence the interpretation of future changes in temperature variability, as illustrated by the projected changes in inter-annual variability in spring for the gridpoint scenarios of daily maximum temperature (figure 6.3). Almost all scenarios based on the A2 emissions scenario project an increase in inter-annual variability throughout the twenty-first century. On the other hand, the majority of the B2 scenarios indicate a decrease in inter-annual variability. Similar differences were found for the other seasons, the three variability measures (i.e., inter-annual, daily, and synoptic variability), the gridpoint and transfer function scenario types, and both the maximum and minimum temperature scenarios.

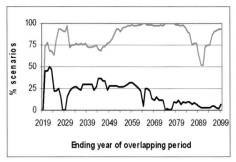

Figure 6.3. Comparison of percent of local maximum temperature scenarios developed using the A2 (red line) and B2 (black line) emissions scenarios that indicate an increase in springtime inter-annual variability compared to a 1990–2019 control period for thirty-year overlapping periods. The scenarios were constructed by interpolating the GCM-simulated temperature series to station locations. Results for a given period are shown for the last year of the thirty-year period.

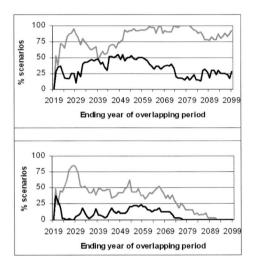

Figure 6.4. Comparison of the percentage of local scenarios of maximum temperature (top panel) and minimum temperature (bottom panel) developed using the A2 (red line) and B2 (black line) emissions scenarios that indicate an increase in springtime synoptic variability compared to a 1990–2019 control period for thirty-year overlapping periods. The scenarios were constructed by interpolating the GCM-simulated temperature series to station locations. Results for a given period are shown for the last year of the thirty-year period.

The percentage of scenarios indicating a variability increase is generally similar for maximum and minimum temperature, suggesting the choice of temperature parameter is not as influential as the choice of time slice and emissions scenario when evaluating variability changes. Nonetheless, some

differences do exist, especially for spring (i.e., March–May), as shown in figure 6.4 using the synoptic variability measure (i.e., the standard deviation of the pre-whitened daily temperature deviations) as an example. An increase in synoptic variability is suggested for maximum temperature developed using the A2 emissions scenario, whereas a decrease in variability is suggested for minimum temperature. In contrast, both maximum and minimum temperature display decreased variability in the future for the B2 emissions scenario.

Previous work for Europe cited above suggests that future changes in temperature variability may vary by season. Similar conclusions can be drawn for the Great Lakes region. The signs of projected changes are particularly divergent for winter and spring, as seen from the projected changes in inter-annual variability for the gridpoint scenarios of maximum temperature developed using the A2 emissions scenario (figure 6.5a). In winter, there is some consensus that inter-annual variability will decrease, whereas in spring the scenarios suggest that inter-annual variability should increase. There is less consistency for summer, although for most of the twenty-first century 50–75% of the scenarios suggest an increase in inter-annual variability. For fall, an increase in inter-annual variability is evident only toward the end of the century. These seasonal variations are reflected in the annual pattern, with greater agreement for increased inter-annual variability in mid-century and again at the very end of the century.

Although there are time periods and scenarios for which the three types of variability (inter-annual, daily, and synoptic) display similar patterns, some differences

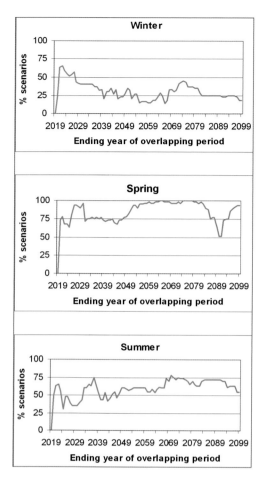

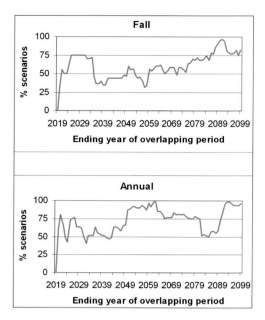

Figure 6.5. Comparison by season of the percentage of local scenarios of maximum temperature developed using the A2 greenhouse gas emissions scenario that indicate an increase in inter-annual variability compared to a 1990–2019 control period for thirty-year overlapping periods. The scenarios were constructed by interpolating the GCM-simulated temperature series to station locations. Results for a given period are shown for the last year of the thirty-year period.

are evident, as illustrated by the variability measures for the A2 gridpoint scenarios at the annual aggregation level (figure 6.6). For maximum temperature, there is considerable agreement among the scenarios that inter-annual variability will increase in the latter two-thirds of the century. On the other hand, there is little or no agreement for daily variability regarding the sign of the projected changes of maximum temperature variability in the middle of the century, and only about 60% of the scenarios suggest an increase in daily variability by the end of the century. For synoptic variability, the majority of the scenarios indicate a decrease in the latter two-thirds of the century. Differences

are also evident for minimum temperature. There is strong agreement that daily and synoptic variability will decrease in the latter two-thirds of the century, whereas inter-annual variability increases in the middle of the century.

A comparison of projected changes in temperature variability for the gridpoint and transfer function scenarios suggests that the variability of free atmosphere variables simulated by GCMs is not always consistent with that of the surface temperature series. For example, the overall impression of the wintertime change in variability for the scenarios developed using the A2 emissions scenario differs for all three variability mea-

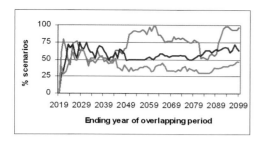

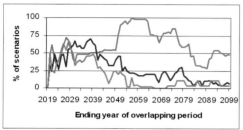

Figure 6.6. Comparison by variability measure (inter-annual, daily, and synoptic) of the percentage of local scenarios of maximum (top panel) and minimum (bottom panel) temperature developed using the A2 greenhouse gas emissions scenario that indicate an increase in variability compared to a 1990–2019 control period for thirty-year overlapping periods. The scenarios were constructed by interpolating the GCM-simulated temperature series to station locations. Results for a given period are shown for the last year of the thirty-year period. The red line shows the percentage of scenarios indicating an increase in inter-annual variability, the blue line represents the increase in daily variability, and the gray line represents the increase in synoptic variability. Please see the text for definitions of the three variability measures.

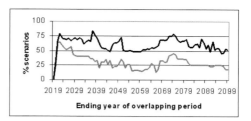

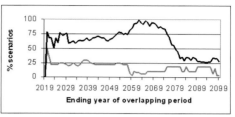

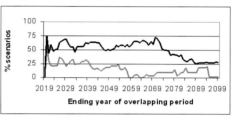

Figure 6.7. Comparison of the gridpoint (GP) and transfer function (TF) scenarios of local maximum temperature developed using the A2 greenhouse gas emissions scenario that indicate an increase in inter-annual (top panel), daily (middle panel), and synoptic variability (bottom panel) compared to a 1990–2019 control period for thirty-year overlapping periods. Results for a given period are shown for the last year of the thirty-year period. The red line shows the percentage of scenarios indicating an increase in variability for the gridpoint scenarios, and the black line represents the percentages for the transfer function scenarios. Please see the text for a description of how the gridpoint and transfer function scenarios were developed.

sures, depending on whether the gridpoint or transfer function scenarios are considered (figure 6.7). All the gridpoint scenarios indicate a decrease in temperature variability, while the transfer function scenarios display an increase for all variability measures. This discrepancy likely reflects errors in the GCM simulations of surface temperature. For example, CGCM1, which is the earlier version of the CGCM2 model used in this analysis, poorly simulates surface temperature variability for mid- and high-latitude locations in spring and fall, when the model overpredicts the occurrence of near-freezing temperatures (Palutikof et al. 1997). A similar error is evident for the CGCM2 simulations

of surface temperature for the gridpoints in the Great Lakes region (data not shown).

Conclusions

Local climate scenarios developed for the Great Lakes region were used to evaluate future changes in temperature variability. The analyses indicate that the sign of projected changes varies with time, and interpreting future temperature variability from only one or a few time slices may be misleading. Rather, it is essential to understand the time-varying nature of projected temperature variability. In addition, the choice of greenhouse emissions scenarios also influences the interpretation of future temperature variability. For the Great Lakes region, the temperature scenarios based on A2 simulations were more likely to project an increase in temperature variability compared to those based on B2 simulations. The choice of temperature parameter—maximum or minimum temperature—appears to have only a modest impact on the interpretation of future changes in temperature variability. In agreement with earlier studies for Europe, the sign of projected changes in temperature variability differs seasonally.

An important finding is that the definition of variability needs to be considered. Most previous studies employed inter-annual variability, but the sign of the projected changes for daily and synoptic variability often differs from that of inter-annual variability, at least for the Great Lakes region. We recommend that daily and synoptic variability be included in addition to inter-annual variability in analyses of climate variability, as these measures are more closely aligned with changes in climate extremes. Finally, the variability changes for local scenarios developed by spatially interpolating GCM temperature simulations (i.e., the gridpoint scenarios) are not always in agreement with the local scenarios developed empirically from free atmosphere variables (i.e., transfer function scenarios). This finding suggests possible inconsistencies in the variability of the GCM simulated time series of temperature and free atmosphere variables.

ACKNOWLEDGMENTS

This work was completed as part of the Pileus Project and was funded by the Environmental Protection Agency—grant #R83081401-0. We thank Jeff Andresen and Aaron Pollyea from the Michigan State Climatologist's Office for providing the historical temperature series used in the analysis. We also express our appreciation to Hannes Thiemann and Hans Luthardt from the Max Planck Institute of Meteorology for their help in obtaining the daily ECHAM4 simulations, to Lawrence Buja and Gary Strand from the National Center for Atmospheric Research, who provided the NCAR CSM1.x daily simulations, and to David Viner at the Climate Research Unit in East Anglia, who made available the daily HadCM3 simulations. The simulations for the CGCM2 were obtained from the CCCma web server.

REFERENCES

Beniston, M., and S. Goyette. 2007. "Changes in Variability and Persistence of Climate in Switzerland: Exploring 20th Century Observations and 21st Century Simulations." *Global and Planetary Change* 57: 1–15.

Flato, G. M., and G. J. Boer. 2001. "Warming Asymmetry in Climate Change Experiments." *Geophysical Research Letters* 28: 195–198.

Gordon, C., et al. 2000. "The Simulation of SST, Sea Ice Extents and Ocean Heat Transports in a Version of the Hadley Centre Coupled Model without Flux Adjustments." *Climate Dynamics* 16: 147–168.

Goubanova, K., and L. Li. 2007. "Extremes in Temperature and Precipitation around the Mediterranean Basin in an Ensemble of Future Climate Scenario Simulations." *Global and Planetary Change* 57: 27–42.

Gregory, J. M., and J. F. B. Mitchell. 1995. "Simulation of Daily Variability of Surface Temperature and Precipitation over Europe in the Current and 2×CO$_2$ Climates using the UKMO Climate Model." *Quarterly Journal of the Royal Meteorological Society* 121: 1451–1476.

Kalnay, E., et al. 1996. "The NCEP/NCAR 40-year Reanalysis Project." *Bulletin of the American Meteorological Society* 77: 437–471.

Katz, R. W., and B. G. Brown. 1992. "Extreme Events in a Changing Climate: Variability Is More Important than Averages." *Climatic Change* 21: 289–302.

Meehl, G. A., and C. Tebaldi. 2004. "More Intense, More Frequent, and Longer Lasting Heat Waves in the 21st Century." *Science* 305: 994–997.

Meehl, G. A., and W. M. Washington. 1995. "Cloud Albedo Feedback and the Super Greenhouse Effect in a Global Coupled GCM." *Climate Dynamics* 11: 399–411.

Nakicenovic, N., et al. 2000. *Special Report on Emissions Scenarios: A Special Report of Working Group III of the Intergovernmental Panel on Climate Change.* New York: Cambridge University Press.

Palutikof, J. P., et al. 1997. "The Simulation of Daily Temperature Series from GCM Output. Part 1: Comparison of Model Data with Observations." *Journal of Climate* 10: 2497–2513.

Raible, C. C., et al. 2006. "Climate Variability— Observations, Reconstructions, and Model Simulations for the Atlantic-European and Alpine Region from 1500–2100 AD." *Climatic Change Climate Variability, Predictability and Climate Risks: A European Perspective* 79: 9–29.

Räisänen, J. 2002. "CO_2-Induced Changes in Interannual Temperature and Precipitation Variability in 19 CMIP2 Experiments." *Journal of Climate* 15: 2395–2411.

Roeckner, E., et al. 1996. "ENSO Variability and Atmospheric Response in a Global Coupled Atmosphere-Ocean GCM." *Climate Dynamics* 12: 737–754.

Rowell, D. P. 2005. "A Scenario of European Climate Change for the Late Twenty-First Century: Seasonal Means and Interannual Variability." *Climate Dynamics* 25: 837–849.

Santos, J., and J. Corte-Real. 2006. "Temperature Extremes in Europe and Wintertime Large-Scale Atmospheric Circulation: HadCM3 Future Scenarios." *Climate Research* 31: 3–18.

Scherrer, S. C., et al. 2005. "European Temperature Distribution Changes in Observations and Climate Change Scenarios." *Geophysical Research Letters* 32: 1–5.

Weisheimer, A., and T. N. Palmer. 2005. "Changing Frequency of Occurrence of Extreme Seasonal Temperatures under Global Warming." *Geophysical Research Letters* 32, doi:10.1029/2005GL023365.

Wigley, T. M. L. 1988. "The Effect of Changing Climate on the Frequency of Absolute Extreme Events." *Climate Monitor* 17: 44–55.

Winkler, J. A. 2009. "Philosophy, Development, Application, and Communication of Future Climate Scenarios for the Pileus Project." *Proceedings of the International Symposium on Climate Change in the Great Lakes Region: Decision Making under Uncertainty,* ed. Thomas Dietz and David Bidwell. East Lansing: Michigan State University Press.

Winkler, J. A., et al. 1997. "The Simulation of Daily Time Series from GCM Output. Part 2: A Sensitivity Analysis of Empirical Transfer Functions for Downscaling GCM Simulations." *Journal of Climate* 10: 2514–2532.

Winkler, J. A., et al. 2000. "Climate Change and Fruit Production: An Exercise in Downscaling." In *Preparing for a Changing Climate: Great Lakes Overview.* A Summary of the Great Lakes Regional Assessment Group for the U.S. Global Change Research Program, pp. 19–24. [Ed. P. Sousounis and J. Bisanz.] Ann Arbor, Mich.: Great Lakes Regional Assessment, University of Michigan, Atmospheric, Oceanic and Space Sciences Department. Also available at: http://www.geo.msu.edu/glra/assessment/assessment.html.

Winkler, J. A., et al. 2002. "Possible Impacts of Projected Climate Change on Specialized Agriculture in the Great Lakes Region." *Journal of Great Lakes Research* 28: 608–625.

7. Wisconsin's Changing Climate: *Temperature*

D. J. LORENZ, S. J. VAVRUS, D. J. VIMONT, J. W. WILLIAMS,
M. NOTARO, J. A. YOUNG, E. T. DeWEAVER, AND E. J. HOPKINS

Introduction

This chapter provides a case study of changes in temperature for the state of Wisconsin using both the observed instrumental record over the past one hundred years and output from coupled Atmosphere-Ocean General Circulation Models (AOGCMs) through the end of the twenty-first century. While, as described in the previous chapters, there are spatial variations across the United States and the Midwest in terms of local manifestations of global climate change, many of the results presented here are also relevant for other states in the Midwestern USA and even the United States as a whole (figure 7.1).

Historical Variability and Trends

OBSERVED TEMPERATURE TRENDS

Annual mean temperature for the last one hundred years from the National Climatic Data Center (NCDC) divisional data indicate large inter-annual variability in the Wisconsin mean temperature, although long-term trends and patterns are still clear (figure 7.1). The period centered on the 1930s and the last few decades are the warmest in

the entire record, while the period before 1920 and the 1960s and '70s were cooler than average. The 1930s' warmth corresponds to the Dust Bowl drought, and the warm temperatures in this period might in part be caused by the reduction in evaporative cooling over the dry soils of this time period.

Superimposed on the Wisconsin mean temperature are spatial means over progressively larger areas (figure 7.1). The mean temperature of the East-North-Central United States and the continental United States correlate well with the mean temperature of Wisconsin. The correlations are as follows:

- Wisconsin and East-North-Central United States = 0.985
- Wisconsin and the Continental United States = 0.789
- Wisconsin and the Northern Hemisphere = 0.442
- Wisconsin and the Globe= 0.4092

As the size of the averaging area increases to the Northern Hemisphere and the global scale, the year-to-year variability in temperature decreases, so that the background trends become more evident. This area averaging reduces the year-to-year variability be-

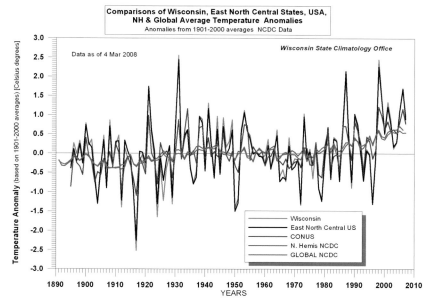

Figure 7.1. Time series of annual-mean temperature anomalies in Celsius averaged over different spatial domains: Wisconsin (red line), East North Central States (Iowa, Michigan, Minnesota, and Wisconsin, black line), continental United States (CONUS) (blue line), Northern Hemisphere (green line), and global (purple line). Data from the National Climatic Data Center (NCDC).

cause it preferentially filters out the natural variability, which tends to vary on smaller spatial scales than the externally forced greenhouse gas signal. As the averaging area increases some of the year-to-year variability in temperature differs from that of Wisconsin. Nevertheless, the long-term trends in temperature are quite similar among all averaging areas, especially the increasing trend over the last thirty years, which has been attributed at least in part to increasing greenhouse gases (IPCC 2007).

Trends in Wisconsin temperature over the last thirty years are not consistent across seasons. Indeed, linear least squares trends in winter are significantly larger than in any other season (figure 7.2). Looking in more detail at the calculated linear least squares trends for the last 112 years, we see that the trend in spring temperature in Wisconsin is about half that of the trend winter temperature (figure 7.2). The trends in summer

and fall, on the other hand, are close to zero. Like Wisconsin, all the other states in the Midwest have the largest positive trends in winter and the second largest positive trends in spring. The trends in summer and fall are more varied among the states. The summer trends are slightly positive in the northwest part of the region and slightly negative elsewhere (see chapter 3 of this volume for a detailed exploration of the negative trends in summer maximum daily temperatures in states such as Missouri).

OBSERVED TRENDS IN TEMPERATURE-RELATED CLIMATE INDICATORS

The seasonality of temperature trends results in large changes in winter-related climate indicators. For example, the duration of ice on Lake Mendota, Wisconsin, has fallen dramatically over the 154-year period of record (figure 7.3). The longest duration was 161

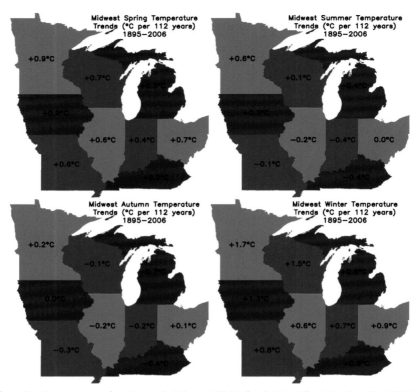

Figure 7.2. Seasonal trends in air temperature for each state in the Midwestern USA. Data from the National Climatic Data Center Climate Division Dataset.

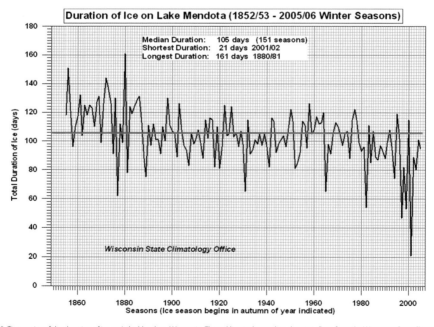

Figure 7.3. Time series of the duration of ice on Lake Mendota, Wisconsin. The red line is the median duration. Data from the Wisconsin State Climatology Office.

days during the winter of 1880/81, while the shortest duration was only 21 days during the winter of 2001/02! From 1980/81 to 2005/06 only five years have been above the median lake ice duration (19%). The changes in mean winter temperatures are also reflected in the extreme temperatures. For example, since the late 1970s, when the strong winter temperature trends began, there have been substantial reductions in the number of days with minimum temperature below 0°F at Green Bay, Wisconsin (figure 7.4).

Trends in extreme winter temperatures are also evident in plant hardiness zones (figure 7.5). The zones correspond to the average annual minimum temperature in degrees Fahrenheit in increments of 10 degrees (Cathey 1990) (see figure 7.5 caption). Over the Midwest plant hardiness zones have shifted northward by over half the width of a hardiness zone in sixteen years.

In contrast to changes in mean winter temperature, the trends in mean summer temperatures in Wisconsin are close to zero. Hence, unlike the winter extremes described above, one might expect little change in the frequency or magnitude of summer extremes. Looking at trends in extremely hot days, however, we do see changes toward fewer extremely hot days. Since the 1940s and '50s there have been substantial decreases in days above 90°F (32.2°C) in Madison, Wisconsin (figure 7.6). Such decreases in summer temperature in response to greenhouse gases have been seen in some regional climate model simulations associated with increases in low-level jet frequency and associated increases in soil moisture (Pan et al. 2004). It is expected that the warming component of the greenhouse gas response will eventually overwhelm the initial summer cooling after a few decades (IPCC 2007).

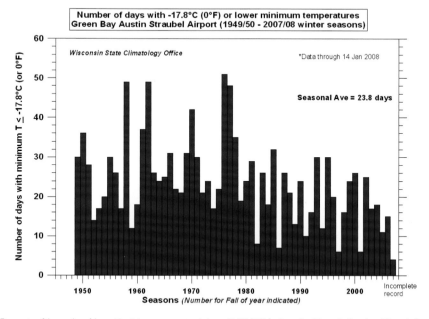

Figure 7.4. Time series of the number of days with minimum temperature below −17.8°C (0°F) for Green Bay, Wisconsin. Data from Wisconsin State Climatology Office.

Differences between 1990 USDA hardiness zones and 2006 arborday.org hardiness zones reflect warmer climate

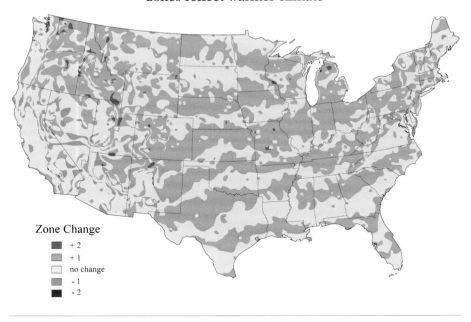

Zone Change

- ■ + 2
- ▨ + 1
- ☐ no change
- ▨ - 1
- ■ - 2

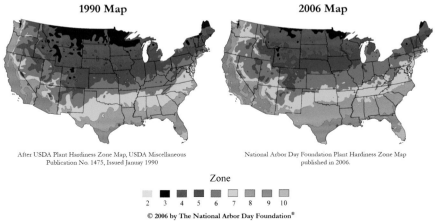

1990 Map	2006 Map

After USDA Plant Hardiness Zone Map, USDA Miscellaneous
Publication No. 1475, Issued January 1990

National Arbor Day Foundation Plant Hardiness Zone Map
published in 2006.

Zone

2 3 4 5 6 7 8 9 10

© 2006 by The National Arbor Day Foundation®

Figure 7.5. Change in plant hardiness zones between 1990 and 2006 (top), the 1990 plant hardiness zones (bottom left), and the 2006 plant hardiness zones (bottom right). The zones corresponds to average annual minimum temperature: zone 2 = −40 to −50°F, zone 3 = −30 to −40°F, zone 4 = −20 to −30°F, zone 5 = −10 to −20°F, zone 6 = 0 to −10°F, zone 7 = 10 to 0°F, zone 8 = 20 to 10°F, zone 9 = 30 to 20°F, zone 10 = 40 to 30°F. Figure courtesy of the National Arbor Day Foundation.

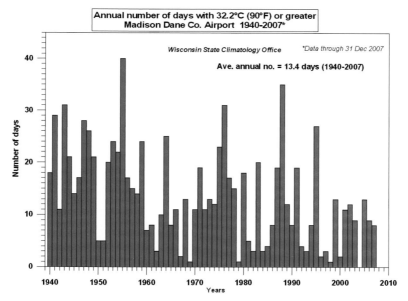

Figure 7.6. Time series of the number of days with maximum temperature above 32.2°C (90°F) for Madison, Wisconsin. Data from the Wisconsin State Climatology Office.

Climate Model Projections for Twenty-First Century

TEMPERATURE

In developing future projections of temperature, monthly mean output from fifteen AOGCMs associated with the IPCC Fourth Assessment Report (IPCC 2007) was first examined in terms of simulation of the present-day annual cycle of Wisconsin mean temperature (figure 7.7a). Overall the AOGCMs tend to be biased on average 2–3°C too warm during the warm half of the annual cycle; although the observed temperature always falls within the envelope of AOGCM simulations, the AOGCM spread in temperature at any given time of year is quite large (as much as 10°C in some seasons).

The ensemble mean projected change in temperature (i.e., the average over the fifteen AOGCMs) between years 2080–2099 and 1980–1999 from the A2 emission scenario (IPCC 2001) is about 4 to 5°C (thick black line in figure 7.7b) and is relatively independent of season. The individual model projections, however, vary significantly from the ensemble mean change and from season to season. For example, two AOGCMs (GFDL-CM2.0 and GFDL-CM2.1) have a very large temperature increase in summer (up to 10°C) while in the remaining months these same two models have a temperature increase of about 4 to 5°C. Some possible reasons for the large spread in summertime climate projections are discussed in chapter 12 of this volume.

It is reasonable to question what is the source of the large variability in temperature response. From a global mean perspective, model projections of cloud cover, amount, and type are the largest source of uncertainty in temperature response to

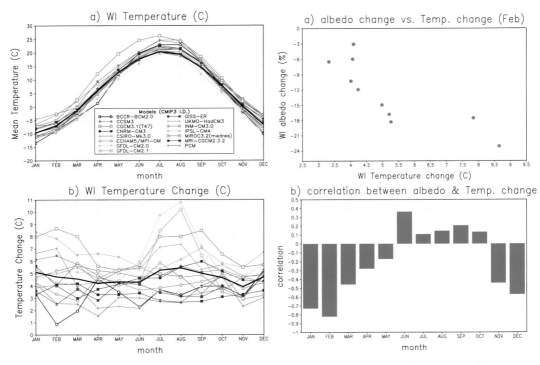

Figure 7.7. a) Wisconsin mean temperature for fifteen IPCC AR4 climate models (thin colored lines) and observations (thick black line). b) Wisconsin mean temperature change for fifteen IPCC AR4 climate models (thin colored lines). Temperature change is defined as the difference between years 2080–2099 and 1980–1999. The ensemble mean change is the thick black line. The observational data are from Maurer et al. (2002).

Figure 7.8. a) Scatter plot of Wisconsin albedo change versus Wisconsin temperature change for ten IPCC AR4 climate models. b) Correlation between Wisconsin albedo change and Wisconsin temperature change as a function of month. Change is defined as the difference between the years 2080–2099 and 1980–1999.

global warming (Soden and Held 2006). However, clouds do not necessarily dominate the spread in temperature response at any particular location. In polar regions, the snow/albedo feedback dominates the spread in temperature response (Qu and Hall 2006). It works as follows: Increasing temperatures lead to reduced snow or ice cover, which then leads to even higher temperatures due to reduction in surface albedo, or reflectivity, from the melting snow or ice. The higher temperatures cause more snow and ice to melt, which amplifies the initial change. If the sensitivity of surface albedo to temperature varies among AOGCMs,

then the snow/albedo feedback can potentially be an important source of variability in temperature response. Figure 7.8 provides evidence that the snow/albedo feedback might be an important contributor to the spread in Wisconsin temperature change. In February, AOGCMs with larger temperature change also have larger decreases in surface albedo. The correlation between temperature change and albedo change varies strongly through the annual cycle. As expected from the snow/albedo feedback, the correlation is strongest in winter and not statistically different to zero in summer (figure 7.8b).

PROJECTED CHANGES IN EXTREME TEMPERATURES

We assessed changes in the frequency of extreme temperature events (heat waves and cold waves) using output of seven AOGCMs included in the World Climate Research Programme's Coupled Model Intercomparison Project phase 3 (CMIP3): CCSM3, GISS-AOM, GFDL CM2.0, IPSL CM4, MIROC3.2, MRI CGCM2.3.2a, and MPI-ECHAM5. We calculated the number of extremely cold days during winter (Dec.–Feb.) and extremely hot days during summer (Jun.–Aug.) for the grid point in each model closest to Madison, Wisconsin (43°N, 89°W), for the late twentieth century (1980–1999) and late twenty-first century (2080–2099). The modern simulation (1980–1999) was forced with observed concentrations of greenhouse gases, while the future run was driven by enhanced greenhouse gas concentrations based on the SRES A1B emission scenario. An extreme temperature day is defined here as one on which the daily mean temperature exceeds by at least two standard deviations the simulated seasonal mean temperature of the late twentieth century. All AOGCMs suggest that future warming will cause an increase (decrease) in the number of extremely hot (cold) days, but the changes are asymmetric. A much larger change is projected for the frequency of extreme heat events than extreme cold: thirteen more extremely hot days and three fewer extremely cold days per year (inter-model mean) (figure 7.9). Although this suggests that the *net* effect is for more frequent temperature extremes, the percent drop in extreme cold events is large (more than 90%) but the absolute decrease is rela-

tively low because there are so few extremely cold days in the late-twentieth-century simulation. Thus, there is a greater potential in the future for more extremely hot days, whereas the reduction in the frequency of extremely cold days is limited by zero. The cause of the large inter-model variation (ranging from a two- to forty-day increase) is uncertain, although some of this difference may be attributable to how the models simulate heat waves in the modern climate. While some of the climate models produce a realistic frequency of extreme heat events, others simulate an excessive number of hot days in their twentieth-century runs. The CCSM3 model, for example, produces several days with daily *mean* temperature over 100°F in Madison during the late twentieth century, a bias that could limit the rise in extremely hot days simulated in a warmer climate and might explain why this model projects the smallest future increase in extreme heat among the models presented here.

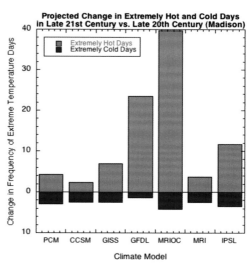

Figure 7.9. Changes in the frequency of extreme temperature days (heat waves and cold waves) assessed from the output of seven IPCC AR4 climate models. See text for the definition of an extreme temperature day.

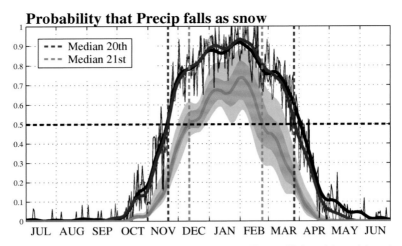

Probability that Precip falls as snow

Figure 7.10. Probability that precipitation falls as snow instead of rain for Madison, Wisconsin. Shown are (black curve) the actual observed probability based on 1950–2002 data from Madison; (blue curve) estimated probability based on precipitation occurring (rain OR snow) and minimum temperature below 30F; (red curve) estimated probability based on adding the 2080–2099 mean temperature change to the temperature used for the blue curve. Thick curves denote smoothed estimates. The shaded light red (darker red) regions indicate estimates based on two (one) standard deviations of the projected temperature change. Vertical dashed lines highlight the present day (blue dashed) and future (red dashed) 50% days, defined as the days where, given a precipitation event, there is a 50% chance that it is snow (and equivalently a 50% chance that it is rain).

PROJECTED CHANGES IN PROBABILITY OF SNOW VERSUS RAIN

Increases in temperature due to global warming have a direct effect on the type of precipitation in winter. To estimate the changes in the probability of snow versus rain at the end of the twenty-first century, however, we do not use the direct output of AOGCMs. The reason is that although AOGCMs represent the main large-scale features of the climate reasonably well, their performance in reproducing regional climatic details is rather poor (Giorgi 2005). This is particularly true for variables such as precipitation. Therefore, we estimate the changes in probability of snow versus rain based solely on the change in temperature from AOGCMs, using a threshold minimum temperature of 30°F to define when precipitation is "snow." As seen in figure 7.10, using this temperature threshold to distinguish rain from snow provides a good fit to the actual probability of snow versus rain. For the A2 emission scenario the projected probability of snow (from the fifteen AOGCMs as shown in figure 7.7) decreases by over 20% in midwinter in Madison, Wisconsin. Moreover, the duration of the snow season (defined as the period where snow is more likely than rain) decreases by over a month and a half.

PROJECTED CHANGES IN WISCONSIN CLIMATE DERIVED USING ANALOGUES

Using output of nine AOGCMs from the IPCC 4AR (CCSM3, CSIRO-Mk3.0, ECHAM5/MPI-OM, GFDL-CM2.1, GISS-ER, IPSL-CM4, MRI-CGCM2.3.2, PCM, UKMO-HadCM3) for two IPCC SRES (A2 and B1), we consider winter and summer temperature and precipitation statistics to identify present-day geographical analogs that most resemble Wisconsin's climate

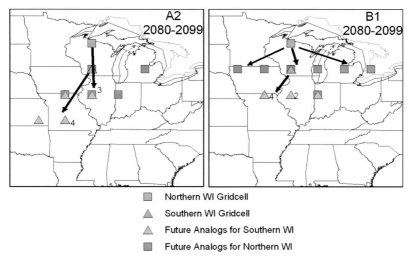

Figure 7.11. Map showing analogs in present climate that most resemble northern Wisconsin (squares) or southern Wisconsin (triangles) at the end of the twenty-first century. The colored symbols show locations of possible present-day analogs (each analog is based on a different AOGCM) for future climate at the corresponding gray symbol. The number by some of the symbols means that multiple models obtained the best analog at this grid point, and the arrows correspond to either the most likely analog or multiple analogs if no particular analog dominates. The left plot is for the A2 emission scenario and the right plot is for the B1 emission scenario.

at the end of the twenty-first century. To identify twentieth-century matches to the twenty-first-century climate projections for Wisconsin, we first calculated standardized Euclidean distances (SEDs) between the late-twentieth-century (1980–1999) mean climates for the two grid cells overlaying Wisconsin (at T42 resolution) and the global set of late-twenty-first-century (2080–2099) mean climates for terrestrial grid cells. The SED index (see Williams, Jackson, and Kutzbach 2007) integrates changes in mean June–August (JJA) temperature, mean December–February (DJF) temperature, mean JJA precipitation, and mean DJF precipitation, and the twenty-first-century changes in all four variables are standardized against inter-annual climate variance for 2080–2099.

Increases in temperature were found to bring about a southward shift in Wisconsin's climate, while increases (decreases) in precipitation were found to bring about an eastward (westward) shift in Wisconsin's climate.

Under the A2 emission scenario, northern Wisconsin's climate will most likely become like that of Illinois, although smaller southward shifts corresponding to southern Wisconsin or Michigan are also possible. Such a southward shift in northern Wisconsin's climate may lead to the reduction or elimination of Wisconsin's Boreal Forest and Northern Mixed Forest communities, which now dominate this region. Under the A2 emission scenario, southern Wisconsin's climate will most likely become like that of southern Missouri, although smaller southward shifts corresponding to Illinois are also possible.

IMPLICATIONS OF POSSIBLE CHANGES IN WISCONSIN'S CLIMATE

Twenty-first-century climate projections from seventeen AOGCMs in the IPCC Fourth Assessment, under scenario A2, were used to drive an offline biosphere simulation model, LPJ-DGVM (Lund-Potsdam-Jena

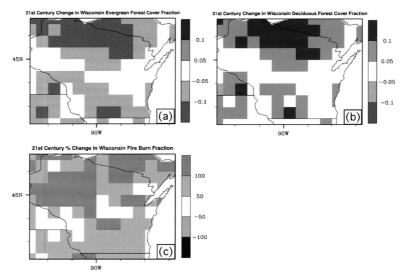

Figure 7.12. Projected mean change in Wisconsin's (a) evergreen forest cover fraction and (b) deciduous forest cover fraction during the twenty-first century, based on LPJ-DGVM (Lund-Potsdam-Jena Dynamic Global Vegetation Model) driven by the climate and carbon dioxide projections of seventeen CMIP3 climate models under A2 scenario. (c) Projected mean percent change in Wisconsin's natural fire burn fraction for the twenty-first century.

Dynamic Global Vegetation Model) (Sitch et al., 2003). From the output of these biosphere simulations, two very distinct patterns emerge over Wisconsin, related to forest distributions and natural fires. The extended growing season leads to a significant replacement of evergreen forests over northern Wisconsin with deciduous forests (figure 7.12). Also, the increase in atmospheric carbon dioxide concentration leads to a build-up of vegetation litter, which serves as fuel for fires. This, combined with more frequent summer droughts, leads to a substantial percent increase (over 50–100%) during the twenty-first century in the area burnt by natural fires.

Concluding Remarks

In summary, we find that over the state of Wisconsin:

1. The observed temperature changes in the historical record are largest in winter and nearly zero in summer.

2. The seasonality of historical temperature trends is associated with changes in winter-related indicators such as lake ice and plant hardiness zones, and little or even opposite sign trends in summer climate indicators.

3. At the end of the next century, ensemble mean AOGCM warming in Wisconsin is about 5°C (A2 emission scenario) and is basically independent of season.

4. There is substantial model-to-model variability, with some AOGCMs predicting nearly 11°C warming in the summer.

5. There is considerable model-to-model variability in the prediction of extreme summertime temperatures: some models predict forty more extremely hot days per summer than present, while other models predict less than two more extremely hot days per summer.

6. In Madison, Wisconsin, the probability of snow in mid-winter drops by over 20% by the end of the century (A2 scenario).

7. Under the A2 emission scenario, the climate of Wisconsin becomes like the climate of present-day Missouri and Illinois.

REFERENCES

Cathey, H. M. 1990. "USDA Plant Hardiness Zone Map." Washington, D.C.: U.S. National Arboretum, Agricultural Research Services, U.S. Dept. of Agriculture. 2002, Misc. Publ. No. 1475. Available at: http://www.usna.usda.gov/Hardzone.

Giorgi, F. 2005. "Climate Change Prediction." *Climate Change* 73: 239–265.

IPCC. 2001. *Climate Change 2001: The Scientific Basis.* Contribution of Working Group I to the Third Assessment Report of the Intergovernmental Panel on Climate Change. [Ed. J. T. Houghton, et al.] New York: Cambridge University Press.

IPCC. 2007. *Climate Change 2007: The Physical Science Basis.* Contribution of Working Group I to the Fourth Assessment Report of the Intergovernmental Panel on Climate Change. [Ed. S. Solomon, et al.] New York: Cambridge University Press.

Maurer, E. P., et al. 2002. "A Long-Term Hydrologically Based Dataset of Land Surface Fluxes and States for the Conterminous United States." *Journal of Climate* 15: 3237–3251.

Pan, Z. T., et al. 2004. "Altered Hydrologic Feedback in a Warming Climate Introduces a 'Warming Hole.'" *Geophysical Research Letters* 31, doi:10.1029/2004GL02528.

Qu, X., and A. Hall. 2006. "Assessing Snow Albedo Feedback in Simulated Climate Change." *Journal of Climate* 19: 2617–2630.

Sitch, S., et al. 2003. "Evaluation of Ecosystem Dynamics, Plant Geography and Terrestrial Carbon Cycling in the LPJ Dynamic Global Vegetation Model." *Global Change Biology* 9: 161–185.

Soden, B. J., and I. M. Held. 2006. "An Assessment of Climate Feedbacks in Coupled Ocean-Atmosphere Models." *Journal of Climate* 19: 3354–3360.

Williams, J. W., S. T. Jackson, and J. E. Kutzbach. 2007. "Projected Distributions of Novel and Disappearing Climates by 2100 AD." *Proceedings of the National Academy of Science* 104: 5738–5742.

8. Overview: *Hydrologic Regimes*

Z. PAN AND S. C. PRYOR

Introduction

As discussed in multiple chapters of this volume, water availability is a key component of multiple economic sectors, including, but certainly not limited to, electricity generation and agricultural activities. Great Lakes water-level fluctuations are an important factor in both the economy and the environment of the upper Midwest. Significant drops in water level during the late 1990s affected both the tourism and shipping industries by isolating shallow inlets, bays, and marinas, cutting off access to ports for commercial and recreational activity, and forcing local communities to undertake expensive dredging operations. However, low lake levels also permit the reestablishment of diverse wetlands and expand fish spawning grounds. In contrast, abnormally high lake levels pose a risk to sensitive shoreline environments by increasing their susceptibility to erosion from higher wave energy during strong storms (Meadows et al. 1997). The impact of climate change on water levels in the Great Lakes is uncertain. Impact assessments show changes in water levels of between –1.38 m and +0.35 m by the end of the twenty-first century (Lofgren et al. 2002; Schwartz et al. 2004).

In this chapter we use the agricultural sector as an example of the key role that hydrology plays in the economy of the Midwestern USA. Agricultural crops contribute annually more than $110 billion to the U.S. economy, a large fraction of which originates in the intensely cultivated Midwest (USGCRP 2001; Rosenzweig et al. 2000). Sixty million acres of corn and soybeans planted in the Midwest represent 60% of all U.S. cultivated cropland (USDA 2004). The combination of favorable climate and fertile soil makes the Midwest one of the most productive agricultural areas in the world.

Inter-annual variability of the climate (specifically precipitation and temperature) in the Midwestern USA is modulated by a number of low-frequency atmosphere-ocean oscillations (e.g., El Niño, the North Atlantic and Arctic oscillations, and the Pacific-North American teleconnection, see chapter 14 of this volume) which in turn influence agricultural yields. The hydrological cycle—its inter-annual variability and possible evolution—thus plays an essential role in the regional and national economy.

Overview of the Hydrological Cycle

While complicated numerical models can simulate more realistically conditions of the climate system and changes therein, simple analytical solutions can illustrate causal relations more clearly under some idealized cases. Before discussing detailed model simulations, we first introduce some simple functional relations among climate and hydrological variables (Held and Soden 2006). Thus in this section we conduct a "thought experiment" that highlights some of the scaling relationships within the hydrological cycle and how they might influence a range of parameters such as: water availability for precipitation (precipitable water, i.e., the total amount of water present in a column of the atmosphere), soil water content, and the atmospheric water vapor content, expressed as a water vapor pressure (e) or water vapor mixing ratio (q). Let's start with the Clausius-Clapeyron (C-C) equation that describes saturation vapor pressure (e_s) changes as a function of air temperature (T):

$$\frac{d \ln e_s}{dT} = -\frac{L}{R_v T^2} \equiv \alpha(T) \qquad (8.1)$$

where L is latent heat of vaporization and R_v is the gas constant for water vapor.

This equation states that saturation vapor pressure increases exponentially with temperature. For a temperature of 288K, $\alpha \cong 0.07°C^{-1}$, implying that the relative e_s increase ($\frac{de_s}{e_s}$) is about 7% for 1°C warming of the air (i.e., the water-holding capacity of the air increases by 7% for a dT=1°C).

Both observations and model simulations indicate that relative humidity (ratio of actual vapor pressure to saturation vapor pressure, i.e., e/e_s) remains fairly constant despite glob-

al warming. Assuming this to be the case, a projected 3°C warming in the entire lower troposphere, where most moisture resides, would lead to a 21% increase in total precipitable water (TPW). Support for the validity of this assumption is derived from observations that columnar atmospheric moisture content (i.e., TPW) does indeed follow the Clausius-Clapeyron scaling reasonably well on a global scale (Held and Soden 2006). More precipitable water in the atmosphere can lead to larger amounts of precipitation. By ignoring small compensating flow in moisture transport, it can be shown that the change in global mean precipitation (dP) associated with a change in temperature and hence atmospheric moisture content is given by:

$$dP=Mdq$$

where M is the mass exchanged per unit time, and q is water vapor mixing ratio (Held and Soden 2006).

Observed precipitation change (dP) does not scale directly with C-C, and only increased by about 2% per °C warming over the historical period. The larger increase in atmospheric vapor than precipitation indicates longer water vapor residence time under warmer climates. Naturally, this "thought exercise" neglects many subtleties of the atmospheric water vapor circulation, but in accord with these inferences hydrological intensity (commonly measured by precipitation minus evaporation) increases under a warmer climate (Roads, Kanamitsu, and Stewart 2002; Bosilovich and Chern 2006).

Daily precipitation accumulations in a range of climate change simulations (using emission scenarios B1, A1B, and A2) conducted with the IPCC AR4 AOGCMs (Intergovernmental Panel on Climate Change

Fourth Assessment Report coupled Atmosphere-Ocean General Circulation Models) consistently show a shift toward more intense and extreme precipitation for the globe as a whole over the twenty-first century (Meehl et al. 2007). Expressed as a percentage of the mean simulated change for 1980 to 1999 (2.83 mm day^{-1}), the rate varies from about 1.4 % °C^{-1} in A2 emission scenario to 2.3 % °C^{-1} in the constant composition commitment experiment (Meehl et al. 2007). At the global scale, the multi-model averaged percentage increase in annual precipitation amount is 2.0% per °C and is larger than decrease in precipitation frequency (of −0.7% per °C) (Sun et al. 2007). This implies an increase in the intensity of precipitation events at the global scale and in some regions of the globe. It may also result in a "tendency for drying of the mid-continental areas during summer, indicating a greater risk of droughts in those regions" (Meehl et al. 2007).

In the following sections we discuss some of the regional-scale manifestations of possible climate states, although it is important to recall that the climate projections for changes

in precipitation regimes are considerably less certain than those for temperature, particularly at the regional scale (Meehl et al. 2007).

Observed Changes in Hydrological Components

PRECIPITATION

Precipitation exhibits greater spatial-temporal variability than temperature, but some long-term trends have been identified. For example, global annual precipitation over land increased by 2.4 mm per decade during 1900–1988 (i.e., an increase of more than 2%) (Dai, Fung, and Del Genio 1997). At a regional scale, the annual total precipitation exhibits a strong southeast to northwest gradient across the Midwestern USA, with annual accumulations exceeding 1200 mm in Kentucky, while those in Minnesota are below 700 mm (see chapter 9 of this volume). Peak precipitation accumulation typically occurs in spring or early summer, when crops need rainfall most. As discussed in chapter 9, trend detection in precipitation

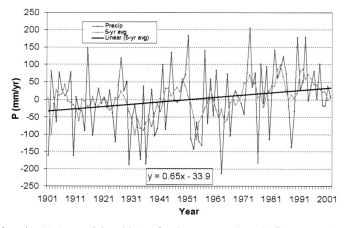

Figure 8.1. Time series of annual precipitation anomaly (annual deviation from the mean computed using data from 1901–2002) averaged over region 35–45°N and 103–83°W, showing the long-term trend (deviation from the century mean) over the twentieth century. Data were obtained from the Climate Research Unit (CRU) of the University of East Anglia (Mitchell and Jones 2005).

time series is very challenging, and hence data from a large proportion of stations examined exhibit no statistically significant trend. However, generally stations in the Midwest that exhibit significant trends (at the 90% confidence level) in annual total precipitation indicate increasing trends of approximately the same fractional magnitude as those reported by Dai, Fung, and Del Genio (1997) and Folland et al. (2001). These trends are, however, of smaller magnitude than those for the central U.S. (figure 8.1).

SOIL MOISTURE

Land surface heterogeneity and its effects on surface exchange processes present a serious challenge to both observationalists and climate modelers who need to reconcile and parameterize the fine spatial scales at which land surface heterogeneity is significant (1 km and finer) and the coarser scales at which most climate simulations are generated (tens to hundreds of kilometers). These influences include, but are not limited to, partitioning incident precipitation between evapo-transpiration, runoff and infiltration/percolation, partitioning of precipitation between rainfall and snow, plus accuracy in simulating annual retreats of snow and ice cover. Soil moisture is one of the most uncertain components of the climate system, in part because it exhibits spatial variability at multiple scales and is not routinely observed on large scales. However, it plays a critical role in surface energy partitioning and influences the structure of the atmospheric boundary layer (Reen et al. 2006).

In the absence of robust observations, our discussion focuses on the direct driver of soil moisture measured by the difference between precipitation and evapo-transpiration (P-E). Following the Clausius-Clapeyron scaling procedure used for atmospheric water vapor and precipitation, we can have:

$$d(P - E) = \alpha dT (P - E) \qquad (8.3)$$

The above equation suggests that if P-E is already positive it will increase, whereas if P-E is negative it will become more negative, exacerbating extremes (Held and Soden 2006). During the cold season (October–April) in the Midwestern USA, current P-E > 0, and thus it would become more positive. However, during the warm season (May–September) when P-E < 0 in some parts of the Midwestern USA, warming may cause a further depletion of soil moisture. This simplified analytical solution is in agreement with most AOGCM projections of drying soil in the region during the summer.

Of course in the real climate system, the factors that dictate soil moisture (e.g., transpiration from vegetation) are much more complicated. The rise of carbon dioxide (CO_2) concentrations in the atmosphere has two potential competing effects on transpiration, and hence water balance and soil moisture (Betts et al. 1997). First, higher CO_2 concentrations can lead to reduced evaporation, as the stomata through which transpiration from plants takes place conduct less water. Second, higher CO_2 concentrations can lead to increased plant growth and thus more leaf area, and hence greater total transpiration from the same ground area. The relative magnitudes of these two effects, however, vary between plant types and also depend on other influences such as the availability of nutrients and the effects of changes in temperature and water availability (Kundzewicz et al. 2007). These competing responses of transpiration to raised CO_2 concentrations in the atmosphere under a warmer climate may partly explain some of the inconsistencies in soil moisture trends among different global and regional climate models, which have varying levels of sophistication in treating vegetation evapo-transpiration.

STREAMFLOW

Streamflow varies seasonally and regionally, depending on the precipitation regime, evapo-transpiration rate, and snowmelt. Temporal lags between peaks in precipitation receipt and streamflow (discharge) vary from long delays in high latitudes and mountainous regions to short delays in relative warm regimes (Dettinger and Diaz 2000). Over North America, streamflow is most variable from year to year in dry regions of the Southwest United States and Mexico, and it varies more (relatively) than precipitation in the same regions.

Streamflow has been increasing in the United States since at least 1940 (USGS 2005), although the increases are not uniform across the range of annual streamflow, geographically or seasonally. Between 1940 and 1999, 40% of tested stations experienced an increase in the annual minimum streamflow, 43% in the annual median streamflow, but only 10% in the annual maximum streamflow (Lins and Slack 1999; Slack and Landwehr 1992; USGS 2005). In contrast, only 8% of the stations had decreases in the annual minimum flow, less than 1% in the annual median flow, and 3% in the annual maximum flow (Lins and Slack 1999). In summary, low-to-moderate streamflow volumes have been increasing at many loca-

tions, but high streamflow volumes have increased at relatively few locations. Embedded in these changes are a number of time scales of variability. Observed streamflow anomalies shifted from generally negative to positive in early 1970s (Lins and Slack 1999). This shift could well be related to inter-annual and decadal climatic modes and specifically to the phase swing of the PDO (Pacific Decadal Oscillation) that occurred around 1976 (Miller et al. 1994; Hu and Feng 2001).

An important signal in streamflow records over the western United States is a shift toward earlier occurrence of peak spring flows by one to four weeks (Stewart, Cayan, and Dettinger 2005). Both snowmelt- and non-snowmelt-dominated streams exhibited earlier streamflow timing, although non-snowmelt streams in the far western United States tend to display later timing. The earlier timing of snowmelt-derived streamflow in the western United States and western Canada was most strongly connected with warmer winter and spring temperatures.

Over the eastern United States, precipitation and streamflow are more evenly distributed throughout the water year. In this part of the country streamflow data from 1971 to 2000 exhibit increased annual minimum and median daily streamflow, but little change the peak discharge (McCabe and Wolock 2002).

Projected Hydrological Regimes in Scenario Climates

PRECIPITATION

As discussed in chapter 1 of this volume, the IPCC 4AR did not present a specific analysis of conditions over the Midwestern USA. However, one of the focal regions of their analysis was central North America (CNA, Christensen et al. 2007). In this region, the projected precipitation by the IPCC AR4 AOGCMs show a winter increase to the north and summer decrease to the south, with the zero-change line migrating from the northern-central United States in winter up to the southern Canada in summer (Christensen et al. 2007). The summer decrease is believed to be related to enhanced subsidence in the southwest United States and northern Mexico associated with amplification of the subtropical high (Mote and Mantua 2002). With northward displacement of the westerlies and intensification of the Aleutian low, annual precipitation in northern North America is projected to increase by 20%, with the largest increase (30%) in winter. The model ensemble median change over the twenty-first century in the central U.S. is 3%, with a maximum (minimum) projected annual change of 18% (−16%). The model consensus is greatest in spring, with a median projected change of +7% (table 8.1). For this season, the middle half (25–75% percentile) of the twenty-one models projected the same sign of precipitation trend.

Under the general influence of the subtropical high, the southwestern United States is very arid. Further drying would result in a deeper atmospheric boundary layer (Paegle, Mo, and Nogues-Paegle 1996; Blackadar 1957), which in turn supports a stronger low-level jet (LLJ) over the Great Plains and thus greater moisture convergence and precipitation in the Midwestern USA. During spring and summer months the LLJ over the Great Plains plays a key role in the atmospheric moisture budget of the region. A number of studies have examined the connection between the LLJ and rainfall in the region

Table 8.1. Precipitation change (%) from 1988–99 to 2088–99 averaged over north-central America, defined as the region bounded by 103–85°W and 20–50°N (Christensen et al. 2007). The values are ensemble means of the IPCC AR4 AOGCMs. The shading in MAM indicates that the signs of middle half (25–75 percentile) models are the same among the twenty-one AOGCMs.

	Min	Median	Max
DJF	−18	5	14
MAM	−17	7	17
JJA	−31	−3	20
SON	−17	4	24
Annual	−16	3	18

(Wallace 1975, Helfand and Schubert 1995). They found almost one-third of the moisture that enters the continental United States is transported by the LLJ (see chapter 3 of this volume).

SOIL MOISTURE AS VIEWED FROM P-E

In accord with earlier work (Manabe and Wetherald 1987), the majority of IPCC 4AR AOGCMs project a decrease in P-E in the twenty-first century, suggesting depletion of soil moisture. For example, "CNA, 15% of the summers in 2080 to 2099 in the A1B scenario are projected to be extremely dry, corresponding to a factor of three increase in the frequency of these events" (Christensen et al. 2007). However, the uncertainty associated with the change is large for P-E, because it is the residual between two large terms (precipitation and evapo-transpiration). Some Regional Climate Models (RCMs) indicated different signs of precipitation and soil moisture trends from those of AOGCMs. While the majority of AOGCM simulations implying drying of the soils under future climate scenarios, this is not observed in RCM simulations. For example, Pan et al. (2004) simulated both precipitation and soil moisture increase toward the middle of the twenty-first century over the central United States. Using a different RCM forced by two AOGCMs, Liang et al. (2006) also noticed the similar moistening of the soil in the region. Aside from apparent differences in horizontal resolutions between regional and global models, different individual model parameterization schemes could contribute to the inconsistency between RCM and AOGCM results (Liang et al. 2006). The difference in hydrological cycles between RCM and AOGCM might partly arise from the disparity in the amount of regional warming projected. Most AOGCMs projected more warming in the central United States than those RCMs, resulting in larger evapo-transpiration. While it is impossible to judge which group of models is more realistic, the southwestern portion of the Midwestern USA is experiencing summer cooling (see chapter 3), especially during the peak global warming period (1975–2005). The simulated twentieth-century climate by the AOGCMs hardly captured the cooling or lack of warming (Kunkel et al. 2006). Thus, future trends of soil moisture over the Midwestern USA remain quite uncertain because even a moderate bias in either precipitation or evapo-transpiration can change the sign of the trend.

STREAMFLOW

Hydrological extremes (floods—river floods, flash floods, and urban floods) and droughts (meteorological drought, hydrological drought, and agricultural drought) depend on precipitation intensity and timing and antecedent conditions of rivers and their drainage basins (Kundzewicz et al. 2007). As discussed above, a robust feature consistent across AOGCM projections is greater annual mean (and possibly extreme) precipitation in the warmer climate (Meehl et al. 2005). However, projected increases in variability may also increase the likelihood of droughts (see the previous section). Thus a warmer climate may increase the risk of both floods and droughts (Wetherald and Manabe 2002; Meehl et al. 2007; Trenberth et al. 2003).

According to the IPCC 4AR, an increase of droughts over low latitudes and mid-latitude continental interiors in summer is likely (IPCC 2007). Projections for the 2090s show regions of strong wetting and drying with a net overall global drying trend (Burke, Brown, and Christidis 2006). The proportion of land surface in extreme drought, globally, is predicted to increase by a factor of 10–30 for the present day to 30% by the 2090s (Kundzewicz et al. 2007). The number of extreme drought events and mean drought duration are likely to increase by factors of two and six, respectively, by the 2090s (Burke et al. 2006).

The response of streamflow to precipitation is referred to as the precipitation elasticity (ε_p):

$$\varepsilon_p = \frac{dQ}{dP} \frac{P}{Q} \qquad (8.4)$$

where P is precipitation and Q is discharge. Precipitation elasticity provides a quantitative measure of the streamflow sensitivity to precipitation changes (Sankarasubramanian, Vogel, and Limbrunner 2001). It depends on many factors, including evapo-transpiration, soil moisture, and water holding capacity, and thus exhibits tremendous spatial variability (Sankarasubramanian,

Vogel, and Limbrunner 2001). Whether or not the elasticity is greater than 1 depends on how close the soil is to saturation. The closer to saturation the soil is, the larger ε_p would be. If the soil is saturated, all the increased rainfall would translate into runoff, resulting in large relative increase in streamflow. On the other hand, if the soil is dry, all increased rainfall penetrates soil, yielding no increase in streamflow. The ε_p also depends on rainfall intensity. The ε_p value in the continental United States ranges from 1.2 to 2.5, except for some areas in western mountainous regions where $\varepsilon_p<1$, meaning that a 1% increase in precipitation would result in 1.2–2.5% gain in streamflow (Sankarasubramanian, Vogel, and Limbrunner 2001). Over the relatively moist Midwestern USA, the value of ε_p is approximately 2 (Jha et al. 2004).

Most climate models agree on the sign of future runoff trends for northern North America, with projected increases of 10 to 40% (Kundzewicz et al. 2007; Milly, Dunne, and Vecchia 2005). Using SWAT (Soil Water Assessment Tool) models forced by RCM projections, Jha et al. (2004) simulated that streamflow in the Upper Mississippi River Basin would increase 51% from the 1990s to the middle of the twenty-first century, although the corresponding precipitation increase was only 21%. These disproportional increases between precipitation and streamflow imply that the increased precipitation tends to fall in areas of near-saturated soils, which translates into greater runoff. These projections illustrate the importance of understanding compounding impacts of climate change on the hydrological cycle.

THE COUPLED HYDROLOGICAL CYCLE IN THE SOUTHWESTERN MIDWESTERN USA

Historical precipitation elasticity with respect to streamflow is highly variable across the Midwest; ranging from values at or below 1 over the upper Midwest, to 2 over much of the southern portions of the region, and exceeding 2.5 in over half of Illinois. Hence it is possible, indeed even likely, that changes in the hydrological cycle in the Midwestern USA will exhibit spatial variability.

As an example of the challenges in making detailed and robust examples of possible future hydrologic regimes, we use the example of the southwestern Midwestern USA (i.e., the region characterized by the "warming hole," see chapter 3 of this volume). Based on the prior discussion, one can envisage the following scenario of the hydrologic cycle during the growing season under a warmer climate: The southwestern United States and northern Mexico are projected to be noticeably warmer, which would enhance the Great Plains nocturnal LLJ through stronger PBL (planetary boundary layer) development (Paegle, Mo, and Nogues-Paegle 1996). At the same time, the northerly flow in the north-central United States becomes stronger, possibly because of greater warming in the western mountain region. The projected LLJ would occur more frequently in the south-central United States and less frequently in the north-central United States, creating a net convergence zone over the Midwestern USA (figure 8.2a). Greater moisture convergence implied by the north-south gradient in changes of LLJ

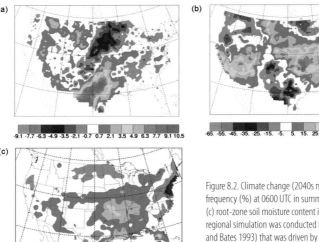

Figure 8.2. Climate change (2040s minus 1990s) in (a) low-level jet frequency (%) at 0600 UTC in summer, (b) summer precipitation, and (c) root-zone soil moisture content in mm. This dynamically downscaled regional simulation was conducted by using RegCM2 (Giorgi, Marinucci, and Bates 1993) that was driven by a HadCM2 A2 scenario (Johns et al. 1997) as initial and lateral boundary conditions. Figures 8.2a and 8.2c are reproduced from Pan et al. (2004) and figure 8.2b is adapted from Pan et al. (2001) with permission of the American Geophysical Union.

frequency produces increased precipitation (figure 8.2b), which in turn leads to increased summer soil moisture (figure 8.2c), enhanced evapo-transpiration, and reduced surface warming compared to surrounding regions. We emphasize the role of the soil moisture reservoir in providing additional "climate memory" that extends the regional reduction in warming and moistening beyond the period of increased precipitation.

In the above chain of processes, both increased precipitation and soil moisture are in close agreement with observations in the twentieth century. The increased soil wetness, however, is somewhat inconsistent with ensemble IPCC AR4 AOGCM projections, where future drying of soil occurs even though precipitation is increased. In fact, the majority of these AOGCMs did not reproduce the lack of warming or "warming hole" observed in the twentieth century. It is perhaps not so surprising that the low-resolution AOGCMs do not project future moistening of soil, as simulated by some regional climate models and suggested by

past observations, given that the majority of AOGCMs are even unable to reproduce the observed summer cooling and moistening in the peak warming of the twentieth century. The discrepancy may partly be amplified by the fact that the driver of the soil moisture, P-E, is a small residual of the two large terms. Small errors in either of the terms in the model can produce large errors in soil moisture.

Concluding Remarks and Summary of the Individual Chapters in the Hydrological Theme

Possible changes in the hydrological cycle both globally and at the regional scale are controlled in part by fundamental consequences of warmer temperatures and the increase in water vapor in the atmosphere, but are also dependent on moisture convergence and atmosphere-surface interactions. Past changes and possible future states are known with much less certainty than thermal regimes. For example, in the

IPCC AR4 the majority of AOGCM simulations indicate "over North America and Europe, the pattern of subpolar moistening and subtropical drying dominates the 21st-century projections" (Christensen et al. 2007). However, as acknowledged in the IPCC 4AR, "large-area and grid-box average projections for precipitation are often very different from local changes within the area" (Christensen et al. 2007).

In the chapters that follow this synthesis we present a range of research focused on the hydrologic cycle over the Midwestern USA. We document observational studies designed to better quantify regional and sub-regional scale trends in historical components of the hydrologic regime (e.g., chapters 9 and 13) and some possible future projections of precipitation under a range of climate scenarios (chapter 10). We also present examples of the tools that are being employed to examine possible impacts of changing precipitation on streamflow (using a SWAT model forced by a subset of IPCC AR4 models [chapter 12]) and soil moisture (chapter 11).

ACKNOWLEDGMENTS

This research was partly supported by the U.S. Department of Energy's Office of Science (BER) through the Midwestern Regional Center of the National Institute for Climatic Change Research at Michigan Technological University. SP acknowledges additional financial support from the NSF Geography and Regional Science program (grants #0618364 and 0647868).

REFERENCES

Betts, R. A., et al. 1997. "Contrasting Physiological and Structural Vegetation Feedbacks in Climate Change Simulations." *Nature* 387: 796–799.

Blackadar, A. K. 1957. "Boundary Layer Wind Maxima and Their Significance for Growth of Nocturnal Inversions." *Bulletin of the American Meteorological Society* 38: 283–290.

Bosilovich, M. G., and J. Chern. 2006. "Simulation of Water Sources and Precipitation Recycling for the MacKenzie, Mississippi, and Amazon River Basins." *Journal of Hydrometeorology* 7: 312–329.

Burke, E. J., S. J. Brown, and N. Christidis. 2006. "Modelling the Recent Evolution of Global Drought and Projections for the 21st Century with the Hadley Centre Climate Model." *Journal of Hydrometeorology* 7: 1113–1125.

Christensen, J. H., et al. 2007. "Regional Climate Projections." In *Climate Change 2007: The Physical Science Basis.* Contribution of Working Group I to the Fourth Assessment Report of the Intergovernmental Panel on Climate Change. [Ed. S. Solomon, et al.] New York: Cambridge University Press.

Dai, A., I. Y. Fung, and A. D. Del Genio. 1997. "Surface Observed Global Land Precipitation Variations during 1900–1988." *Journal of Climate* 10: 2943–2962.

Dettinger, M. D, and H. F. Diaz. 2000. "Global Characteristics of Stream Flow Seasonality and Variability." *Journal of Hydrometeorology* 1: 289–310.

Giorgi F., M. R. Marinucci, and G. T. Bates. 1993. "Development of a Second-Generation Regional Climate Model (RegCM2) I: Boundary-Layer and Radiative Transfer Processes." *Monthly Weather Review* 121: 2794–2813 .

Folland, C. K., et al. 2001. "Observed Climate Variability and Change." In *Climate Change 2001: The Scientific Basis,* [ed. J. H. Houghton et al.], 99–182. Cambridge: Cambridge University Press.

Held, I. M., and B. J. Soden. 2006. "Robust Responses of the Hydrological Cycle to Global Warming." *Journal of Climate* 19: 5686–5699.

Helfand, H. M., and S. D. Schubert. 1995. "Climatology of the Simulated Great Plains Low-Level Jet and Its Contribution to the Continental Moisture Budget of the United States." *Journal of Climate* 8: 784–806.

Hu, Q., and S. Feng. 2001. "Variation of ENSO and Interannual Variation in Summer in the

Central United State. *Journal of Climate* 14: 2469–2480.

IPCC. 2007. *Climate Change 2007: The Physical Science Basis.* Contribution of Working Group I to the Fourth Assessment Report of the Intergovernmental Panel on Climate Change. [Ed. S. Solomon, et al.] New York: Cambridge University Press.

Jha, M., et al. 2004. "Impacts of Climate Change on Stream Flow in the Upper Mississippi River Basin: A Regional Climate Model Perspective." *Journal of Geophysical Research* 109, doi:10.1029/2003JD003686.

Johns, T. C., et al. 1997. "The Second Hadley Centre Coupled Ocean-Atmosphere GCM: Model Description, Spinup and Validation." *Climate Dynamics* 13: 103–134.

Kundzewicz, Z. W., et al. 2007. "Freshwater Resources and Their Management." In *Climate Change 2007: Impacts, Adaptation and Vulnerability.* Contribution of Working Group II to the Fourth Assessment Report of the Intergovernmental Panel on Climate Change, ed. M. L. Parry, et al., 173–210. Cambridge: Cambridge University Press.

Kunkel, K. E., et al. 2006. "Can CGCMs Simulate the Twentieth-Century 'Warming Hole' in the Central United States?" *Journal of Climate* 19: 4137–4153.

Liang, X.-Z., et al. 2006. "Regional Climate Model Downscaling of the U.S. Summer Climate and Future Change." *Journal of Geophysical Research* 111, doi:10.1029/2005JD006685.

Lins, H. F., and J. R. Slack. 1999. "Streamflow Trends in the United States." *Geophysical Research Letters* 26: 227–230.

Lofgren, B. M., et al. 2002. "Evaluation of Potential Impacts on Great Lakes Water Resources Based on Climate Scenarios of Two GCMs." *Journal of Great Lakes Research* 28: 537–554.

Manabe, S., and R. T. Wetherald. 1987. "Large-Scale Changes of Soil Wetness Induced by an Increase in Atmospheric Carbon-Dioxide." *Journal of the Atmospheric Sciences* 44: 1211–1235.

McCabe, G. J., and D. M. Wolock. 2002. "A Step Increase in Streamflow in the Conterminous United States." *Geophysical Research Letters* 29: 2185–2188.

Meadows, G. A., et al. 1997. "The Relationship between Great Lakes Water Levels, Wave Energies, and Shoreline Damage." *Bulletin of the American Meteorological Society* 78: 675–683.

Meehl, G. A., et al. 2005. "Overview of the Coupled Model Intercomparison Project." *Bulletin of the American Meteorological Society* 86: 89–93.

Meehl, G. A., et al. 2007. "Global Climate Projections." In *Climate Change 2007: The Physical Science Basis.* Contribution of Working Group I to the Fourth Assessment Report of the Intergovernmental Panel on Climate Change. [Ed. S. Solomon, et al.] New York: Cambridge University Press.

Miller, A. J., et al. 1994. "The 1976–77 Climate Shift of the Pacific Ocean." *Oceanography* 7: 21–26.

Milly, P. C. D., K. A. Dunne, and A. V. Vecchia. 2005. "Global Pattern of Trends in Streamflow and Water Availability in a Changing Climate." *Nature* 438: 347–350.

Mitchell, T. D., and P. D. Jones. 2005. "An Improved Method of Constructing a Database of Monthly Climate Observations and Associated High-Resolution Grids." *International Journal of Climatology* 25: 693–712.

Mote, P. W., and N. J. Mantua. 2002. "Coastal Upwelling in a Warming Future." *Geophysical Research Letter* 29, doi:10.1029/2002GL016086.

Paegle, J., K. C. Mo, and J. Nogues-Paegle. 1996. "Dependence of Simulated Precipitation on Surface Evaporation during the 1993 United States Summer Floods." *Monthly Weather Review* 124: 345–361.

Pan, Z., et al. 2001. "Evaluation of Uncertainty in Regional Climate Change Simulations." *Journal of Geophysical Research* 106: 17,735–17,751.

Pan, Z., et al. 2004. "Altered Hydrologic Feedback in a Warming Climate Introduces a 'Warming Hole.'" *Geophysical Research Letters* 31, doi:10.1029/2004GL02528.

Reen, B. P., et al. 2006. "A Case Study on the Effects of Heterogeneous Soil Moisture on Mesoscale Boundary-Layer Structure in the Southern Great Plains, USA, Part II:

Mesoscale Modeling." *Boundary-Layer Meteorology* 120: 275–314.

Roads, J., M. Kanamitsu, and R. Stewart. 2002. "CSE Water and Energy Budgets in the NCEP-DOE Reanalyses." *Journal of Hydrometeorology* 3: 227–248.

Rosenzweig, C., et al. 2000. "Climate Change and U.S. Agriculture: The Impacts of Warming and Extreme Weather Events on Productivity, Plant Diseases and Pests." Center for Health and the Environment, Harvard Medical School. Available at: http://forum.decvar.org/documents/rosen.pdf.

Sankarasubramanian, A., R. M. Vogel, and J. F. Limbrunner. 2001. "Climate Elasticity of Streamflow in the United States." *Water Resources Research* 37: 1771–1781.

Schwartz, R. C., et al. 2004. "Modeling the Impacts of Water Level Changes on Great Lakes Community." *Journal of the American Water Resources Association* 40: 647–662.

Slack, J. R., and J. M. Landwehr. 1992. "Hydro-Climatic Data Network: A U.S. Geological Survey Streamflow Data Set for the United States for the Study of Climate Variations, 1874–1988." U.S. Geol. Surv. Open-File Rept. 92–129.

Stewart, I. T., D. R. Cayan, and M. D. Dettinger. 2005. "Changes toward Earlier Streamflow Timing across Western North America." *Journal of Climate* 18: 1136–1155.

Sun, Y., et al. 2007. "How Often Will It Rain?" *Journal of Climate* 20: 4801–4818.

Trenberth, K. E., et al. 2003. "The Changing Character of Precipitation." *Bulletin of the American Meteorological Society* 84: 1205–1217.

USDA. 2004. *2002 Census of Agriculture, Farm and Ranch Irrigation Survey (2003)*. U.S. Department of Agriculture Special Studies Volume 3, Part 1, AC-02–SS-1. Washington, D.C.: USDA National Ag. Statistics Service.

USGCRP. 2001. *U.S. National Assessment of the Potential Consequences of Climate Variability and Change.* Available at: http://www.usgcrp.gov/usgcrp.

USGS. 2005. *Streamflow Trends in the United States.* Available at: http://pubs.usgs.gov/fs/2005/3017/#fig3.

Wallace, J. M. 1975. "Diurnal Variations in Precipitation and Thunderstorm Frequency over the Conterminous United States." *Monthly Weather Review* 103: 585–598.

Wetherald, R. T., and S. Manabe. 2002. "Simulation of Hydrologic Changes Associated with Global Warming." *Journal of Geophysical Research* 107, doi:10.1029/2001JD001195.

9. Did Precipitation Regimes Change during the Twentieth Century?

S. C. PRYOR, K. E. KUNKEL, AND J. T. SCHOOF

Introduction

MOTIVATION

As discussed in chapter 1 of this volume, agricultural yields in the Midwestern USA are critical to provision of food for the nation and are critically linked to precipitation receipt (Rosenzweig and Hillel 2008). For example, the drought of 1988 resulted in estimated economic losses of $56 billion (Rosenzweig et al. 2000). Equally, the possibility of an increase in extreme precipitation events over the contiguous United States is of great interest to a nation where flood-related annual economic losses increased from $1 billion in the 1940s to $6 billion annually in the 1990s (both figures adjusted to 1997 $) (Easterling et al. 2000). Flash floods currently account for an average of two hundred deaths a year (Winkler 1988), and the floods of 1993 in the Mississippi River Basin caused over $23 billion in agricultural losses (Rosenzweig et al. 2000). Further, prior research has indicated increases in "great floods" under climate change scenarios (Milly et al. 2002).

OBJECTIVES

The precipitation climate of a station/region may be described using multiple descriptors, for example: (1) the probability of precipitation (i.e., the likelihood that any day chosen at random will exhibit precipitation or not); (2) the time sequence of precipitation and non-precipitation days (often simulated using Markov chain models (Wilks 1992, Schoof and Pryor 2008), where the order of the model reflects the number of days in the sequence under consideration); (3) the precipitation amount on "wet days" (e.g., daily intensity, percentiles of precipitation accumulation, or the sum of precipitation accumulated over a sequence of days or on intense precipitation days); (4) total accumulation of precipitation on "wet days" (annual total precipitation); and/or (5) the seasonal distribution of precipitation (i.e., the seasonality). Here we present an overview of precipitation climates in the Midwestern USA and quantify historical evolution using the following descriptors (see figure 9.1 for examples from two stations analyzed herein):

1. Annual total precipitation.

2. Number of rain days and daily intensity, where the latter is computed as annual total/number of precipitation days (i.e., the mean precipitation accumulated on a "wet day").

3. Extreme precipitation. Changes in extreme precipitation events are of great interest given the economic and agricultural losses associated with floods (Easterling et al. 2000). Herein we use two descriptors—the sum of precipitation accumulated on the top-10 wettest days of the year and the total precipitation accumulated during the wettest pentad. These metrics are insensitive to under-reporting of light precipitation early in the time series (see below).

4. Duration of consecutive days without precipitation. Drought duration/intensity has been linked to changes in global sea surface temperature fields (particularly ENSO conditions) (Seager 2007); see chapter 1 of this volume. Given the importance of Midwest agricultural production to the United States, the possibility for changes in drought frequency/intensity is of great consequence. The mean duration of consecutive days without precipitation is used herein as a simple metric of this phenomenon.

5. Seasonality of precipitation receipt. The consequences of possible changes in precipitation receipt are dictated, in part, by the timing of precipitation events and by changes therein that determine partitioning

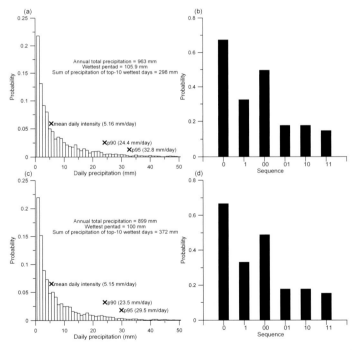

Figure 9.1. a) Histogram of daily precipitation amounts at West Lafayette, Indiana (40.42°N, −86.92°E) during 1971–2000, where p90 and p95 indicate the 90th and 95th percentile values and values of other descriptors of the probability distribution as presented herein are given by the embedded text. b) Sequencing of wet (1) and dry (0) days for this station for a 0th-order process (i.e., probability that an individual day will be a dry day [0] or wet day [1]), and for 1st-order process (i.e., the probability that a dry day will be followed by a dry day [00], that a dry day will be followed by a wet day [01], that a wet day will be followed by a dry day [10], and that a wet day will be followed by a wet day [11]). c), d) as a), b) but for Dubuque, Iowa (42.40°N, −90.67°E). Both stations occasionally exhibit daily accumulations >50 mm; the axes in a) and c) are truncated to 50 mm to increase legibility.

of the water into runoff/evapo-transpiration/infiltration and thus flood forecasting and ecosystem responses (Groisman, Knight, and Karl 2001; Epstein et al. 2002; Rosenberg et al. 2003). Here we quantify seasonality using the index (SI) of Walsh and Lawler (1981):

$$SI = \frac{1}{R} \sum_{n=1}^{12} \left| X_n - \frac{R}{12} \right| \qquad (9.1)$$

where R is the annual total precipitation and X_n = monthly total precipitation in month n.

An alternative metric of seasonality is also presented. This metric is based on the day of year at which the 10th, 25th, 50th, 75th and 90th percentiles of the annual total precipitation are achieved (referred to here as percentile day-of-year).

Data and Methods

Analyses presented herein are conducted using metrics of daily precipitation computed at the annual time scale using time series of data records from individual stations across the Midwestern USA. These time series are derived from the data set described by Kunkel et al. (1998, 2005). They are drawn principally from the National Weather Service's (NWS's) cooperative observer network (COOP), and have been extended to the pre-digital era and subject to homogenization efforts to correct for observer error, station discontinuity, and digitization errors. The resulting data set was pre-screened to select stations that have at least eighty years of data between 1901 and 2000 during which >360 days of valid data per year are available, and which have the first year of valid

data prior to 1910 and the last year subsequent to 1990, thus ensuring the trends derived reflect changes over the entire course of the twentieth century. We analyze station-specific daily precipitation time series at the annual time scale and include data from years where valid data are available for >360 days. Hence, if the data record from a given station fulfills the first data selection criteria (eighty "complete" years and start and end dates), but does not have >360 days of valid observations in 1995, the trend is computed excluding 1995.

There is evidence that very light precipitation amounts were underreported in the early part of the observation record (Groisman and Knight 2009). This is a potential source of bias in the number of days with precipitation and other metrics of the precipitation probability distribution, particularly the daily intensity. Hence, a precipitation threshold of 0.05 inch (1.27 mm) was used to define a rain day (see Pryor, Howe, and Kunkel 2009). The resulting daily intensity from the later analysis will be biased high, but the trends may be a more accurate portrayal of the behavior of the physical system.

Precipitation time series exhibit greater temporal variability than temperature, which confounds analysis of long-term trends. As one example, data from Dubuque (figure 9.1) indicate that the standard deviation of annual total precipitation is 23% of the mean annual total precipitation computed using data from 1901–2000. Comparable values for the top-10 wettest days and the wettest pentad are 24% and 36%, respectively. Because our focus is on the mean rate of change of precipitation metrics during the twentieth century, trends are computed using linear analysis of the annual time

Table 9.1. Fraction of stations (in percent) by metric that exhibit a statistically significant (positive or negative) trend between 1901 and 2000

Metric\Number of stations that exhibit specified trend	Negative trend	Positive trend
Annual total precipitation	2	24
Sum of precipitation accumulated on the top-10 wettest days	8	22
Wettest pentad	2	11
Daily intensity	17	12
Number of rain days	5	43
Mean number of consecutive days without precipitation	33	3
Seasonal index	10	4

series. Application of OLSR (ordinary least squares regression) leads to an (erroneous) increase in the probability of detecting a "statistically significant" trend (Power 1993; Pryor, Howe, and Kunkel 2008). Hence the trend analysis is conducted using application of the nonparametric Kendall's tau-based slope estimator (Alexander et al. 2006) to annual values of each of the metrics. This technique does not assume a distribution for the residuals and is relatively robust to outliers (Sen 1968). The linear trend is deemed statistically significant if it differs from 0 at the 90% confidence level, and the magnitude of the trend is given by the median of the series of slopes ($\frac{Y_j - Y_i}{t_j - t_i}$), where Y_x is the value of the metric at a given point t in time, and $1 \le i < j \le n$, where n is the total number of data points).

Because of the high inter-annual variability of percentile day-of-year estimates, these values are analyzed in terms of the difference between two thirty-year periods: 1941–1970 and 1971–2000. The significance of differences in mean percentile day-of-year in the two time periods is quantified using confidence intervals derived using bootstrapping of the annual values during 1971–2000 (i.e., an assessment of the differences between the time periods is conducted relative to inter-

annual variability in the 1971–2000 period). If the mean percentile day-of-year from 1941–1970 period lies within the middle 900 values in an ordered sequence of the distribution of 1000 realizations from the 1971–2000 period, the percentile day-of-year values in the two time windows are deemed to be not significantly different at the 90% confidence level.

Results

Results from analysis of the metrics described above in terms of the current climatology (derived using data from 1971–2000) and temporal trends over the period 1901–2000 are provided in table 9.1 and the following sections.

ANNUAL TOTAL PRECIPITATION

Annual total precipitation exhibits strong northwest to southeast gradients across the Midwest (figure 9.2a), with annual accumulations over 1200 mm in Kentucky and below 700 mm in Minnesota. While data from a large proportion of stations exhibit no trend, in accord with prior research (Groisman et al. 2004), generally the stations that exhibit significant trends in an-

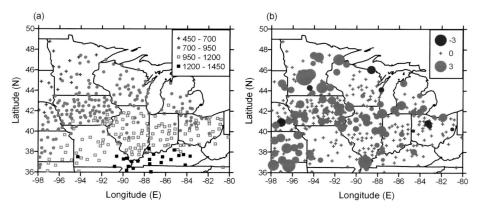

Figure 9.2. a) Mean annual total precipitation (mm) during 1971–2000, and b) the trend in annual total precipitation 1901–2000 expressed in percent per decade. Red circle indicates a statistically significant increase through time; blue circle indicates a statistically significant decline. Plus symbol indicates trend was not significant (shown as 0 in the legend). The diameter of the dot scales linearly with trend magnitude.

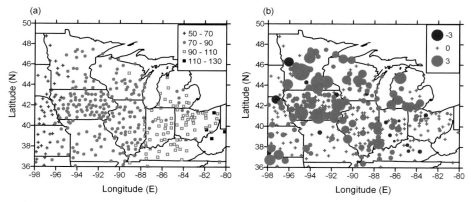

Figure 9.3. a) Mean annual number of rain days (1971–2000). b) Trend in annual number of rain days 1901–2000 expressed in percent per decade. Symbols are as figure 9.2.

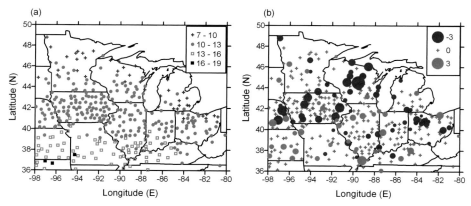

Figure 9.4. a) Mean daily intensity (mm per day) (1971–2000). b) Trend in mean daily intensity 1901–2000 expressed in percent per decade. Symbols are as figure 9.2.

nual total precipitation indicate an evolution toward higher annual accumulation (figure 9.2b, table 9.1). Percentage changes are of the order of 1–3% per decade, which if sustained implies an increase of 10–31% by 2100 and a doubling of annual total precipitation in 240–710 years.

PRECIPITATION FREQUENCY AND MEAN INTENSITY

Unlike annual total precipitation, the mean number of rain days primarily exhibits variability with longitude, from approximately fifty to seventy (i.e., ~1 day in 7) in the western portion of the domain, to over a hundred in the east (i.e., ~1 day in 3 or 4) (figure 9.3a). Daily intensity varies more with latitude from 7–10 mm/day in the north to over 15 mm/day in the south (figure 9.4a). Trends in mean annual number of rain days are overwhelmingly toward increased precipitation frequency (figure 9.3b, table 9.1). Almost half of stations indicate statistically significant increases in number of rain days over the twentieth century, with rates of increase of >1.5% per decade. By comparison, few stations exhibit significant trends in mean daily intensity (figure 9.4b, table 9.1), and they do not exhibit substantial spatial clustering.

METRICS OF EXTREME PRECIPITATION

Over 30% of annual total precipitation at most stations in the Midwestern USA is obtained in the ten wettest days of the year (figure 9.5). In the west of the region, as much as 50% of annual accumulated pre-

cipitation is attributable to as few as ten days in some years. This emphasizes the importance of intense precipitation events in the Midwest, and is evident from analysis of the precipitation received during the top-10 wettest days (figure 9.6a) and the wettest pentad (five-day period, figure 9.7a). Spatial patterns in both metrics closely mirror those present in the annual total precipitation, with highest values in the south of the domain and lowest values in the north (figure 9.2). In general, stations that exhibit significant changes in the metrics of extreme precipitation indicate trends toward increased values. Twenty-two percent of stations exhibit significant increases in the total accumulated precipitation during the top-10 wettest days of the year, while 11% exhibit significant increases in accumulated precipitation during the wettest pentad. In both cases the increases are up to 3% per decade. However, the majority of stations exhibit no change at the 90% confidence level (table 9.1, figure 9.6b and figure 9.7b).

DURATION WITHOUT PRECIPITATION

The mean number of consecutive days without precipitation, like the number of rain days per year, indicates variability principally with variations in longitude and declines from over 6.5 days in the west to fewer than 4.8 days in the east. Time series from approximately one-third of stations analyzed indicated a significant trend toward declining values of this metric, consistent with an increase in rain days (table 9.1 and figure 9.8b), which implies that the change in rain day frequency has not been associated with major changes in the order-1 Markov chain

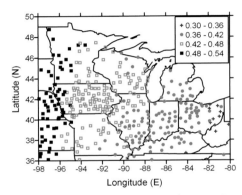

Figure 9.5. Fraction of the mean annual total precipitation that derives from the top-10 wettest days in a year during 1971–2000.

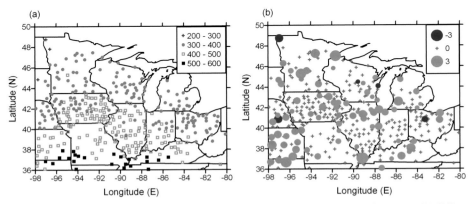

Figure 9.6. a) Mean sum of the top-10 wettest days in a year (mm) (1971–2000). b) Trend in sum of the top-10 wettest days in a year 1901–2000 expressed in percent per decade. Symbols are as figure 9.2.

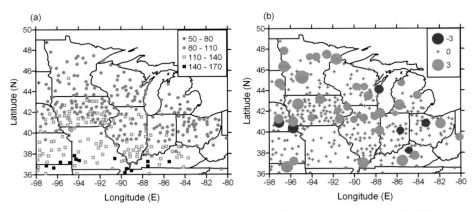

Figure 9.7. a) Mean precipitation accumulated in the annual wettest pentad (mm) (1971–2000). b) Trend in the wettest pentad 1901–2000 expressed in percent per decade. Symbols are as figure 9.2.

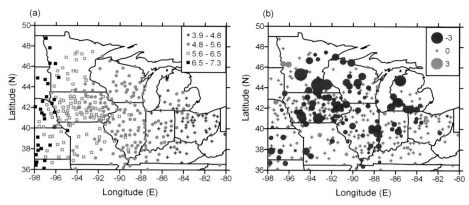

Figure 9.8. a) Mean number of consecutive days without precipitation (1971–2000). b) Trend in number of consecutive days without precipitation 1901–2000 expressed in percent per decade. Symbols are as figure 9.2.

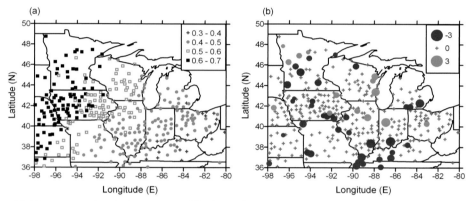

Figure 9.9. a) The mean seasonality index during 1971–2000. b) Trend in annual SI 1901–2000 expressed in percent per decade. Symbols are as figure 9.2.

transition probabilities (see examples in figure 9.1) (Schoof and Pryor 2008).

SEASONALITY

Values of mean SI decline from west to east (figure 9.9a), ranging from over 0.6 to under 0.4. These values are generally higher than values computed for the majority of the United Kingdom (Walsh and Lawler 1981), and also the northeast United States (Hsu and Wallace 1976), but they are considerably lower than is observed in southern California, where values exceed 1.0 (Pryor and Schoof 2008). Indeed, values computed

for the Midwestern USA generally fall into the SI classes defined by Walsh and Lawler (1981) as "equitable but with a definite wetter season" and "rather seasonal with a short drier season." Several stations across the Midwest exhibit increased annual total precipitation over the twentieth century, increased frequency of precipitation, increases in metrics of extreme precipitation, and a decline in the mean number of consecutive days without precipitation; however, these changes do not appear to have been manifest as widespread changes in SI (figure 9.9b, table 9.1), with relatively few stations exhibiting increases or decreases.

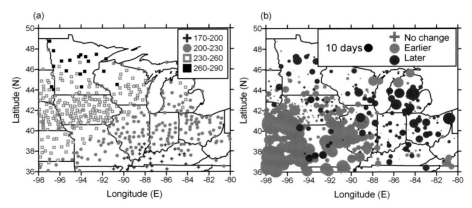

Figure 9.10. a) Mean percentile day-of-year for the 50th percentile accumulation as computed for water years 1971–2000. b) Difference in mean 50th percentile day-of-year values computed for 1971–2000 v 1941–1970. If the 1971–2000 period exhibits significantly later achievement of the 50th percentile than the earlier time period, the station is shown by a filled red circle. If the 1971–2000 period exhibits significantly earlier achievement of the 50th percentile than the earlier time period, the station is shown by a blue circle. In each case the diameter of the circle scales linearly with the difference in mean percentile day-of-year. If the mean percentile day-of-year from 1941–1970 lies within the middle 900 values in an ordered sequence of the distribution of 1000 bootstrap realizations from the 1971–2000 period, the mean percentile day-of-year values are deemed to be not significantly different at the 90% confidence level and the station is shown by a + symbol.

However, an analysis of the date on which the 50th percentile of annual total accumulation is achieved does indicate some shifts (figure 9.10). During 1971–2000 the water year day (recall the water year starts on 1 October), on which half of the annual total precipitation is achieved, ranged from approximately day 180 in the south (i.e., during April), to approximately day 260 in the northwest (i.e., during June), emphasizing the relative importance of summer precipitation in the northwest of the region. When the 1971–2000 period is compared with 1941–1970 we observe that there has been a shift toward earlier attainment of the 50th percentile day-of-year in the southwest of the region, indicating relative moistening of the winter and spring period. Conversely parts of Indiana, the Michigan Peninsula, and Ohio indicate later attainment of the 50th percentile day-of-year in the 1971–2000 time-frame than during 1941–1970. These results indicate substantial sub-regional

variability in changes to the precipitation seasonality (Pryor and Schoof 2008).

CHANGES IN THE RELATIVE IMPORTANCE OF EXTREME EVENTS?

Karl and Knight (1998) analyzed gridded station data and concluded that "since 1910, precipitation has increased by about 10% across the contiguous United States." The increase in annual total precipitation across the Midwest was equal to, or greater than, this change at approximately a quarter of stations studied herein. Karl and Knight (1998) further stated that when data are averaged over the contiguous United States, over half of the increase in total precipitation between 1910 and 1995 "is due to positive trends in the upper 10 percentiles of the precipitation distribution." Use of percentile-based metrics in precipitation analysis is confounded by reporting bias (Pryor, Howe, and Kunkel 2009), so herein we use two alternative metrics of extreme precipitation. The results indicate very similar rates of change in annual total precipitation and the sum of precipitation received in the top-10 wettest days in the year (figure 9.11) when expressed in percent change per decade. This implies that some of the increase in annual total precipitation has derived from intensification of extreme events, but this is not the sole driver of the change in annual total accumulation. This is also evidenced in the increase in annual number of rain days. As an example, returning to one of the stations depicted in figure 9.1, data from Dubuque indicate the sum of precipitation received on the top-10 wettest days of the year increased by approximately 1.6%/decade, while annual total precipitation increased by 2.3%/decade, which implies the top-10 wettest days, while increasing in intensity, actually represented a declining fraction of the annual accumulated precipitation. Thus there is no compelling evidence of systematically larger trends in the extreme metrics than in annual total precipitation.

Concluding Remarks

Several stations across the Midwest experienced increased annual total precipitation over the twentieth century, with trends of over 2% per decade at some stations (figure 9.2 and figure 9.11). Seventy percent of stations analyzed exhibit no trend in annual total precipitation at the 90% confidence level, but these results are broadly consistent with findings of prior research in that they indicate a greater prevalence of positive than negative long-term trends in annual total precipitation (Groisman et al. 2004). These magnitudes of change are also consistent with the 10–20% per century increase in annual precipitation reported by Groisman et al. (2004), the greater than 10% precipitable water increase between 1973 and 1995 (Ross and Elliott 1996), and increased low-level moisture availability and horizontal moisture convergence (Trenberth et al. 2003). An even larger fraction of stations considered exhibit evidence of an increased frequency of precipitation in the Midwestern USA, even when a threshold of 1.27 mm is used to

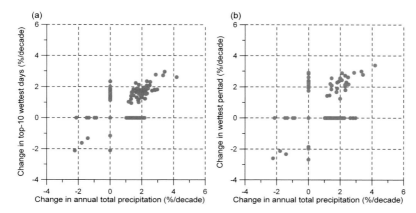

Figure 9.11. a) Scatter-plot of the change in annual total precipitation versus change in top-10 wettest day precipitation, and b) Scatter-plot of the change in annual total precipitation versus change in the wettest pentad. If the change is not statistically significant a value of 0 is plotted for the trend.

define a precipitation day. Time series from a substantial number of stations also exhibit evidence of increased intensity of precipitation on the top-10 wettest days of the year, and a few stations exhibit evidence of intensification of the wettest pentad. Approximately one-third of stations studied showed evidence of a decline in the mean number of consecutive days without precipitation, but results for daily intensity and the seasonality of precipitation receipt as measured using a standard seasonality index may be summarized as indicating "no change."

It is important to recall that precipitation time series exhibit tremendous inter-annual variability, and thus it is difficult to identify and quantify trends. Nevertheless, returning to the example of the Dubuque station, the linear trend in annual total precipitation is 2.3% per decade, which means the linear trend over the century is approximately equal to the inter-annual variability as measured using the ratio of the standard deviation of annual values to the mean of the annual values. The temporal trends in wettest pentad and sum of precipitation accumulated on the top-10 wettest days of the year are 2.0%/decade and 1.6%/decade, which means that when scaled to the century time step they too are of comparable magnitude to the current inter-annual variability.

Returning to the original motivation for this research—the possibility for changes in flood frequency—these analyses imply:

1. The Midwest is highly sensitive to changes in extreme precipitation events. Over much of the domain 40% of annual total precipitation derives from the top-10 wettest days, despite the fact that over much of the domain precipitation is observed on approximately one day in three to seven days.

2. Although the majority of station time series studied exhibit no change in the precipitation metrics, possibly due to the very high inter-annual variability, between one-quarter and one-half of all stations across the Midwestern USA exhibited an increase in precipitation frequency, annual accumulated precipitation, and the intensity of extreme events (as manifest in the sum of the top-10 wettest days). The increase in annual total precipitation, where observed, has not been associated with large-scale changes in seasonality as measured using a standard seasonality index. However, analysis of the day of the year on which certain percentiles of annual total precipitation were achieved indicates spatially coherent patterns of change. In some regions the mean day of the year on which the 50th percentile of annual precipitation was achieved differed by over twenty days between 1971–2000 and 1941–1970. Additionally, linear trends in annual total precipitation and some other metrics of the precipitation regime when expressed as percent per century are of comparable magnitude to current inter-annual variability. This implies the evolution of these parameters during the twentieth century is substantial.

3. The increase in annual total precipitation, where observed, is not disproportionately due to increases in extreme events, though intensification of high-magnitude precipitation has caused some fraction of the increase in annual total precipitation. While floods are not solely a product of heavy precipitation, there is evidence that the most common cause of flooding is

"intense and/or prolonged storm precipitation" (Nott 2006). Given the increase in intensity of extreme precipitation events, an increased risk of flooding seems likely.

ACKNOWLEDGMENTS

Financial support from the NSF Geography and Regional Science program (grants #0618364 and 0647868) and the Office of the Dean of the College of Arts and Science of Indiana University is gratefully acknowledged.

REFERENCES

Alexander, L. V., et al. 2006. "Global Observed Changes in Daily Climate Extremes of Temperature and Precipitation." *Journal of Geophysical Research* 111, doi:10.1029/2005JD006290

Easterling, D. R., et al. 2000. "Climate Extremes: Observations, Modeling, and Impacts." *Science* 289: 2068–2074.

Epstein, H. E., et al. 2002. "The Relative Abundance of Three Plant Functional Types in Temperate Grasslands and Shrublands of North and South America: Effects of Projected Climate Change." *Journal of Biogeography* 29: 875–888.

Groisman, P. Y., and R. W. Knight. 2008. "Prolonged Dry Episodes over the Conterminous United States: New Tendencies Emerged during the last 40 Years. *Journal of Climate* 21: 1850–1862.

Groisman, P. Y., R. W. Knight, and T. Karl. 2001. "Heavy Precipitation and High Streamflow in the Contiguous United States: Trends in the Twentieth Century." *Bulletin of the American Meteorological Society* 82: 219–246.

Groisman, P. Y., et al. 2004. "Contemporary Change of the Hydrological Cycle over the Contiguous United States: Trends Derived from in situ Observations." *Journal of Hydrometeorology* 5: 64–85.

Hsu, C.-P., and J. M. Wallace. 1976. "The Global Distribution of the Annual and Semiannual Cycles in Precipitation." *Monthly Weather Review* 104: 1093–1101.

Karl, T. R., and R. W. Knight. 1998. "Secular Trends of Precipitation Amount, Frequency, and Intensity in the United States." *Bulletin of the American Meteorological Society* 79: 231–241.

Kunkel, K. E., et al. 1998. "An Expanded Digital Daily Database for Climatic Resources Applications in the Midwestern United States." *Bulletin of the American Meteorological Society* 79: 1357–1366.

Kunkel, K. E., et al. 2005. "Quality Control of Pre-1948 Cooperative Observer Network Data." *Journal of Atmospheric and Oceanic Technology* 22: 1691–1705.

Milly, P. C. D., et al. 2002. "Increasing Risk of Great Floods in a Changing Climate." *Nature* 415: 514–517.

Nott, J. 2006. *Extreme Events: A Physical Reconstruction and Risk Assessment.* Cambridge: Cambridge University Press.

Power, M. 1993. "The Statistical Implications of Autocorrelation for Detection in Environmental Health Assessment." *Journal of Aquatic Ecosystem Health* 2: 197–204.

Pryor, S. C., J. A. Howe, and K. E. Kunkel. 2009. "How Spatially Coherent and Statistically Robust Are Temporal Changes in Extreme Precipitation across the Contiguous USA?" *International Journal of Climatology* 29: 31–45.

Pryor, S. C., and J. T. Schoof. 2008. "Changes in Precipitation Seasonality over the Contiguous USA." *Journal of Geophysical Research* 113, doi:10.1029/2008JD010251.

Rosenberg, N. J., et al. 2003. "Integrated Assessment of Hadley Centre (Hadcm2) Climate Change Projections on Agricultural Productivity and Irrigation Water Supply in the Conterminous United States—I. Climate Change Scenarios and Impacts on Irrigation Water Supply Simulated with the HUMUS Model." *Agricultural and Forest Meteorology* 117: 73–96.

Rosenzweig, C., and D. Hillel. 2008. *Climate Variability and the Global Harvest: Impacts of El Niño and Other Oscillations on Agroecosystems.* New York: Oxford University Press.

Rosenzweig, C., et al. 2000. *Climate Change and U.S. Agriculture: The Impacts of Warming* and *Extreme Weather Events on Productivity, Plant Diseases and Pests.* Boston: Center for Health and the Environment, Harvard Medical School.

Ross, R. J., and W. P. Elliott. 1996. "Tropospheric Water Vapor Climatology and Trends over North America: 1973–93." *Journal of Climate* 9: 3561–3574.

Schoof, J. T., and S. C. Pryor. 2008. "On the Proper Order of Markov Chain Model for Daily Precipitation Occurrence in the Contiguous United States." *Journal of Applied Meteorology and Climatology* 47: 2477–2486.

Seager, R. 2007. "The Turn of the Century North American Drought: Global Context, Dynamics and Past Analogs." *Journal of Climate* 20: 5527–5552.

Sen, P. K. 1968. "Estimates of the Regression Coefficient Based on Kendall's Tau." *Journal of the American Statistical Association* 63: 1379–1389.

Trenberth, K. E., et al. 2003. "The Changing Character of Precipitation." *Bulletin of the American Meteorological Society* 84: 1205–1217.

Walsh, P. D., and D. M. Lawler. 1981. "Rainfall Seasonality: Description, Spatial Patterns and Changes through Time." *Weather* 36: 201–208.

Wilks, D. S. 1992. "Adapting Stochastic Weather Generator Algorithms for Climate Change Studies." *Climatic Change* 22: 67–84.

Winkler, J. A. 1988. "Climatological Characteristics of Summertime Extreme Rainstorms in Minnesota." *Annals of the Association of American Geographers* 78: 57–73.

10. Climate Change and Streamflow in the Upper Mississippi River Basin

E. S. TAKLE, M. JHA, AND E. LU

Introduction

Recent analyses of observations and modeling report acceleration of the hydrological cycle at high latitudes in the Northern Hemisphere during the twentieth century (Stocker and Raible 2005; Wu, Wood, and Stott 2005). Detailed evaluation of the spectrum of precipitation events for the central United States (Groisman et al. 2005) reveal that the occurrence of extreme intense precipitation events has increased over the twentieth century (see also chapter 9 of this volume). Most notably, however, all of this increase (20% increase, statistically significant at the 0.01 level) occurred during the last thirty years of the twentieth century. Assessments of local and regional impacts of changes in the hydrological cycle in future climates call for improved capabilities for modeling the hydrological cycle and its individual components at the sub-watershed level.

Publication of the Fourth Assessment Report (AR4) by the Intergovernmental Panel on Climate Change (IPCC 2007) has brought increased attention to climate change at regional scales. Although extensive simulations of climate change by Regional Climate Models (RCMs) have yet to be reported, the AR4 offers more interpretation of coupled Atmosphere-Ocean General Circulation Model (AOGCM, also known as Global Climate Model [GCM]) results at regional scales than previous reports. Analysis from multiple AOGCMs reveals that while all models project the Midwestern USA to warm with increasing greenhouse gas concentrations, there is less agreement among models on whether precipitation will increase or decrease. As discussed in chapter 8, increases in temperature will markedly affect streamflow even if precipitation is unchanged due to (1) higher potential evaporation and hence higher atmospheric water-holding capacity, (2) altered rain-snow events, and (3) changed timing of snowmelt. Items (2) and (3) primarily affect seasonality of runoff and hence streamflow with less impact on annual total, whereas (1) can be expected to have a larger impact on annual total streamflow. When precipitation changes are superposed on temperature changes simplistic analyses quickly fail, underscoring the need for detailed modeling of complex and interacting physical processes.

Takle, Jha, and Anderson (2005) and Takle et al. (2006) report preliminary studies to determine the ability of AOGCMs to produce suitable input for the Soil and Water Assessment Tool (SWAT) watershed model (Arnold and Fohrer, 2005) to simulate components of the hydrological cycle in the Upper Mississippi River Basin (UMRB). In observations of streamflow at Grafton, Illinois, with simulated results, we found that no individual low-resolution AOGCM was able to give a distribution of annual flows that was not statistically different from the mean of observed values. However, the ensemble of nine models did produce credible distribution that was statistically significant. Also, the one model for which we had both high-resolution and low-resolution versions produced annual streamflows not statistically different from observations at high resolution even though results from its low-resolution sister model were statistically different from observations.

Streamflow models such as SWAT accept a wide range of meteorological datasets and use internal weather generators to fill in key missing values and create refined details, such as the partitioning of daily precipita-

tion between rain and snow. Because Jha et al. (2004) showed that for the UMRB SWAT provided good results for annual streamflow, it is not clear whether spatial refinement of AOGCM results is warranted for simulating streamflow for this watershed. Use of data from AOGCMs directly is an alternative to using RCMs or statistical models to downscale global results. AOGCM have improved physical process models and process resolution since the third assessment report of the Intergovernmental Panel on Climate Change (2001). Furthermore, advances in computing capabilities now permit the use of multi-model ensembles, which we have shown (at least for one region and one period) to provide a reliable source of weather data for assessing streamflow (Takle, Jha, and Anderson 2005). We report herein an extension of the studies reported by Takle, Jha, and Anderson (2005). We use twenty-first-century results of ten AOGCM being made available for the IPCC Fourth Assessment Report (PCMDI 2005) directly as input to SWAT for calculating components of the hydrological cycle in the region experiencing changes in

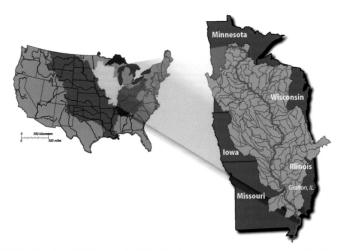

Figure 10.1. The Upper Mississippi River Basin (UMRB) and the 119 USGS eight-digit sub-watersheds upstream from Grafton, Illinois.

hydrological regime over the last thirty years (Pan et al. 2004; Groisman et al. 2005).

The UMRB has a drainage area of 447,500 km² up to the point just before the confluence of the Missouri and Mississippi Rivers near Grafton, Illinois (figure 10.1). Land cover in the basin is diverse, but the category having the largest percentage is cropland (43%) used for row crop and small-grain agriculture. This creates challenges for simulating the hydrological cycle of the basin because of the important role of land management, including tile drainage. Other forms of land use include: land devoted to forages (pasture, hay, and rangeland), conservation reserve plantings, forests, wetlands, lakes, prairies, and urban areas.

For modeling with SWAT, the basin is divided into 119 sub-watersheds that are consistent with U.S. Geological Survey (USGS) eight-digit hydrologic cataloging unit (HCU) watershed boundaries as described by Seaber, Kapinos, and Knapp (1987). Each of these sub-watersheds is then further subdivided into hydrological response units (HRUs), creating a total of nearly 2,800 HRUs. Meteorological data from approximately 550 observing stations in the UMRB were averaged to provide precipitation and maximum and minimum temperature input for each eight-digit sub-watershed. Details of land use, soils, and topography data for the UMRB are described by Gassman et al. (2006). Our study domain lacks fine-scale orographic features that otherwise would surely compromise the ability of AOGCMs to describe the spatial distribution of hydrological processes over a region containing only a few AOGCM gridpoints.

Models

SWAT MODEL

The SWAT model (Arnold and Fohrer 2005) is a physically based, continuous time, long-term watershed scale hydrology and water quality model. SWAT version 2005 was used for this analysis. For simulation of streamflow by use of SWAT, the UMRB is divided into multiple sub-watersheds, which are further subdivided into HRUs, over which we assume land use, management, and soil characteristics are homogeneous. SWAT has routines to simulate components of the hydrological cycle, plant growth, and management practices as well as water-quality-related factors not considered in this report. Contributions to streamflow from each HRU in a sub-watershed are summed and then routed through channels, ponds, and/ or reservoirs to the watershed outlet. The role of subsurface tile drainage is included but likely underestimated due to poor data on the amount of tile installed in the last fifteen years.

Meteorological input to SWAT includes daily values of maximum and minimum temperature, total precipitation, mean wind speed, total solar radiation, and mean relative humidity. These data can be provided from observing stations or climate models. SWAT has an internal weather generator that fills in missing values from the observation files. The hydrologic cycle as simulated by SWAT at the HRU level is based on the balance of precipitation, surface runoff, percolation, evapo-transpiration, and soil water storage. SWAT takes total daily precipitation from models or observations and classifies it as rain or snow using the average daily temperature. When climate model output is provided, SWAT uses only total liquid precipitation and does its own partitioning to rain or snow. Snow cover is allowed to be non-uniform cover due to shading, drifting, topography, and land cover and is allowed to decline non-linearly based on an areal depletion curve. Snowmelt, a critical factor in partitioning between runoff and baseflow, is controlled by the air and snow pack temperature, the melting rate, and the areal coverage of snow. On days when the maximum temperature exceeds 0°C, snow melts according to a linear relationship of the difference between the average snow pack maximum temperature and the base or threshold temperature for snowmelt. The melt factor varies seasonally, and melted snow is treated the same as rainfall for estimating runoff and percolation. SWAT simulates surface runoff volumes for each HRU using a modified SCS curve number method (USDA Soil Conservation Service 1972). Further details can also be found in the SWAT user's manual (Neitsch et al. 2002).

AOGCMS

AOGCM results that included daily values needed for our simulations of current and future scenario climates were available from ten models (see table 10.1) in the IPCC Data Archive, including two versions of models from the Geophysical Fluid Dynamics Laboratory. The results provide a useful preliminary view of the hydrologic cycle components resulting from direct use of data generated by multiple AOGCMs. Takle, Jha, and Anderson (2005) found that streamflow data resulting from an AOGCM ensemble were serially uncorrelated at all lags and formed unimodal distributions, suggesting that the data may be modeled as independent samples from an identical normal distribution. The test of the hypothesis of zero difference between mean annual streamflow of the pooled AOGCM/SWAT and observed (OBS)/SWAT results gave a p-value of 0.5979, suggesting that use of AOGCM ensemble results may provide a valid approach for assessing annual streamflow in the UMRB. We use model output from runs of

Table 10.1. AOGCMs used in the SWAT-UMRB simulations

Institution	Model Name	Lon x Lat Resolution
Bjerknes Centre for Climate Research (Norway)	BCCR_BCM2.0	2.8° x 2.8°
Canadian Centre for Climate Modelling and Analysis	CCCMA_CGCM3.1	3.8° x 3.7°
Météo-France / Centre National de Recherches Météorologiques (France)	CNRM_CM3	2.8° x 2.8°
CSIRO Atmospheric Research (Australia)	CSIRO_MK3.0	2.8° x 2.8°
NOAA Geophysical Fluid Dynamics Laboratory (USA)	GFDL_CM2.0	2.5° x 2.0°
NOAA Geophysical Fluid Dynamics Laboratory (USA)	GFDL_CM2.1	2.5° x 2.0°
Center for Climate System Research (Japan)	MIROC3.2_MEDRES	2.8° x 2.8°
Meteorological Institute of the University of Bonn (Germany)	MIUB_ECHO_G	3.8° x 3.7°
Max Planck Institute for Meteorology (Germany)	MPI_ECHAM5	1.9° x 1.9°
Meteorological Research Institute (Japan)	MRI_CGCM2.3.2A	2.8° x 2.8°

the ten models listed in table 10.1 that have output for the twenty-first century (21C) A1B emission scenario (IPCC 2001) for the period 2046–2065.

Results

We have shown in previous studies (Jha et al. 2004; Takle, Jha, and Anderson 2005; Takle et al. 2006) that SWAT simulates annual total streamflow very well in validation experiments conducted using high-resolution meteorological data outside the calibration period, including years having droughts and floods. The seasonal cycle is less well simulated, particularly during years with precipitation extremes, but does capture quite well the main seasonal characteristics. This can be seen in figure 10.2, where the streamflow simulated by SWAT with data from all observing stations used as input (dark blue line) matches the measured streamflow at Grafton, Illinois (purple line) well except during June, July, and September, when errors of about 20% are common.

Accurate simulations of streamflow by SWAT for future scenario climates will re-quire input from AOGCMs that represent true features of these climates. Two features of AOGCMs that impede their ability to represent climate as required by SWAT are coarse resolution and model bias. Fortunately we can use global model simulations of the contemporary climate to assess the impact of these two features, as will be discussed in the following sub-sections.

INFLUENCE OF MODEL RESOLUTION

AOGCMs have equivalent grid spacing far too coarse to allow for model gridpoints in individual eight-digit watersheds. The UMRB has nominal dimension of 7° N-S by 5° E-W. A comparison of these dimensions with the model resolutions in table 10.1 shows that the number of gridpoints "representing" the basin for the low-resolution models ranges from about 5 for the Canadian Centre model to about 19 for the MPI model. The GFDL model grid spacing, which we use for the "degrading" the observations, has about 17 gridpoints (see figure 10.3). By contrast, 131 weather stations were used to represent baseline climate in

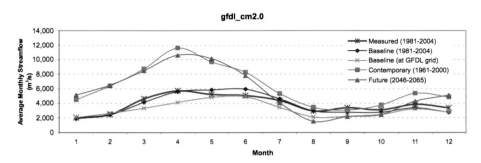

Figure 10.2. Mean monthly UMRB streamflow measured at Grafton, Illinois. Also plotted are UMRB streamflow simulated by SWAT for (a) the baseline period for which observed monthly streamflow data at Grafton, Illinois, are available (blue line), (b) the baseline period but allowing SWAT to use observed meteorological variables only from observing stations nearest the GFDL grid points (green line), (c) a forty-year period (1961–2000) where meteorological input values are simulated by the GFDL CM2.0 model (pink line), and (4) a future scenario (2046–2065) climate as simulated by the same GFDL model (green line).

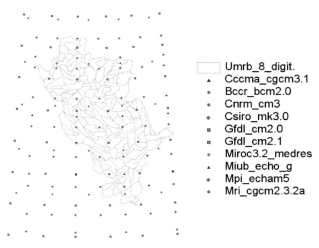

Umrb_8_digit.
▲ Cccma_cgcm3.1
• Bccr_bcm2.0
• Cnrm_cm3
+ Csiro_mk3.0
▪ Gfdl_cm2.0
▪ Gfdl_cm2.1
• Miroc3.2_medres
▲ Miub_echo_g
+ Mpi_echam5
• Mri_cgcm2.3.2a

Figure 10.3. Distribution of grid points over the Upper Mississippi River Basin from various AOGCMs.

the region (corresponding to a grid spacing of about 0.6°×0.6° if they were uniformly distributed).

Ideally, a simulation with SWAT would have at least one weather station per sub-basin. While this condition is met for the observing network in the UMRB, this requirement needs to be reconsidered when climate model data are used. It is instructive to consider the impact of model resolution on simulated hydrology.

We "degraded" the observations by using only those data points (17) closest to the GFDL model gridpoints as input to SWAT. The results, shown in figure 10.2, reveal that, for this basin, total annual streamflow at Grafton, Illinois, generally is reduced when coarse resolution observed data are used, as compared with results using high-resolution observations. When meteorological data were withheld from all observing locations except those 17 nearest GFDL gridpoints (light blue line), the streamflow is uniformly reduced from March through September. This "degraded" or low-resolution baseline simulated streamflow (light blue line) is a

better comparison for the low-resolution GFDL contemporary simulations (pink line). With GFDL model meteorological input, SWAT produces excessive stream-flow in all months, with maximum error of about 300% in April. And the curve for the degraded observations is even further from the model results, suggesting that a major source of error in the GFDL model is due to something other than resolution.

Streamflow produced with GFDL data is extreme when compared to other models in the list. As shown in figure 10.4, the mean of all ten models (pink line) produces a seasonal cycle much closer to the observations (purple line) and the SWAT-simulated streamflow using all station data (dark blue line). Since all models have coarse resolution, it is informative to compare their results with the degraded baseline (light blue line) even though it would be more accurate to create "degraded" observations for each model depending on its resolution. It is noteworthy that the mean streamflow simulated by the model ensemble matches very well the streamflow represented by the degraded observations for June–October and November. Largest errors occur in March–May.

From figure 10.4 we can conclude that compensating for model resolution produces a model ensemble simulation that matches the "degraded" streamflow remarkably well from July through December but leaves large absolute errors in other months and erroneously simulates peak flow in April rather than June.

INFLUENCE OF MODEL BIASES

It is noteworthy that precipitation, snowfall, and runoff are "events" whereas snowmelt, baseflow, evapo-transpiration (ET), potential evapo-transpiration (PET), and total water yield are continuous values. Snowfall (partitioning of precipitation to snow fraction) depends on temperature on the day of snowfall. ET and PET depend on temperature every day (more strongly in the warm season). Other components are not directly (although they are indirectly) dependent on temperature.

AOGCMs generally produce too many light rain events and too few intense events (Gutowski et al. 2003) even if rainfall totals are accurate. The impact of this bias, compared to the true intensity spectrum, is to reduce runoff and increase ET and/or baseflow. Low bias on rainfall amount likely would lead to low runoff, baseflow, ET, and hence water yield, while excess rain rate would have the opposite effect.

We compensated for model biases of temperature by computing the differences of monthly means between the model contemporary climate and the observed climate and subtracting these from the daily values produced by the model. We used the same correction for both contemporary and future scenario climate. For precipitation we computed the ratio of model-to-observed monthly mean. Corrected precipitation values were determined by dividing daily model values by the appropriate monthly correction ratio.

Impact of the bias correction for the GFDL CM2.0 model is revealed by comparison of SWAT-simulated streamflow using GFDL bias corrected values, shown in figure 10.5, with streamflow computed with uncorrected values shown in figure 10.2. A major reduction in streamflow is created by this correction so that error is approximately 20% rather than 300% (when compared to

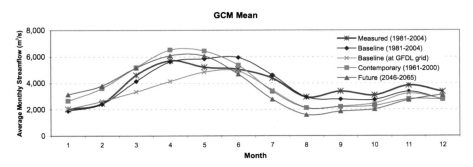

Figure 10.4. As in figure 10.2 except this is the composite of monthly results for all AOGCMs listed in table 10.1.

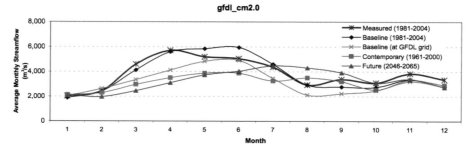

Figure 10.5. As in figure 10.4 except that variables are corrected for model biases.

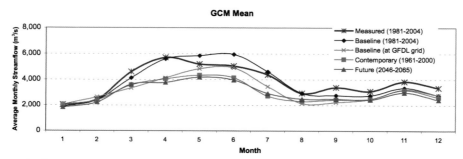

Figure 10.6. As in figure 10.5 except that variables are corrected for model biases.

the streamflow simulated with the degraded observations) in the cool season and almost no change in error in the warm season.

Closer inspection of the hydrological components reveals that annual precipitation was corrected downward by 10%, but snowfall was corrected down by slightly over 50% and runoff downward by slightly under 50%. This suggests that the temperature (the GFDL model is too warm) dominated the correction and underscores the importance of accurate temperature for estimating snowfall and snowmelt.

By performing this bias correction on all models we get the AOGCM model mean streamflow produced by SWAT, as shown in figure 10.6. Comparison of the composite contemporary climate results (pink line) with streamflow produced by the degraded observations (light blue line) reveals excellent agreement except for May–July. Comparison with figure 10.4 shows that the

60%-error maximum in April has been completely eliminated and the maximum error now is in July and has magnitude of about 25%. It is noteworthy that the bias correction had major impact on streamflow for the cold season but had almost no impact on the warm season.

CLIMATE CHANGE

AOGCM simulations of precipitation over the UMRB for the current climate have large errors (31% too high for MPI ECHAM5 to 28% too low for MRI CGCM2.3.2a). For the future climate all models show very modest changes in precipitation, MRI CGCM2.3.2a having the largest change with a 7.5% decrease. We corrected each model for its biases in meteorological variables and used the results to simulate streamflow with SWAT for the future scenario climate of 2046–2065, as shown in figure 10.6. Comparison of the

Table 10.2. Mean annual streamflow for contemporary and future scenario climates for each model compared with baseline at the GFDL grid spacing (a) without and (b) with bias corrections made on temperature and precipitation.

	(a) no bias correction			(b) bias correction		
	Streamflow (m³/s)			Streamflow (m³/s)		
Measured	45,995			45,995		
Baseline	45,060			45,085		
Baseline (at GFDL grid)	38,263			42,403		
BASELINE	*Cont.*	*Future*	*% Change*	*Cont.*	*Future*	*% Change*
BCCR_BCM2.0	23,011	18,378	−20	36,142	36,175	0
CCCMA_CGCM3.1	41,430	46,554	12	34,077	32,483	−5
CNRM_CM3	47,367	46,437	−2	30,685	28,785	−6
CSIRO_MK3.0	27,864	28,891	4	37,585	36,658	−2
GFDL_CM2.0	74,840	68,430	−9	37,422	39,419	5
GFDL_CM2.1	54,158	55,685	3	38,640	41,439	7
MIROC3.2_MEDRES	14,468	9,668	−33	37,130	36,614	−1
MIUB_ECHO_G	55,572	42,827	−23	36,763	35,214	−4
MPI_ECHAM5	99,954	99,004	−1	39,079	34,791	−11
MRI_CGCM2.3.2A	13,709	15,451	13	35,377	32,752	−7

future scenario streamflow (green line) with the contemporary climate (pink line) shows essentially no change in streamflow and retention of the essential seasonal features (e.g., major peak in May, broad minimum in mid to late summer, and minor peak in November) of the current climate. Comparison of table 10.2 (a) and (b) reveals how reduction in bias affects each AOGCM's estimate of change in streamflow due to climate change.

Further examination of all hydrological components simulated by SWAT is in progress to ensure that these streamflow simulations are consistent with other features of the AOGCM future scenarios.

Conclusions

Output from an ensemble of ten AOGCMs was examined for use in driving a regional hydrological model for contemporary and future scenario climates. Examination of the role of resolution of driving meteorological fields for the streamflow model revealed that low-resolution meteorology generally leads to erroneously low streamflow in all months. However, when these biases are eliminated, simulations of streamflow improve dramatically, especially in the cool season (first half of the year). The AOGCMs used in this analysis showed very little change in precipitation due to increase in greenhouse gases, and hence very little change in streamflow.

ACKNOWLEDGMENTS
We acknowledge the international modeling groups for providing their data for analysis, the Program for Climate Model Diagnosis and Intercomparison (PCMDI) for collecting and archiving the model data, the JSC/CLIVAR Working Group on Coupled Modelling (WGCM) and their Coupled Model Intercomparison Project (CMIP) and Climate Simulation Panel for organizing the model data analysis activity, and the IPCC WG1 TSU for technical support.

The IPCC Data Archive at Lawrence Livermore National Laboratory is supported by the Office of Science, U.S. Department of Energy. We thank Z. Pan for his constructive comments on an early draft. Partial support for this work was provided by USDA National Research Initiative Grant #20063561516724.

REFERENCES

Arnold, J. G., and N. Fohrer. 2005. "Current Capabilities and Research Opportunities in Applied Watershed Modeling." *Hydrological Processes* 19: 563–572.

Gassman, P. W., et al. 2006. "Upper Mississippi River Basin Modeling System Part 1: SWAT Input Data Requirements and Issues." In *Coastal Hydrology and Processes,* ed. V. P. Singh and Y. J. Xu, 103–117. Highlands Ranch, Colo.: Water Resources Publications.

Groisman, P. Y., et al. 2005. "Trends in Intense Precipitation in the Climate Record." *Journal of Climate* 18: 1326–1350.

Gutowski, W. J., Jr., et al. 2003. "Temporal-Spatial Scale of Precipitation Errors in Central U.S. Climate Simulation." *Journal of Climate* 16: 3841–3847.

IPCC. 2001. *Climate Change 2001: The Scientific Basis.* Contribution of Working Group I to the Third Assessment Report of the Intergovernmental Panel on Climate Change. [Ed. J. T. Houghton, et al.] New York: Cambridge University Press.

IPCC. 2007. *Climate Change 2007: The Physical Science Basis.* Contribution of Working Group I to the Fourth Assessment Report of the Intergovernmental Panel on Climate Change. [Ed. S. Solomon, et al.] New York: Cambridge University Press.

Jha, M., et al. 2004. "Impact of Climate Change on Stream Flow in the Upper Mississippi River Basin: A Regional Climate Model Perspective." *Journal of Geophysical Research* 109, doi:10.1029/2003JD003686.

Neitsch, S. L., et al. 2002. *Soil and Water Assessment Tool: User Manual, Version 2000.* Texas Water Resour. Inst., TR-192. GSWRL 02-02, BRC 02-06.

Pan, Z., et al. 2004. "Altered Hydrologic Feedback in a Warming Climate Introduces a 'Warming Hole.'" *Geophysical Research Letters* 31, doi:10.1029/2004GL020528.

Program for Climate Model Diagnosis and Intercomparison (PCMDI). 2007. Available at: http://www-pcmdi.llnl.gov/ (accessed 06/25/08).

Seaber, P. R., F. P. Kapinos, and G. L. Knapp. 1987. *Hydrologic Units Maps.* U.S. Geological Survey, Water-Supply Paper 2294.

Stocker, T. F., and C. C. Raible. 2005. "Water Cycle Shifts Gear." *Nature* 434: 830–833.

Takle, E. S., M. Jha, and C. J. Anderson. 2005. "Hydrological Cycle in the Upper Mississippi River Basin: 20th Century Simulations by Multiple GCMs." *Geophysical Research Letters* 32, doi:10.1029/2005GL023630.

Takle, E. S., et al. 2006. "Upper Mississippi River Basin Modeling Systems, Part 4: Climate Change Impacts on Flow and Water Quality." In *Coastal Hydrology and Processes,* ed. V. P. Singh and Y. J. Xu, 135–142. Highlands Ranch, Colo.: Water Resources Publications.

USDA Soil Conservation Service. 1972. *National Engineering Handbook.* Section 4, "Hydrology," chapters 4–10.

Wu, P., R. Wood, and P. Stott. 2005. "Human Influence on Increasing Arctic River Discharges." *Geophysical Research Letters* 32, doi:10.1029/2004GL021570.

11. The Influence of Land Cover Type on Surface Hydrology in Michigan

J. A. ANDRESEN, W. J. NORTHCOTT,
H. PRAWIRANATA, AND S. A. MILLER

Introduction

The hydrologic behavior of a given area is dependent on a number of physical factors including climate, geology, and topographical setting. It is also dependent on the land cover type, which in turn influences surface energy and water balances. The interrelationship and interaction between these factors may be complex. In general, the greater the surface albedo, the lower the net radiation of the surface, which in turn impacts the rate of potential evapo-transpiration, available soil moisture, and so on. The amount and type of vegetation on the surface can influence the albedo, the rate and amount of precipitation reaching the soil, and rate of evapo-transpiration. For example, decreases in vegetation leaf area index and forest maturity as a result of logging and fire were found to have been associated with increases in surface runoff and decreases in evapo-transpiration on an annual basis in several previous studies (Bosch and Hewlitt 1982; VanShaar, Haddeland, and Lettenmaier 2002; Matheussen et al. 2000).

In Michigan and the Great Lakes region, land cover type has varied considerably over time. In a study of land cover change since 1850, Ramankutty and Foley (1999) concluded that cropland area in the Midwestern USA increased steadily until approximately 1940, but has remained relatively stable since. Much of the historical conversion to cropland in Michigan and the Great Lakes region was associated with logging/clear-cutting of forests or removal of savannah/grassland areas. At the same time, climate in the region has also changed. According to a recent study (Andresen 2009), major trends across the state since the beginning of the twentieth century are somewhat similar to larger-scale global and hemisphere trends, including warming temperatures from approximately 1900 to 1940, followed by a cooling trend from the early 1940s to the late 1970s, which was in turn followed by a second warming trend that began around 1980 and has continued to the present. Much of the warming trend during the past two to three decades was associated with warmer minimum temperatures during the winter and spring seasons. Another important regional trend identified in that study was an increase of precipitation since approximately 1940, which has leveled off somewhat during the past decade (Andresen et al. 2001). Seasonal snowfall totals have increased significantly during the past several decades in areas of the region frequented by lake-effect snowfall, while totals in non-lake-effect areas have remained steady or decreased slightly. Some of these climate trends may be related to long-term land use changes, especially the conversion of natural land-scapes to agriculture and urban/suburban development (Lamptey, Barron, and Pollard 2005; Bonan 1999). Given the many linkages between land cover, surface hydrology, and climate, the major objectives in this study were an examination of the effect of major land cover types on surface water balance in Michigan and identification of any water balance component trends during the historical period of record.

Methodology

Estimation of the surface water balance in this study was carried out with a process-based simulation model that accounts for the major hydrologic processes and their interactions. The Soil and Water Assessment Tool (SWAT), a continuous, process-based daily time-step simulation model, was used to estimate the long-term water balance across several different treatment parameters including climate zone, soil type, land use class, and land management practice. The hydrologic modeling work was carried out in two separate phases: (1) a case study simulation for validation of the model and (2) surface water balance simulations at a number of individual sites representative of differing climates in the state over the historical period of record. For the validation phase, the Augusta Creek watershed of southwest Lower Michigan was used.

SWAT MODEL

The Soil and Water Assessment Tool (SWAT) is a process-based, continuous, watershed-scale hydrology and water quality simulation model developed by the USDA Agricultural Research Service's Grassland, Soil and Water Research Lab (Neitsch et al. 2002). The model was developed for the purpose of assisting water resource managers in predicting and assessing the impact of management on water, sediment, and agricultural chemical yields in large, ungaged watersheds. The hydrologic components of the model have been rigorously tested on watersheds of varying size (Gassman et al. 2007; Arnold et al. 2000; Srinivasan et al. 1998). The basic model operates on a daily time step and allows continuous simulation over many years. Recent additions allow for simulating surface runoff and infiltration using the Green and Ampt (1911) approach using rainfall data of anytime increment and hourly channel routing. The SWAT model has eight major system components: hydrology, weather, erosion and sedimentation, soil temperature, plant/crop growth, nutrients, pesticides, and agricultural management.

To simulate the spatial heterogeneity of land cover, topography, soil type, and climate, watersheds may be divided into a number of user-delineated sub-basins. Each sub-basin is then further subdivided into individual hydrological response units (HRUs); an HRU is an individual combination of land use/cover/management, soil type, and meteorological data.

The daily or sub-daily water budget is computed for each HRU in the watershed. Daily surface runoff is calculated using the SCS curve number approach or on an hourly basis using the Green and Ampt method (USDA/Soil Conservation Service 1972). Peak runoff rate is calculated using a modified rational formula, and the routing of in-channel flow between sub-basins is computed with Manning's equation and the Muskingum or variable storage method.

The meteorological input to the model consists of up to five parameters: rainfall (daily or sub-daily), daily maximum and minimum temperatures, mean daily relative humidity, total daily solar radiation, and average daily wind speed. The model is able to estimate missing data values in the United States from a spatial reference meteorological database consisting of data obtained from the National Weather Service. Potential evapo-transpiration data can be provided by the user or estimated in the model with one of a number of standard methods, including Penman-Monteith, Priestly-Taylor or Hargreaves (Neitsch et al. 2002).

CLIMATE DATA

Due to its unique location in the Great Lakes Basin, Michigan has a diverse climate with respect to precipitation and temperature (see chapters 1, 2, and 9 of this volume). The climate in Michigan is described generically as humid continental, with hot summers, cold winters, and precipitation occurring each month. The proximity of the Great Lakes plays an important role in modifying regional climate (see chapter 21), with an increase in cloudiness and precipitation in areas downwind of the lakes during the fall and winter months due to the passage of relatively cooler and drier air (typically from continental source regions)

across open water. Spatially averaged across the state, Michigan averages 830 mm of precipitation per year, with annual totals varying from less than 720 mm in east central and northeastern sections of the Lower Peninsula to just above 960 mm in the extreme western Upper and southwestern Lower Peninsulas (figure 11.1). In terms of seasonality, February is climatologically the driest month, while August and September are the wettest. About 60% of the annual total is recorded during the May–October growing season. Summer precipitation falls primarily in the form of showers or thunderstorms, while a more steady type of precipitation of lighter intensity dominates the winter months. Seasonal snowfall varies from less than 100 cm in the southeastern Lower Peninsula to more than 550 cm in sections of the northern Upper and northwestern Lower Peninsulas.

To determine the impact of climate on surface hydrology, we chose five locations across the state with representative climates and high quality, long-term daily climate data (at least sixty years). The five locations selected were Coldwater, Bay City, Big Rapids, East Jordan, and Chatham (figure 11.1).

Daily series of precipitation and maximum and minimum temperatures for each study site were obtained from the NOAA Summary of the Day (NOAA/NCDC 1897–2004). The individual lengths of the data series ranged from 68 to 108 years, and all but one of the series began in 1895–1900. Missing temperature and precipitation data within the observed period of record were estimated by spatial interpolation of data from neighboring stations and were provided by the Midwest Regional Climate Center in Champaign, Illinois (S. Hilberg, personal

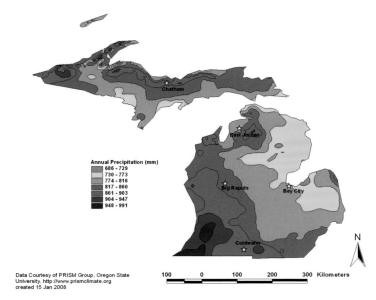

Annual Precipitation (mm)
686 - 729
730 - 773
774 - 816
817 - 860
861 - 903
904 - 947
948 - 991

Data Courtesy of PRISM Group, Oregon State
University, http://www.prismclimate.org
created 15 Jan 2008

100   0   100   200   300 Kilometers

Figure 11.1. Average annual precipitation totals in Michigan, 1971–2000. Data courtesy of PRISM Group, Oregon State University.

communication). Daily solar radiation data for the entire study period were synthetically generated on the basis of the observed precipitation and temperature data with the Weather Generator (WGEN) methodology of Richardson and Wright (1984).

SOILS DATA

Soils across Michigan are highly heterogeneous and range in texture from sandy to clayey, each with varying hydraulic properties that affect infiltration, runoff, and percolation. To simplify the hydrologic characteristics of given soil, each soil has been classified by the USDA-NRCS into one of four hydrologic soil groups (HSG) designated A, B, C, or D. Soils are grouped primarily by their infiltration capacity, with A soils being sandy or sandy loam with low runoff potential, while D soils are clayey with much less infiltration capacity and high runoff potential. County-level or SSURGO-level soils mapping has been produced in

digital form on seventy-four of Michigan's eighty-three counties (USDA/NRCS 1995). We assembled individual county data layers into a single GIS coverage. The majority of soils within the state fall into coarser-textured, more well-drained A and B soil hydrologic soils groups. Of the soils that are digitally mapped and classified, 28% are A soils and 30% are B soils. SWAT contains a nationwide database of the physical, chemical, and biological properties of NRCS classified soils. Each of the Michigan soils used in this study was cross-referenced with this database, and the resulting properties were used in the hydrologic simulations. With the SWAT model, we simulated the groundwater recharge from our different land uses across 265 individual soil types that represent the vast majority of soils in the state (504 total). These 265 soil types were segregated by hydrologic group, with totals of 78 A, 120 B, 30 C, and 37 D soils. We did not consider soils that have dual classification (e.g., A/D) soils as these categories represent soils that in

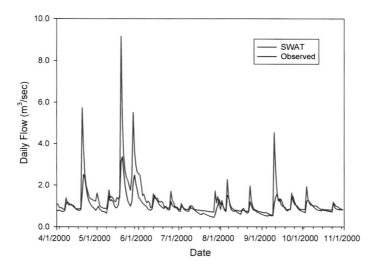

Figure 11.2. SWAT-simulated versus observed daily streamflow for Augusta Creek at Augusta, Michigan, April–October, 2000.

their natural state respond as a D soil (e.g., an undrained wetland), but if they are artificially drained will respond as an A soil from a surface runoff standpoint. Due to the site-specific drainage management aspect, these types of soils were deemed too complex to model in a general sense across the state. For simplicity, the slope of the landscape was assumed to be near flat, with a slope of 0.5 degrees for all simulations.

LAND COVER/LAND USE DATA

There are multiple land cover/land use classes for the state of Michigan, ranging from forest and wetland to agricultural row crop and urban. A geospatial data set from the Michigan Department of Natural Resources (MDNR 2001) was used for all land use classification in the study. This data set was derived from 30-meter spatial resolution LANDSAT and Thematic Mapper (TM) satellite imagery taken between 1997 and 2001. These image data were assembled and clas-

sified into thirty individual land use classes. Land use classes from the MDNR scheme were matched with representative land cover type parameterizations provided in the SWAT system for all simulations.

SWAT MODEL VALIDATION IN AUGUSTA CREEK WATERSHED

In a test validation of the SWAT model in representing surface water hydrology of a given sub-region, we chose the Augusta Creek watershed of southwestern Lower Michigan. This 9,590 ha watershed is highly diverse and contains a mix of soil and land cover types representative of the state as a whole. The model was parameterized using 30 m digital elevation data and 2001 land use data. Soils data were obtained from the SSURGO soils database. Climatological data (precipitation and temperature) were obtained from a NOAA Climatological Cooperative Network station at the Kellogg Biological Station (Hickory Corners,

Michigan), which lies within the watershed. Observed daily surface water flow data were obtained for Augusta Creek from the U.S. Geological Survey and were used to evaluate the efficiency of the model. Daily simulations from the period 1981–2000 were used to calibrate and validate the model, with odd years used for calibration and even years for validation. An example of model output from 2000 is given in figure 11.2. The model was calibrated primarily by adjusting the curve number of the various land use soil type combinations, which partitions rainfall between water that runs off into the stream and water that infiltrates into the soil profile for plant use and ground water recharge. The curve numbers were adjusted based on a calibration scheme using a Shuffled Complex Evolution algorithm to achieve the best fit of the estimates with the observed data.

In general, the simulated estimates follow the daily hydrology of the watershed, with an overall r^2 value between observed and simulated flow of 0.67. The overall mean difference between simulated and observed flow was 0.266 m^3/s, suggesting a positive bias in the SWAT output, especially in the overestimation of peak flow during large flow events (figure 11.2). This is most likely due to the hydraulics of the Augusta Creek floodplain. The Augusta Creek floodplain is relatively unaltered from its natural state, which gives it a large buffering capacity during high-flow events. Streamside wetlands are abundant along the creek and absorb much of the flow, lowering the flood peak. The SWAT stream routing model was designed for simulating the hydraulics of channelized streams and hence tends to overestimate the peak flows of streams with significant floodplain storage. Even with

these over-predictions, the overall mean absolute difference between simulated and observed flow was only 0.299 m^{3}s^{-1}. Given the strong correlation between SWAT and the observed data and the relatively low bias and differences, the model was deemed appropriate for the set of historical simulations.

Attempting to simulate recharge across a multitude of soil types, land cover, and climate zones requires numerous computer simulations and analyses. Our scenario simulations consisted of running the SWAT model over all of the combinations of 5 locations, 9 land covers, and 265 soil types. This resulted in 11,925 individual simulations. The water balance components were averaged over the 1971–2000 simulation period and by their soil hydrologic group. So for each location, land cover, and soil hydrologic group, there is an individual groundwater recharge/percolation value.

Results and Discussion

Averages of the surface water balance components at the five sites across four representative land cover types from 1971–2000 are given in table 11.1.

Differences between the land cover types are striking, especially for the urban cover, for which the relatively low surface permeability resulted in runoff and percolation (shallow groundwater recharge) values that are significantly higher and lower, respectively, than the other three vegetated surfaces. A graphical illustration of the differences across land cover types is given in figure 11.3 for the water balance terms at Coldwater for type B hydrologic group soils. Mean evapotranspiration, the largest sink for surface water, ranges in this case from 376 mm over

Table 11.1. Simulated annual mean water balance components (mm) by soil hydrologic group at each study location, 1971–2000

| | | Forest | | | | Agricultural row crops | | | |
| | | Soil hydrologic group | | | | | | | |
Location	Water Balance Component (mm)	A	B	C	D	A	B	C	D
Coldwater	Precipitation	862	862	862	862	862	862	862	862
	Evapo-transpiration	533	575	608	600	399	422	426	436
	Percolation	318	272	225	223	320	233	157	132
	Surface Runoff	10	14	28	38	142	206	278	292
Big Rapids	Precipitation	832	832	832	832	832	832	832	832
	Evapo-transpiration	496	538	572	565	367	391	397	405
	Percolation	318	270	219	213	296	214	142	119
	Surface Runoff	16	21	39	52	166	225	292	305
Chatham	Precipitation	827	827	827	827	827	827	827	827
	Evapo-transpiration	481	517	541	532	356	377	382	393
	Percolation	320	276	227	221	290	209	137	118
	Surface Runoff	23	32	57	72	179	240	306	314
East Jordan	Precipitation	800	800	800	800	800	800	800	800
	Evapo-transpiration	489	530	563	556	364	385	390	399
	Percolation	299	252	203	200	294	217	148	126
	Surface Runoff	9	14	31	41	140	196	260	273
Bay City	Precipitation	736	736	736	736	736	736	736	736
	Evapo-transpiration	478	521	556	550	370	393	397	410
	Percolation	250	204	162	159	248	172	110	91
	Surface Runoff	6	9	17	26	117	170	227	234
		Grassland				Urban high density			
		Soil hydrologic group							
Location	Water Balance Component (mm)	A	B	C	D	A	B	C	D
Coldwater	Precipitation	862	862	862	862	862	862	862	862
	Evapo-transpiration	341	376	391	402	454	471	479	456
	Percolation	517	408	295	207	58	12	2	4
	Surface Runoff	2	77	174	252	349	378	380	402
Big Rapids	Precipitation	832	832	832	832	832	832	832	832
	Evapo-transpiration	316	346	360	369	413	435	445	424
	Percolation	510	384	270	190	71	21	6	7
	Surface Runoff	4	100	199	271	346	375	380	399
Chatham	Precipitation	827	827	827	827	827	827	827	827
	Evapo-transpiration	301	330	343	352	418	413	416	398
	Percolation	513	370	257	180	82	32	13	11
	Surface Runoff	10	125	225	293	355	388	398	411
East Jordan	Precipitation	800	800	800	800	800	800	800	800
	Evapo-transpiration	310	340	353	361	402	426	437	419
	Percolation	486	376	271	195	81	27	10	11
	Surface Runoff	2	82	173	241	316	345	351	369
Bay City	Precipitation	736	736	736	736	736	736	736	736
	Evapo-transpiration	312	343	357	365	402	416	422	401
	Percolation	421	323	228	157	46	11	3	4
	Surface Runoff	2	69	150	212	287	308	311	331

a grassland surface to 575 mm for forest. This is physically plausible, as a forest (for the simulation, a mixed, northern deciduous forest was assumed) transpires over the entire growing season, has potentially more plant extractable water availability due to a deeper rooting system, and also intercepts a relatively larger fraction of precipitation in the canopy, which may subsequently evaporate. The grassland surface transpires similarly to the other vegetated surfaces during the growing season until mid-late summer, when early senescence rapidly reduces transpiration rates, resulting in a lower seasonal total. The mean runoff totals increase from 14 mm for the forest cover to 77 mm over grassland to 206 mm for the agricultural row crop to 378 mm for the urban surface, the result of both decreasing surface permeability and canopy interception of precipitation. The percolation or shallow groundwater recharge term is almost nil for the urban surface at 12 mm, increasing to 233 mm for row crop to 272 mm for forest to 408 mm for grassland.

Considering the overall proportions of the water sink terms of evapo-transpiration, runoff, and percolation with the precipitation source term, the results are similar, with the surface cover type having a large impact. For evapo-transpiration, the relative proportion of mean annual precipitation ranges from 36.4% for grassland on an A soil at Chatham to 75.5% for forest on a D soil at Bay City. For runoff, the proportions range from 0.2% on grassland A soils at Coldwater to 49.7% on the urban surface over a D soil at Chatham. The percolation term was found to be most variable of the balance terms, ranging from only 0.2% under an urban surface with a C soil to as much as 62.0% under grassland with an A soil at Chatham. Overall, these patterns are consistent with earlier results of Twine, Kucharik, and Foley (2004), who investigated changes in pre- and post-settlement vegetation types on surface energy balance in the Upper Midwest. Four other vegetated surface cover types were also considered in the study that are typical of the landscape in Michigan and the Upper Midwest region: alfalfa (a perennial crop), scrub vegetation (typically secondary growth on former agricultural lands), and both marsh-covered and forested wetlands. The water balance proportions and patterns

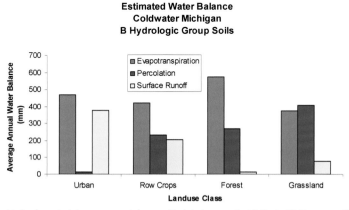

Figure 11.3. SWAT-simulated surface water balance components for group B hydrologic group soils at Coldwater, Michigan, across different land use categories, 1971–2000.

for these surfaces were very similar to those of the grassland and forest, with some relative increase in evapo-transpiration rates for the marsh surfaces (data not shown). A second, low-density urban cover was also considered, and the results were very similar to but less extreme than those of the high-density cover (data not shown).

An overall influence of soil hydrologic group is also apparent in the water balance results (table 11.1), but is less pronounced than for the differing land cover types. In general, the more coarse-textured and well-drained the soil (e.g., type A and B soils), the less the annual runoff, evapo-transpiration, and percolation. One interesting result was an increase in evapo-transpiration of approximately 5–10% from the well-drained A soils to the poorly drained D soils. This is a result of the relatively greater water-holding capacities of the finer-textured C and D type soils, which typically contain clay as a major component. This suggests that soil type (and texture) are relatively less important than the surface cover type in determining the long-term water balance. Predictably, the greatest relative differences in annual evapo-transpiration across soil hydrologic groups were found for the forested surface and the least for the urban surface.

In selecting our study locations, we also considered the form of precipitation, that is, rainfall versus snow or other frozen precipitation types, as the form of precipitation should tend to modify the timing of precipitation runoff and percolation (due to surface storage in a snowpack, melting, etc.), and this issue is taken into account in the SWAT simulation model. In our simulations, there were no apparent differences in the balance components at Chatham and

East Jordan, with mean annual snowfall totals 449.3 cm and 270.5 cm, respectively, and a relatively higher fraction of the annual precipitation occurring as snow. Mean annual snowfall totals at the other three sites are less than 180 cm. This suggests that while the intra-seasonal rates of runoff and percolation may differ due to frozen versus liquid precipitation, the ultimate total over a year varies little. We also investigated the impact of surface slope on the balance components, with a range of surface slopes from 0.5 to 15° (typical of the region) considered. In general, there was a tradeoff of runoff and percolation. The greater the slope, the greater the runoff and the less the percolation. The evapo-transpiration rates decreased by a small amount at higher slopes, but did not differ greatly across the slope treatments (data not shown).

As mentioned earlier, climate in Michigan and the Upper Midwest region has become wetter and warmer in recent decades, and these trends translated into changes in water balance components over time. Overall temporal variability of the components was greatest for precipitation, followed by percolation, runoff, and finally evapo-transpiration. A representative time series of the water balance components from Coldwater is given in figure 11.4, in which the data are plotted as nine-year moving averages to identify longer, decadal-scale temporal changes. Variations in the association between precipitation and the various water balance sink components is evident. Simple correlation coefficients in this case between precipitation and the components range from 0.46 for evapo-transpiration to 0.81 for runoff to 0.90 for percolation. The most significant long-term trend in this

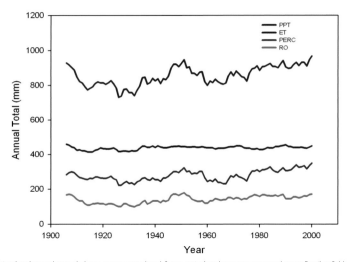

Figure 11.4. Trends of simulated annual water balance components (mm) for an agricultural row crop cover and group B soil at Coldwater, Michigan, 1902–2004. All series are plotted as nine-year moving averages.

case (and in virtually all of the vegetated location/cover type/soil group combinations) is an increase in precipitation from approximately 725 mm during the mid 1930s to 975 mm during 1995–2004. Some of this more than 20% increase in annual precipitation is reflected in increases in runoff values (approx. 110 mm to 190 mm) and evapotranspiration rates (approx. 410 mm to 460 mm). The largest individual portion of the increase was associated with an increase in percolation (approx. 220 to 340 mm). The relatively small change in evapo-transpiration suggests that plants are unable to take full advantage of the increasing water in the soil before it drains below the rooting zone as shallow groundwater recharge. It also suggests relative increases of groundwater over time as baseflow into rivers and lakes. This is consistent with significant upward trends in low streamflows in the Upper Midwest regions identified by Douglas, Vogel, and Kroll (2000) and increases in streamflow rates across the Great Lakes region from 1948–1988 (Lettenmaier, Wood,

and Wallis 1994). In a study of changes of streamflow in a projected warmer and wetter climate of the future across the Upper Midwest region, Jha et al. (2003) found increased annual groundwater recharge of 43% resulting from an annual precipitation increase of 21%.

Finally, given increasing concern and new regulations governing water resources in the region, it is interesting to consider the potential impact of agricultural irrigation on groundwater in relation to our results. Seasonal irrigation application totals for the majority of irrigated crops in the region are typically range from 150 to 250 mm/year (Algozin, Bralts, and Ritchie 1988), and almost all of these crops are grown on coarse-textured, well-drained soils (i.e., hydrologic groups A and B). Given mean values of percolation on group A and B soils across our study sites generally in the range of 200–300 mm/year, this suggests that irrigation for agriculture could be at least theoretically sustainable over a long period in terms of groundwater resource usage.

Concluding Remarks

A study to examine the links between ground water recharge, land use/cover type (including agricultural crops), and climate was undertaken in Michigan. The Soil and Water Assessment Tool (SWAT), a continuous, process-based, daily time-step simulation model, was used for all simulations. For the simulations, nine major land use categories were selected from approximately thirty in the state: low- and high-density urban, agricultural row crop (corn), agricultural forage (alfalfa), forest (oak deciduous), natural grassland, shrub/scrub, and forested and marsh wetlands. For soils, 265 individual soil types representing the majority of soils in the state were selected and segregated by hydrologic group (A, B, C, and D). Climate data were obtained from five locations across the state with relatively complete, long-term (one hundred years or greater) series length: Coldwater, Bay City, Big Rapids, East Jordan, and Chatham. Recharge simulations were run for all combinations of soil types, land cover, and location (climate) for a total of 11,925 individual simulations. In terms of the hydrological balance, evapotranspiration was the largest water sink term over all land use and soil types, followed by percolation/shallow groundwater recharge and surface runoff. The latter two variables were found to be inversely proportional to one another, with greatest recharge values and least runoff found on coarse-textured, highly permeable soils (group A). All of the hydrological water sink terms (evapo-transpiration, runoff, and recharge) were found to be positively correlated with precipitation, especially recharge. Several temporal trends were noted in the model output. Precipitation increases across the state during the late 1930s through the late 1990s were associated with increases in all three water balance sink components. The model results suggest that the majority of the additional precipitation of the past few decades has ended up as groundwater recharge.

ACKNOWLEDGMENTS
Financial support from the Great Lakes Protection Fund and the Michigan Agricultural Experiment Station (Project MICL03373) is gratefully acknowledged.

REFERENCES
Algozin, K. A., V. F. Bralts, and J. T. Ritchie. 1988. "Irrigation Strategy Selection Based on Crop Yield, Water, and Energy Use Relationships: A Michigan Example." *Journal of Soil and Water Conservation* 43: 428–431.

Andresen, J. A. 2009. "Historical Climate Trends in Michigan and the Great Lakes Region." In *Proc. Int. Symposium on Climate Change in the Great Lakes Region: Decision Making Under Uncertainty,* 15–16 March 2007. East Lansing: Michigan State University (in press).

Andresen, J. A., et al. 2001. "Assessment of the Impact of Weather on Maize, Soybean, and Alfalfa Production in the Upper Great Lakes Region of the United States, 1895–1996." *Agronomy Journal* 93: 1059–1070.

Arnold, J. G., and N. Fohrer. 2005. "Current Capabilities and Research Opportunities in Applied Watershed Modeling." *Hydrologic Processes* 19: 563–572.

Arnold, J. G., et al. 2000. "Regional Estimation of Base Flow and Groundwater Recharge in the Upper Mississippi River Basin." *Journal of Hydrology* 227: 21–40.

Bonan, G. B. 1999. "Frost Followed the Plow: Impacts of Deforestation on the Climate of the United States." *Ecological Applications* 9: 1305–1315.

Bosch, J. M., and J. D. Hewlitt. 1982. "A Review of Catchment Experiments to Determine the Effect of Vegetation Changes on Water Yield and Evapotranspiration." *Journal of Hydrology* 55: 3–23.

Douglas, E. M., R. M. Vogel, and C. N. Kroll. 2000. "Trends in Floods and Low Flows in the United States: Impact of Spatial Correlation." *Journal of Hydrology* 240: 90–105.

Gassman P. W., et al. 2007. *The Soil and Water Assessment Tool: Historical Development, Applications, and Future Research Directions.* Working Paper 07-WP 443. Ames: Center for Agricultural and Rural Development, Iowa State University.

Green, W., and A. Ampt. 1911. "Studies on Soil Physics: 1. The Flow of Air and Water through Soils." *Journal of Agricultural Science* 4: 1–24.

Jha, M., et al. 2003. *The Impacts of Climate Change on Stream Flow in the Upper Mississippi River Basin: A Regional Climate Model Perspective.* Working paper 03-WP 337. Ames: Center for Agricultural and Rural Development, Iowa State University.

Lamptey, B. L., E. J. Barron, and D. Pollard. 2005. "Impacts of Agriculture and Urbanization on the Climate of the Northeastern United States." *Global and Planetary Change* 49: 203–221.

Lettenmaier, D. P., E. F. Wood, and J. R. Wallis. 1994. "Hydro-Climatological Trends in the Continental United States, 1948–1988." *Journal of Climate* 7: 586–607.

Matheussen, B., et al. 2000. "Effects of Land Cover Change on Streamflow in the Interior Columbia Basin (USA and Canada)." *Hydrological Processes* 14: 867–885.

Michigan Department of Natural Resources. 2001. *Integrated Forest Monitoring Assessment and Prescription* (*IFMAP*). 2nd rev.

Neitsch, S., et al. 2002. *Soil and Water Assessment Tool Theoretical Documentation Version 2000.* Temple, Tex.: Grassland, Soil and Water Research Laboratory, USDA Agricultural Research Station.

NOAA/NCDC. 1897–2004. *Climatological Data.* Asheville, N.C.: National Oceanic and Atmospheric Administration, National Climatic Data Center.

Ramankutty, N., and J. A. Foley. 1999. "Estimating Historical Changes in Land Cover: North American Croplands from 1850 to 1992." *Global Ecology and Biogeography* 8: 381–396.

Richardson, C. W., and D. A. Wright. 1984. *WGEN: A Model for Generating Daily Weather Variables.* Pub. ARS-8. USDA Agricultural Research Service.

Srinivasan, R., et al. 1998. "Large Area Hydrologic Modeling and Assessment Part II: Model Application." *Journal of the American Water Resources Association* 34: 91–101.

Twine, T. E., C. J. Kucharik, and J. A. Foley. 2004. "Effects of Land Cover Change on the Energy and Water Balance of the Mississippi River Basin." *Journal of Hydrometeorology* 5: 640–655.

USDA/NRCS. 1995. *Soil Survey Geographic* (*SSURGO*) *Database.* USDA/NRCS, Misc. Publ. 1527. Lincoln, Neb.: National Soil Survey Center. Available at: http://www.ftw.nrcs.usda.gov/ssur_data.html.

USDA/Soil Conservation Service. 1972. *National Engineering Handbook.* Section 4 (Hydrology). Washington, D.C.: USDA/SCS.

VanShaar, J. R., I. Haddeland, and D. P. Lettenmaier. 2002. "Effects of Land-Cover Changes on the Hydrological Response of Interior Columbia River Basin Forested Catchments." *Hydrological Processes* 16: 2499–2520.

12. Wisconsin's Changing Climate:
Hydrologic Cycle

D. J. LORENZ, S. J. VAVRUS, D. J. VIMONT, J. W. WILLIAMS,
M. NOTARO, J. A. YOUNG, E. T. DeWEAVER, AND E. J. HOPKINS

Introduction

This chapter provides an overview of changes in precipitation in Wisconsin both in the observed record and in climate model projections for the end of the twenty-first century (figure 12.1). We also discuss the role of precipitation in modulating the temperature response to global warming. Many of the results presented here are also relevant for the other states composing the Midwestern USA and the United States as a whole.

Observed Precipitation Trends

Inter-annual variability in the Wisconsin statewide average annual total precipitation (figure 12.1) is very large (like that of much of the Midwestern USA; see chapter 9), and hence long-term trends and patterns can be hard to identify. The most obvious long-term pattern is a relatively dry spell from the 1920s and '30s to the 1960s and '70s and a relatively wet period after this. Qualitatively the data imply a slight positive trend in Wisconsin precipitation over the last one hundred years. Trends in Wisconsin precipitation over the last one hundred years are not distributed equally over the annual

cycle (see also analyses presented in chapters 9 and 10). The largest positive trends in precipitation in Wisconsin have occurred in summer, while basically zero change in precipitation has occurred in winter (figure 12.2). With two exceptions (Missouri in summer and Iowa in autumn), all Midwestern states have experienced increasing precipitation in spring, summer, and autumn. In winter, there is no clear pattern to the sign of the precipitation trends in the states composing the Midwestern USA.

Climate Model Projections of Precipitation for the End of the Twenty-First Century

For future projections of precipitation, we use coupled Atmosphere-Ocean General Circulation Models (AOGCMs) participating in the Intergovernmental Panel on Climate Change (IPCC) Fourth Assessment Report. The AOGCMs capture the major seasonality of Wisconsin-mean precipitation (figure 12.3), but exhibit larger model-to-model variability than for temperature (see chapter 7). The ensemble-mean change in precipitation (i.e., the average over the fifteen AOGCMs) for the end of the twenty-

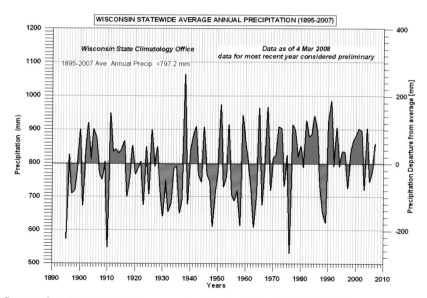

Figure 12.1. Time series of annual-mean precipitation in mm averaged over Wisconsin. Data from the National Climatic Data Center.

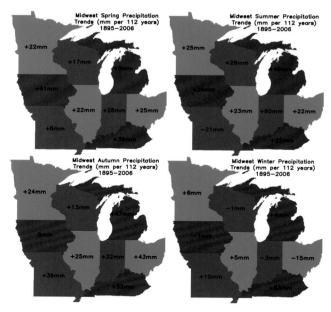

Figure 12.2. Seasonal trends in precipitation for each state in the Midwest (where winter is DJF, spring is MAM, summer is JJA, and autumn/fall is SON). Data from the National Climatic Data Center Climate Division Dataset. The trends were computed using ordinary least squares regression.

first century in the A2 emission scenario (IPCC 2001) is negative from June to September and positive for the rest of the year (thick black line). Individual model projections, however, vary substantially from the ensemble-mean change. Even the sign of precipitation change varies among AOGCMs, particularly in the warmer months.

To gauge the consistency of model projections of precipitation change, we plot the fraction of models with increasing precipitation for summer and for winter (figure 12.4). A value of 1 means that all models predict an increase in precipitation, while a value of 0 means that all models predict a decrease in precipitation (no models give exactly zero change in precipitation). A value of 0.5 signifies the greatest uncertainty in the sign of change since half project increases and half project decreases in precipitation. In the summertime, there is considerable uncertainty in the sign of the precipitation change in the Midwest: most of the region is in the 40% to 60% range. In the wintertime, on the other hand, the models tend to agree that precipitation will increase in the Midwest, with greater certainty in the northern Midwest, where up to 100% of models predict increases in precipitation. The result that changes in winter precipitation are

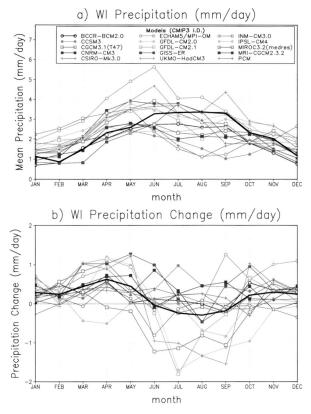

Figure 12.3. a) Wisconsin-mean precipitation for fifteen IPCC AR4 climate models (thin colored lines) and observations (thick black line). b) Wisconsin-mean precipitation change for fifteen IPCC AR4 climate models (thin colored lines). Precipitation change is defined as the difference between years 2080–2099 and 1980–1999. The ensemble-mean change is the thick black line. The observational data are from Maurer et al. (2002).

Projected change in Precipitation (IPCC models)

a) fraction of models with increasing P (June–Aug.)

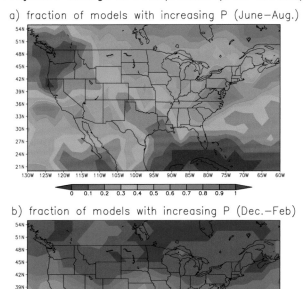

b) fraction of models with increasing P (Dec.–Feb)

Figure 12.4. Fraction of models with increasing precipitation (2080–2099 mean minus 1980–1999 mean) for a) June–August and b) December–February. Fifteen IPCC AR4 climate models (listed in figure 12.3) are used for this calculation.

more certain than summer precipitation also tends to hold elsewhere over the continental United States. In the wintertime, the increases in precipitation are in the northern United States and in Canada, while the precipitation decreases are in the far southern United States and in Mexico.

The climate model results for the precipitation change in the Midwestern USA contrast with the observed trends in the twentieth century. The AOGCMs show a tendency toward wetter winters and inconsistent changes in summer precipitation (figure 12.5), while the historical data tend to indicate stronger trends in the summer.

Relationship between Changes in Precipitation/Evaporation and Changes in Temperature

In chapter 7, we noted that some AOGCMs simulate large increases in summer temperatures over Wisconsin; these same models also have strong decreases in summertime precipitation (figure 12.5). For example, the two models with the largest temperature increase also have the largest precipitation decrease in the month of July in Wisconsin. This association between temperature and precipitation can be seen more easily in a scatter-plot of the fifteen AOGCM simulations (figure 12.6). In addition, the correla-

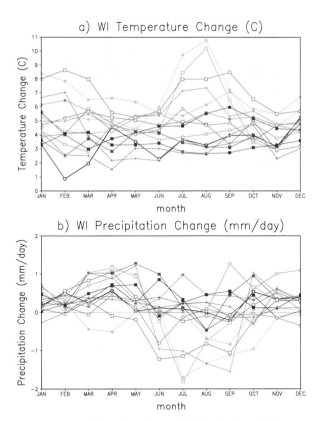

Figure 12.5. a) Wisconsin-mean temperature change for fifteen IPCC AR4 climate models (thin colored lines; AOGCMs are represented by the same colors as in figure 12.3). Temperature change is defined as the difference between years 2080–2099 and 1980–1999. b) Wisconsin-mean precipitation change for fifteen IPCC AR4 climate models (thin colored lines). Precipitation change is defined as the difference between years 2080–2099 and 1980–1999.

tion varies strongly by month, with the largest correlations in July through September (figure 12.6).

This correlation between summertime precipitation change and summertime temperature change occurs over a large region of the United States, as does the expected correlation between summertime evaporation change and summertime temperature change (figure 12.7). The correlations between evaporation and temperature are strongest over the central United States and weaken as one moves into Canada and the western United States. In northern Canada and Alaska the correlations are positive, which presumably means that evaporation in these regions

is limited primarily by energy rather than moisture. These relationships are described further in chapter 8 and, as described in that chapter, imply that in some regions soil moisture is critical to driving changes in temperature. Under this scenario, the chain of causality runs as follows: (1) decreases (increases) in precipitation lead to (2) decreases (increases) in soil moisture, which lead to (3) decreases (increases) in evaporative cooling, which lead to (4) increases in temperature above (below) the background global warming amount. An alternative hypothesis is that (1) decreases (increases) in precipitation are associated with (2) decreases (increases) in cloudiness, which lead to (3) increases

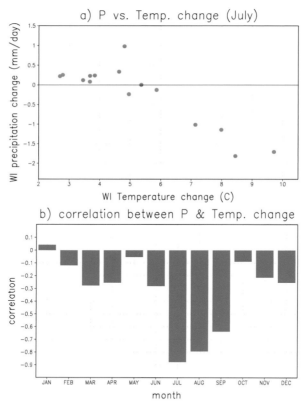

Figure 12.6. a) Scatter-plot of Wisconsin precipitation change versus Wisconsin temperature change for fifteen IPCC AR4 climate models (shown in figure 12.3). b) Correlation between Wisconsin precipitation change and Wisconsin temperature change as a function of month. Change is defined as the difference between the years 2080–2099 and 1980–1999.

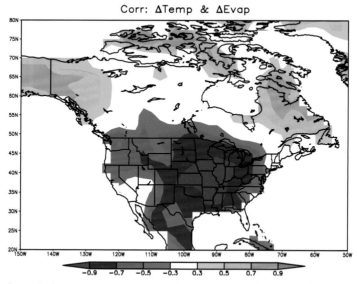

Figure 12.7. Correlation between local evaporation change and local temperature change. The evaporation and temperature change are averaged over the months of July and August. Analysis based on the same fifteen AOGCMs as depicted in figure 12.3.

(decreases) in solar radiation at the surface, which lead to (4) increases in temperature above (below) the background global warming amount. To test this hypothesis, we calculated the local correlation between temperature change and cloud fraction change in the AOGCM output. The temperature and cloud fraction correlation is quite large in some areas of the central United States (<−0.7). Hence, changes in cloudiness also appear to be playing a role.

Relationship between Changes in Precipitation and Changes in Circulation

We hypothesize above that the change in precipitation is impacting not only the availability of moisture but also temperature. Hence, if we better understand the reasons for the model-to-model variability in precipitation change we will also better understand the model-to-model variability temperature

change. To analyze the inter-model variability in precipitation about the ensemble-mean, we use Maximum Covariance Analysis (MCA) (Bretherton, Smith, and Wallace 1992). Given a number, n, of realizations of two spatial fields (i.e., zonal wind and precipitation), MCA finds pairs of spatial patterns that explain the most covariance between the two fields. Or, in other words, MCA finds pairs of spatial patterns (i.e., one zonal wind pattern and one precipitation pattern) that are most synchronous over the n realizations of the two fields. Typically, the multiple realizations in MCA are over time, but in our case the realizations are over both AOGCMs and time. We use MCA to summarize the inter-model co-variability of the 500 mb zonal wind change and precipitation change during the months of July and August over the domain 25 to 80°N and 70 to 130°W. There are therefore thirty realizations for the MCA (fifteen models and two months). We remove the ensemble-mean for

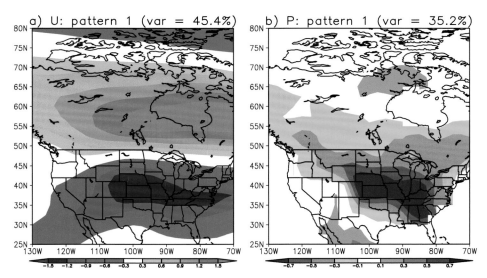

Figure 12.8. The first pair of MCA patterns. a) The 500 mb zonal wind MCA pattern (m/s). b) The precipitation MCA pattern (mm/day). The percent of the inter-model variance explained by the patterns is given in the top right corner of each plot. The patterns represented here are the homogeneous regression patterns (see Bretherton et al. 1992) instead of the normalized patterns so that the actual amplitudes can be discerned.

each month prior to the analysis and thus present anomaly fields in order to focus on the inter-model variability.

The leading pair of MCA patterns for the 500 mb zonal wind anomalies and the land precipitation anomalies are shown in figure 12.8. The positive phase of the zonal wind pattern has increases in Canada and decreases in the US. (The sign of the MCA patterns is arbitrary. Here we designate the positive phase as the phase that is the same sign at the majority of gridpoints as the ensemble-mean change in both zonal wind and precipitation.) The latitude of maximum zonal wind in the twentieth-century climate lies between the main positive and negative centers in figure 12.8a. Therefore, the positive (negative) phase of the zonal wind pattern represents a poleward (equatorward) shift in climatological zonal winds. The precipitation pattern shows strong decreases in precipitation in the central United States with weaker decreases in precipitation over a large portion of North America. The interpretation of the MCA is as follows: models with a strong poleward shift of the climatological zonal winds have strong decreases in precipitation over the central United States. Because a poleward shift of the zonal winds and a decrease in precipitation occur prominently in the ensemble-mean, the reverse of the above statement is as follows: models with little change in zonal wind have small increases in precipitation over the central United States.

The strong relationship between the zonal wind change and precipitation change is further emphasized in figure 12.9. Each point represents an AOGCM in either July or August. The x-axis is the projection of the climate models' zonal wind change on the MCA zonal wind spatial pattern (the zonal

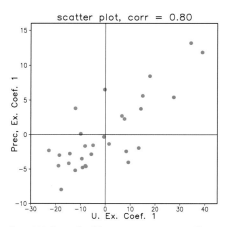

Figure 12.9. Scatter plot of the precipitation expansion coefficient versus the 500mb zonal wind expansion coefficient (from MCA). Each point represents a climate model in either July or August.

wind expansion coefficient, Bretherton et al. 1992). The y-axis is the projection of the climate models' precipitation change on the MCA precipitation pattern (the precipitation expansion coefficient). The correlation between the expansion coefficients is quite large (=0.80). Hence, the positive (negative) phase of the zonal wind pattern is associated with the positive (negative) phase of the precipitation pattern.

In the discussion above, the 500 mb zonal wind should be thought of as a proxy for the jet stream and the storm track. Thus, the MCA implies a decrease in the jet stream, and the storm track over the central United States leads to a decrease in precipitation there. It is not clear why there is not a prominent increase in precipitation in Canada associated with the increase in zonal winds there. The strong relationship between the zonal wind and the precipitation may seem puzzling in light of the strong associations of day-to-day and week-to-week fluctuations in precipitation with the Great Plains low-level jet (chapter 3 of this volume), which almost exclusively projects onto the meridional component of the wind. When we perform the MCA between 850 mb meridional wind and precipitation, however, the correlation between the two expansion coefficients is considerably less (0.65 instead of 0.80). Moreover, we find that increases (decreases) in southerly winds are associated with decreases (increases) in precipitation, which is exactly opposite to the relationship expected based on short-term fluctuations in the current climate! Clearly more research is necessary to better understand the role of circulation in forcing precipitation changes under global warming.

Projected Changes in Precipitation Extremes

We assess changes in the frequency of extreme precipitation events using the output of seven climate models included in the World Climate Research Programme's Coupled Model Intercomparison Project phase 3 (CMIP3): CCSM3, GISS-AOM, GFDL CM2.0, IPSL CM4, MIROC3.2, MRI CGCM2.3.2a, and MPI-ECHAM5. We calculated the simulated precipitation amounts on the ten wettest days for the grid point in each model covering Madison, Wisconsin (43°N, 89°W), for the late twentieth century (1980–1999) and late twenty-first century (2080–2099). The modern simulation was forced with observed concentrations of greenhouse gases, while the future run was driven by enhanced greenhouse gas concentrations based on the SRES A1B emissions scenario. Almost all the AOGCMs project greater intensity of extreme precipitation, with the very heaviest events (wettest and second wettest days) showing the most pronounced enhancement (figure 12.10). The

inter-model mean increase in intensity is about 10% for all but the two wettest days, which become 19% and 37% wetter. The PCM is the outlier model, projecting less extreme heavy precipitation for all but the wettest day. The increase of extreme precipitation among these models over southern Wisconsin is consistent with the results of other climate models and the theoretical discussion presented in chapter 8. However, the magnitude of extreme precipitation events simulated by AOGCMs (including those analyzed here) in the modern climate is usually far less than observed. For example, the rainfall on all ten of the wettest days recorded in Madison exceeds 80 mm, whereas the corresponding *simulated* precipitation amounts range from just 35–50 mm (inter-model mean). Some of this difference is undoubtedly due to the large spatial scale of AOGCMs and failures of parameterizations (e.g., convection [Dai 2006]), compared with point measurements of precipitation recorded in observations.

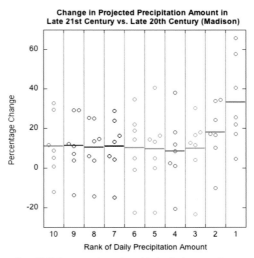

Figure 12.10. Percentage change in precipitation for the wettest day, the second-wettest day, etc., for seven climate models (open circles) for Madison, Wisconsin. The model-mean is shown by the solid line.

Concluding Remarks

Historical observations indicate annual-mean precipitation in Wisconsin has increased over the last century, with the largest positive trends occurring in summer (when the climatological precipitation is also largest). AOGCM simulations indicate:

1. There is larger model-to-model variability in both the mean and the change in precipitation than for temperature.

2. In Wisconsin, over 90% of the models predict precipitation increases in winter, while between 50 and 60% of models predict precipitation increases in summer.

3. There is a strong relationship between precipitation change and temperature change in the summer over Wisconsin: models with large decreases in precipitation also have large increases in temperature. Here we argue that the precipitation is driving the local temperature enhancement primarily via changes in evaporative cooling over land and secondarily via decreases in cloudiness.

4. The model-to-model variability in central U.S. precipitation change is strongly linked to the poleward shift of the mid-latitude jet. This suggests that the poleward shift of the jet stream decreases continental precipitation, which in turn dries the soil and decreases cloudiness, which in turn lead to extreme summertime temperatures.

5. Extreme precipitation tends to increase more rapidly than the mean precipitation.

REFERENCES

Bretherton, C. S., C. Smith, and J. M. Wallace. 1992. "An Intercomparison of Methods for Finding Coupled Patterns in Climate Data." *Journal of Climate* 5: 541–560.

Dai, A. 2006. "Precipitation characteristics in Eighteen Coupled Climate Models." *Journal of Climate* 19: 4605–4630.

IPCC. 2001. *Climate Change 2001: The Scientific Basis.* Contribution of Working Group I to the Third Assessment Report of the Intergovernmental Panel on Climate Change. [Ed. J. T. Houghton, et al.] New York: Cambridge University Press.

IPCC. 2007. *Climate Change 2007: The Physical Science Basis.* Contribution of Working Group I to the Fourth Assessment Report of the Intergovernmental Panel on Climate Change. [Ed. S. Solomon, et al.] New York: Cambridge University Press.

Maurer, E. P., et al. 2002. "A Long-Term Hydrologically Based Dataset of Land Surface Fluxes and States for the Conterminous United States." *Journal of Climate* 15: 3237–3251.

Trenberth, K. E. 1999. "Conceptual Framework for Changes of Extremes of the Hydrological Cycle with Climate Change." *Climate Change* 42: 327–339.

13. Spatial and Temporal Dimensions of Extreme Rainfall in the Twin Cities Metropolitan Area

K. A. BLUMENFELD

Introduction

Extreme rainfall events have well-known consequences: flash floods, infrastructure damage, property damage, and even loss of life (Davis 2001). The phenomenon has recently received intense research attention, partly because these events have become and may become both more frequent and more severe in response to anthropogenic warming (Hennessy et al. 1997; Karl and Knight 1998; Huntington 2006; see also chapters 8 and 9 of this volume), and partly because populations have generally become more vulnerable to them (Changnon et al. 2000).

In extreme rainfall research, one consideration often overlooked is the impact changing precipitation regimes will have on our built environment, for example, the transportation infrastructure. Much of our infrastructure has been designed to withstand extraordinarily intense rainfall, often "100-year" events. Specifically, the federally sponsored design values can be found in the (former) U.S. Weather Bureau's *Technical Publication 40* (Hershfield 1961—hereinafter TP40). Huff and Angel (1992—hereinafter HA92) compiled design values for the Mid-

western USA, and critiqued the techniques, record lengths, and spatial density of data used in TP40. Skaggs (1998) and Skaggs and Blumenfeld (2006) analyzed data from Minnesota's high spatial density rain gauge network and have raised questions about (1) the spatial resolution of extreme rainfall climatologies, (2) the stability of design values through time, and (3) statistical techniques used to estimate the low-frequency character of extreme rainfall. They conclude both TP40 and HA92 underestimate the true potential for extreme rainfall in Minnesota and the Twin Cities Metropolitan Area (TCMA), and these authors have initiated a dialogue about exactly what question climatologists mean to ask/answer when estimating design values (e.g. Blumenfeld, Skaggs, and Zandlo 2004; Blumenfeld and Skaggs 2005).

In this chapter, I expand on this research to address two objectives: to estimate the frequencies of extreme rainfall over a variety of spatial scales, and to demonstrate the potential utility of estimating rainfall frequencies for long records with an application of Generalized Extreme Value (GEV) statistical techniques (Faragó and Katz 1990). One goal of this chapter is to understand whether the disastrous, record-shattering rainfalls in southeastern Minnesota during August 2007 could have been anticipated from the data and from GEV approaches.

Data and Methods

EXTREME RAINFALL ESTIMATES IN THE TCMA

This portion of the study uses Minnesota's high spatial density rain gauge network, which comprises NWS COOP data, as well as several other locally and regionally implemented networks (table 13.1). This network is unique both for its high spatial density and for its longevity; the network has had over one thousand observers since the early 1970s. The densest part of the network is centered on the TCMA, and has averaged just over three hundred observers annually since 1970. The present study area consists of nested boxes within a 9600 km² area, centered on a portion of Minneapolis and adjacent suburbs, with the smallest and largest boxes 10 km and 70 km on a side, respectively (figure 13.1). As reported by Blumenfeld, Skaggs, and Zandlo (2004), every daily rainfall report for the period 1970–2002 has

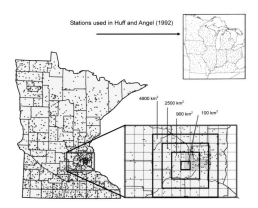

Figure 13.1. The complete Minnesota high spatial density network shown in June 2002 (left), with the 9600 km² TCMA study area and specific nested boxes (right). For comparison, the data density from Huff and Angel's (1992) rainfall frequency atlas is also shown (top).

been assigned to a corresponding 10 km×10 km cell within the 9600 km² area (ninety-six cells total). An intensive data comparison and editing scheme allowed all observers within the same cell to "compete" with one another for each annual maximum daily precipitation (AMP) value (figure 13.2). The editing resulted in an AMP time series of varying length for each cell. Fifty cells had all thirty-three years of data, another twenty-five cells had at least twenty-two years, and only five cells (all on the far eastern domain, in Wisconsin) had no data, leaving a total of 2535 values area-wide.

Table 13.1. Composition of the Minnesota high spatial density rain gauge network

Network	# Stations	Quality	Coverage
NWS COOP	About 150	Good	Obs 12 mo/yr
MN DNR	About 80	Good	Obs 12 mo/yr
Met. Mosquito Dist.	About 60	Good early, declining	Obs 9 mo/yr
"Backyard" Volunteers	About 300	Good	Split: 12 mo/yr; 9 mo/yr
Future Farmers of America	Declining:1000 in 1970s, ~10 at present	Poor	Variable
KSTP	About 30	Good	Discont.; 12 and 9 month obs 1980–1990s
Soil/Water Cons Dists	700 to 1000	Good	Variable, most in summer
Misc., e.g., WCCO	Variable	Variable	Variable

In this study, I analyze the AMP values within areas of 100, 900, 2500, and 4900 km² (i.e., each of the nested boxes) in order to understand how the basic statistics and estimated return period values vary with the size of the area in question. The thirty-three-year record for each area will lead to more uncertainty in the estimates of long return periods (e.g., one hundred years) than if longer records had been used, though it is sufficient for GEV estimates (Faragó and Katz 1990). The spatial density of observers may compensate for the short record length. For instance, the text of TP40 described a substudy in which increasing the spatial density of observers had far more influence over resulting estimates than merely increasing the record lengths. Ideally, we would have both a long record (e.g., >33 yr) and continuous high spatial observer density.

Collecting the annual precipitation maxima from areas rather than fixed points asks, "For a given return period, what is the rainfall threshold expected to be met or exceeded *somewhere* within the specified area?" Note that this is a different question than "For a given return period, what is the rainfall threshold expected to be met or exceeded at a *fixed point*?" Traditional methods involve analyzing an annual maxima time series from a single point (and then interpolating values to areas between points), whereas this method involves collecting the largest annual value from within an area and creating the annual series from those values. Thus, the present method approximates a "moving rain gauge" that samples only the largest daily rainfall amount for each year within the specified area.

I use two GEV-based methods to estimate the 100-year return period for a storm of 24-hour duration.

1. Three-parameter maximum-likelihood method (hereinafter MLE), given by:

$$[R-(P+Q)k^{-1}]k^{-1}=0,$$
$$(P+Q)k^{-1}b^{-1}=0, \quad Qb^{-1}=0 \qquad (13.1)$$

Where $P=n-\Sigma e_j$; $Q=\Sigma e_j f_j-(1-k)\Sigma f_j$; $R=n-\Sigma y_j+\Sigma y_j e_j$; $y_j=(x_j-u)/b$; $e_j=exp(-y_j)$; and $f_j=exp(ky_j)$.

The MLE and other GEV methods are quite sensitive to "extremes of the extreme," with estimates often shaped by only the one or two largest events within a time series. In such instances, $k<0$, and hence the distribution approximates the Frechét form of the GEV. If these sorts of events do not exist within the record, then k will be near,

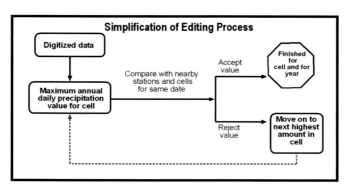

Figure 13.2. Flow diagram of steps within editing environment.

or possibly less than, zero, and hence the distribution will approximate the Gumbel or Weibull GEV forms. In terms of return-period estimates, Frechét distributions will yield the highest values, followed by Gumbel and then Weibull. One potential consequence for the methodology herein is that as one steps from the smallest nested box to progressively larger ones, the lowest series values will be replaced faster than the highest, and thus, the distributions will propagate away from Frechét-like qualities. Even though basic intuition would suggest that a relatively small area's 100-year event cannot exceed the 100-year threshold for a larger area that contains the smaller area, increases in the value of k without sufficiently large increases in u or b could lead to such internal inconsistencies.

2. To account for potential problems with the MLE method, I also use a hybrid GEV method, initializing u and b from the Empirical Moments (EM) method (Gumbel 1958), given by:

$$u = \mu - (\Gamma b), \quad b = \sigma \sqrt{6} / \pi, \qquad (13.2)$$

Where μ and σ are the series mean and standard deviation, respectively, and Γ=Euler's constant (0.57722).

A 100-year event, in theory, has probability p of 0.01 in any given year, with a "reduced design value" or "reduced variate" of $y = \{1 - [-\log(1 - p)]^k\} / k$ (Faragó and Katz 1990). The magnitude of the 100-year, 24-hour event is then estimated as $by + u$. For each box, k is held constant as the median of the MLE estimates for k, -0.25. For comparison I also use a simple two-parameter "Gumbel" technique, which has no shape parameter, k, so the reduced variate is defined as $y = -\log(-\log[1 - p])$.

One problem that arose in earlier research (e.g., Blumenfeld, Skaggs, and Zandlo 2004) is the effect observer density has on series mean AMP values. For a 100 km² area, nine observers is the approximate "saturation" threshold. For example, when N (observers) ≤9, then N explains nearly all of the variance in the mean AMP ($r^2 = 0.94$), with a trendline coefficient of 3.685 mm for each missing observer, significant at the 99% level. Conversely, for N>9, the number of additional observers has a nonlinear, statistically insignificant effect on mean AMP values. As in the previous research, I correct the AMP value by adding the trendline coefficient (3.685 mm) to each "missing" observer for any area with fewer than nine observers per 100 km². In the earlier research, this had the effect of completely rearranging the spatial pattern of mean AMP values over the TCMA (figure 13.3).

LONG-RECORD EXTREME RAINFALL ESTIMATES

Here I estimate 100-year, 24-hour rainfalls by applying the hybrid EM method (equation 13.2) to both the 121-year NWS COOP record at Grand Meadow, and a 116-year composite AMP series for Minnesota. I also estimate the 100-year 24-hour values from consecutive, overlapping periods of thirty and forty years within each of these data sets. Using the short periods illustrates regimes and trends within each record. Grand Meadow is the nearest USHCN station to the catastrophic, record-breaking floods in southeastern Minnesota during August 2007, where 24-hour amounts in excess of 300 mm covered a large area, including a report of 437 mm at La Crescent (figure 13.4). Thus, its complete record can be used to determine the probability and the

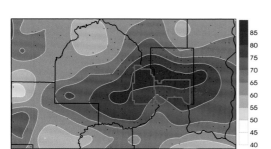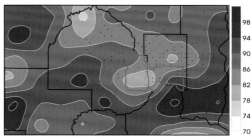

Figure 13.3. Mean annual maximum daily precipitation (AMP) (mm) before (left) and after (right) correcting for observer disparities. The correction scheme suggests that areas in the outer TCMA are generally missing the true maxima in most events, whereas central TCMA areas are not. Note the scale differences between the two maps, despite their employing the same color scheme.

return period of such a storm in that area. The composite record for Minnesota is used to estimate the magnitudes and recurrence intervals for rainfall expected *somewhere* within the state. The 2007 flooding event was much more severe than anything on record, yet one question is, why did we not see it coming?

Results

AREAL EXTREME RAINFALL RETURN PERIOD ESTIMATES IN THE TCMA

As expected, increasing the box size leads to increases in mean AMP values, though the return period estimates for a 100-year storm are neither linear nor predictable (table 13.2). The Gumbel technique yields the smallest

estimates (176.7–238.7 mm), and the hybrid GEV method the largest (373.3–448.8 mm). For comparison, the corresponding point-estimated values from HA92 and TP40 are 143 and 152 mm, respectively. The largest actual value in the thirtythree-year record in the study area is 324 mm, which is closest to the 100-year estimates from the MLE method. The MLE, however, exhibits some curious behavior as area increases. Notably, the 100 km² area has larger values than the larger areas within which it is nested, which is physically impossible. This inconsistency results from the low values in a series being the "easiest" to replace; therefore, the lower values in the AMP distribution increase faster than the higher ones, as box sizes increase. Although the values of the

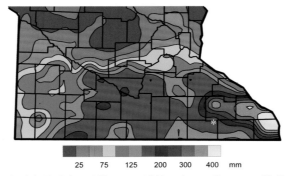

Figure 13.4. Isohyet pattern associated with the 18–19 August 2007 extreme rainfall in southeastern Minnesota, modified from original by Minnesota State Climatology Office (available at http://www.climate.umn.edu/img/flash_floods/ff070818-20.pdf).

Table 13.2. Basic statistical and GEV properties of extreme rainfall in each study box. The three right-hand columns indicate the estimates for "100-year storms" from a two-parameter Gumbel method similar to that in TP40, and from the two GEV methods. The empirical method of HA92 was not feasible, due to the short record lengths.

Area (km²)	Mean AMP (mm)	Mean N obs/100 km²	Adjustment (9-Mean N*3.685) (mm)	Corr AMP (mm)	Gumbel est. (mm)	100y, 24h est. from GEV (MLE)	100y, 24h est. from GEV (hybrid EM)
100	87.59	12.15	0.00	87.59	176.7	309.5	373.3
900	107.83	8.67	1.20	109.03	197.7	262.4	374.4
2500	119.25	5.66	12.31	131.56	218.3	296.7	432.7
4900	128.05	4.24	17.56	145.60	238.7	304.7	448.8

location parameter u increase, compressing the distribution's range leads to reductions in the scale and shape parameters, which ultimately determine the trajectory of the GEV estimates through the return periods beyond the bounds of the data (i.e., beyond thirty-three years). Compared to the other approaches, the hybrid EM method generates very large estimates, though its values are within the range of those measured during the 18–19 August 2007 major flooding event outside of the TCMA in southeastern Minnesota.

LONG-RECORD ANALYSES: GRAND MEADOW

Applying the GEV EM method to the full Grand Meadow record produces a 100-year storm estimate of 237 mm, which is 91 mm greater than the estimate for climate division nine given by HA92 (figure 13.5). One explanation for this discrepancy is that the present study uses data from the period of relatively large annual maxima during the 1990s and 2000s, after HA92 was published. The thirty- and forty-year running hybrid series resolve

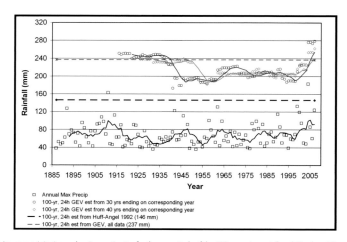

Figure 13.5. Annual maximum precipitation and various estimates for the magnitude of the 100-year storm at Grand Meadow, Minnesota, 1886–2007. Seven-year moving averages of thirty- and forty-year running EM estimates are shown as blue and red traces. Fixed estimates from EM GEV method and from HA92 shown for comparison.

the temporal variability within the long record and similar to observations of Kunkel, Andsager, and Easterling (1999) and Kunkel et al. (2003), resolves a period of relatively low extreme rainfall values from the 1930s to the early 1990s. The recent high annual maxima are certainly influencing EM-based estimates for the whole record. Additionally, the EM method produced extraordinarily large values over the TCMA nested boxes.

Even the very aggressive hybrid EM method, however, does not pick up on the potential for the intensity of rainfall experienced on 18–19 August 2007 in La Crescent, where 437 mm fell in twenty-four hours. Indeed that storm computes to a nearly 1,000-year event when applying the EM method to the Grand Meadow record. It might very well be that at Grand Meadow proper—or more appropriately, at the point occupied by the Grand Meadow NWS COOP rain gauge— such a storm does have an approximate 0.001 probability in any given year. In fact, given the aggressiveness of the EM method, one could possibly argue that Grand Meadow

has *at most* a 0.001 annual probability of experiencing this sort of storm. Considering, however, (1) the evidence from the TCMA study and (2) the reality of the August 2007 floods, it is quite likely that the probability of such a storm is much higher than 0.001 at *some point* within the Grand Meadow area. In the TCMA study area, the La Crescent storm had an approximate 0.01 annual probability of occurring at some point within an area of 2500 km^2 (see table 13.2), which is roughly the average county size in Minnesota. Assuming the TCMA and southeastern Minnesota have similar extreme rainfall climates, the La Crescent storm could be thought of as a 100-year storm at the county level, though perhaps only a 1,000-year storm at any given point. In other words, moving the analysis focus from a fixed point to a moving point within a county increases the probability of a La Crescent–type storm by a factor of ten. This is a distinction that, understandably, has escaped precipitation design research attention.

LONG-RECORD ANALYSES: MINNESOTA

The composite AMP record for Minnesota is an extreme example of the moving rain gauge approach used over the TCMA nested boxes: in this case, the gauge roams over the entire state, in search of each AMP value (figure 13.6). The resulting series mean AMP is 165 mm. This value is not as much larger than the TCMA mean values as one might expect, most likely because of gauge density. The thirty-three years of TCMA data were completely during the high observer density era, whereas the first seventy-nine years of the Minnesota record were not. From 1970 to 2006, the mean AMP for Minnesota was 203 mm, compared with 147 mm for the ear-

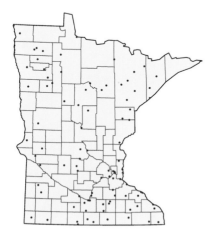

Figure 13.6. Locations of annual maximum precipitation values in Minnesota, 1891–2006. Note that some locations received the annual maximum more than once.

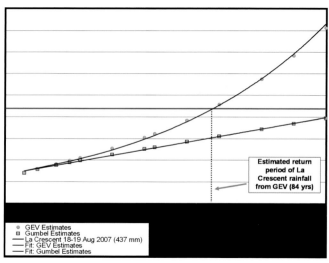

Figure 13.7. Return period estimates for extreme rainfall somewhere in Minnesota, from EM GEV and two-parameter methods.

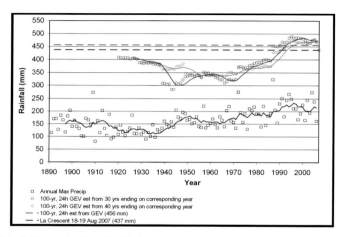

Figure 13.8. As in figure 13.5, but for Minnesota based on 1891–2006 data, and with La Crescent rainfall from 18–19 August 2007 and full-record GEV 100-year estimate shown for comparison.

lier period. Two-sample t-tests indicate these periods were significantly different at all thresholds (pooled variance t=−7.23, p<.001). Thus, it is likely that the stepwise increase in gauge densities led to much of the stepwise increase in AMP values.

The EM-based estimate for a 100-year storm somewhere in Minnesota, based on 1891–2006 data, is 456 mm, and the La Crescent storm equates to an 84-year event (figure 13.7). Using the two-parameter Gumbel approach, the 100-year storm is 310 mm, and the La Crescent storm's periodicity exceeds 1,000 years. These results, especially for the EM-based approach, however, are highly nuanced. On one hand, the annual maxima responded positively and significantly to the tenfold increase in gauge densities in the early 1970s. For instance, before 1970, thirty- and forty-year EM-based estimates statewide averaged 350–360 mm (figure 13.8). Increases in gauge density, however, do not explain all of the increases in extreme rainfall values and return period estimates. Even during the high observer density period, two distinct extreme rainfall regimes emerge. From 1970 to 1986, only five years (30%) had AMP values in excess of 203 mm (8 in.), while from 1987–2006, thirteen years (65%) exceeded that threshold. This positive trend in extreme rainfall was apparent in the Grand Meadow record, and has been observed in other investigations (e.g. Kunkel, Andsager, and Easterling 1999; Kunkel et al. 2003; see also chapter 9 of this volume). Though basic statistics suggest that the La Crescent event was an outlier with respect to the climatic record of Minnesota (the previous high value was 324 mm, or 12.75 in.), in the context of the distribution of recent extreme rainfall values, it was not. Many of the EM-based 100-year rainfall estimates from thirty and forty years of data late in the study period far exceeded the La Crescent threshold (see figure 13.8).

Concluding Remarks

Evaluating return periods for extreme rainfall occurring at some point within an area, rather than at a fixed point, leads to larger values than fixed-point approaches, and the values also increase as the size of the area under consideration increases. In the TCMA, a two-parameter Gumbel method yielded 100-year return values substantially smaller than the largest values observed during the thirty-three-year record in each nested box. The MLE method produced values slightly lower than the empirical maximum for the 100 km² boxes, but failed to produce equal or larger estimates as box size increased—a result that is physically implausible considering that the boxes were

nested. A hybrid GEV method (EM) that held the shape parameter fixed through all box sizes yielded estimates that were much larger than the empirical maximum but were consistent with values reported during the extreme rainfall event of 18–19 August 2007 elsewhere in the state. Considering that the TCMA data set was only thirty-three years long, the hybrid GEV results may be reasonable, especially in light of the August 2007 events. On the other hand, with just a thirty-three-year base period, these estimates will have greater uncertainty than those drawn from a longer record.

Even with the seemingly high-biased hybrid approach, point-based estimates for Grand Meadow did not hint at the potential for rainfalls of the magnitude observed at La Crescent. In fact, the La Crescent storm was estimated to have a 1,000-year return period. Even so, the estimate was substantially larger (by 91 mm) than the HA92 estimates. If the hybrid GEV method was applied to the annual maximum precipitation series for Minnesota, however, the La Crescent storm became an 84-year event—that is, an 84-year storm *somewhere within the state*. Moreover, the latter two decades of the Minnesota record are largely responsible for the high 100-year return estimates and portray the La Crescent storm as a 70-year event. If this intensification of the extreme precipitation regime continues, it would only complicate existing problems with design values and their inability to account for spatial variation of extreme rainfall. There is no guarantee, however, that this regime will continue; Kunkel, Andsager, and Easterling (1999) and Kunkel et al. (2003) have demonstrated considerable temporal variability in extreme precipitation regimes.

The distinction between evaluating return periods at a *fixed point* or *some point* is an important one. The traditional fixed-point approach involves calculating individual return period estimates from single station records and then interpolating values to the areas between the stations. This method does not account for larger rainfall amounts that fall between rain gauges, but is nevertheless accurate for almost any randomly selected point. In other words, a random point in the TCMA is likely to see the HA92 100-year rainfall just once per hundred years. The some-point approach involves calculating a return period estimate from many stations within an area. Given sufficient gauge density (≥9 observers per 100 km²), this method will capture most of the meaningful spatial variability of extreme rainfall, but the resulting return period estimates will be valid for an area, not any random point within that area. As an example, the 100-year rainfall estimate for the 2500 km² box in the TCMA is valid for some (unknown) point in that box, but not for a randomly selected point in that box; in the latter case, the HA92 estimate, or perhaps a more up-to-date point-based one, would hold.

For precipitation design purposes, assuming the data are reasonably up-to-date, fixed-point approaches will yield correct or "protective" values in most places most of the time. Failures, however, will be common at discrete locations fairly regularly and occasionally will be catastrophic. Damages to property, crops, state park land, and infrastructure in Minnesota and Wisconsin resulting from the August 2007 floods may exceed $1 billion. Regulators will need to decide whether we wish to tolerate occasional losses of this magnitude. Also, recall that

the mean AMP in Minnesota is 165 mm, so during an average year, the TP40 and HA92 thresholds are exceeded somewhere in the state at least once. By contrast, the some-point approach to precipitation design will be overprotective in most places most of the time, but will lead to few failures. Monetarily, the difference amounts to high repair and maintenance costs versus high initial investments, and the debate about which is more appropriate is a matter of public policy. If the recent enhancement of our extreme precipitation regime continues, current point-based estimates may be too low, and much of Minnesota may be more susceptible to flooding and damage than previously thought.

ACKNOWLEDGMENTS
I would like to thank Dick Skaggs for his insights, and the entire Minnesota State Climatology Office staff for their help with data-related issues.

REFERENCES
Blumenfeld, K. A., and R. H. Skaggs. 2005. "The Frequency and Spatial Variability of Extreme Rainfall over the Twin Cities Metropolitan Area." *Proceedings: 2nd Midwest Extreme and Hazardous Weather Conference.* Champaign, Ill.: Central Illinois Chapter, American Meteorological Society.

Blumenfeld, K. A., R. H. Skaggs, and J. A. Zandlo. 2004. "Using a Dense Precipitation Gauge Network to Estimate Maximum Daily Precipitation." *Proceedings: 14th Conference on Applied Climatology.* Seattle, Wash.: American Meteorological Society.

Changnon, S. A., et al. 2000. "Human Factors Explain the Increased Losses from Weather and Climate Extremes." *Bulletin of the American Meteorological Society* 81: 437–442.

Davis, R. S. 2001. "Flash Flood Forecast and Detection Methods." In *Severe Convective Storms. Meteorological Monograph, No. 50,* ed. C. A. Doswell III, 481–525. Boston: American Meteorology Society.

Faragó, T., and R. Katz. 1990. "Extremes and Design Values in Climatology." *WMO Technical Document* 386.

Gumbel, E. J. 1958. *Statistics of Extremes.* New York: Columbia University Press.

Hennessy, K. J., J. M. Gregory, and J. F. B. Mitchell. 1997. "Changes in Daily Precipitation under Enhanced Greenhouse Conditions." *Climate Dynamics* 13: 667–80.

Hershfield, D. M. 1961. "Rainfall Frequency Atlas of the United States." *U.S. Weather Bureau Technical Paper* 40.

Huff, F. A., and J. Angel. 1992. "Rainfall Frequency Atlas of the Midwest." *Illinois State Water Survey Bulletin* 71.

Huntington, T. G. 2006. "Evidence for Intensification of the Global Water Cycle: Review and Synthesis." *Journal of Hydrology* 319: 83–95.

Karl, T. R., and R. Knight. 1998. "Secular Trends of Precipitation Amount, Frequency and Intensity in the United States." *Bulletin of the American Meteorological Society* 79: 231–241.

Kunkel, K. E., K. Andsager, and D. Easterling. 1999. "Long Term Trends in Extreme Precipitation Events over the Conterminous United States and Canada." *Journal of Climate* 12: 2515–2527.

Kunkel, K. E., et al. 2003. "Temporal Variations of Extreme Precipitation Events in the United States: 1895–2000." *Geophysical Research Letters* 30, doi:10.1029/2003GL018052.

Skaggs, R. H. 1998. "Intensity of Extreme Rainfall over Minnesota." *Minnesota Department of Transportation Research Report* 1998-09U.

Skaggs, R. H., and K. A. Blumenfeld. 2006. "Precipitation Design Values for Minnesota: Are They Adequate?" *CURA [Center for Urban and Regional Affairs] Reporter* (Winter): 24–29.

14. Overview: *North American Atmospheric Circulation Effects on Midwestern USA Climate*

J. S. M. COLEMAN AND K. KLINK

Introduction

Midwestern climate is mainly governed by fluctuations in hemispheric-scale atmospheric circulation patterns that arise from energy imbalances between low and high latitudes. The seasonal position of longwave ridges and troughs (e.g., Rossby waves) across North America dictates the development and propagation of shortwave (synoptic scale) features. These synoptic scale systems vary in frequency and intensity, depending largely on the baroclinic structure of the Rossby wave configuration within the prevailing upper-level mid-latitude westerlies. Annual atmospheric circulation variability arises from alterations in the strength and arrangement of the semi-permanent pressure systems (i.e., subtropical highs, subpolar lows) that are often identified with a particular teleconnection pattern phase (e.g., the North Atlantic Oscillation).

Daily meteorological conditions at a given location (e.g., temperature, wind speed and direction, cloud cover, precipitation, etc.) are highly dependent on the relative position of synoptic-scale features such as troughs, ridges, cyclones, and anticyclones. Short- and/or long-term deviations from the climatological norm commonly lead to extremes in temperature and precipitation, producing heat waves and cold spells, droughts and floods. Depending on the duration and timing, these atmospheric circulation shifts have important consequences for agricultural productivity (e.g., Robeson 2002; Southworth et al. 2000), human health (e.g., Coleman 2008), the renewable energy economy (e.g., Klink 2007) and recreational activities (e.g., Patten, Smith, and O'Brien 2003) in the Midwestern USA.

Many model-based climate change scenarios, such as those described in the Fourth Assessment Report of the Intergovernmental Panel on Climate Change (IPCC 2007), suggest that there may be hemispheric-scale shifts in the position and/or intensity of both longwave and shortwave circulation features (Christensen et al. 2007; Meehl et al. 2007). Possible changes include a poleward shift of mid-latitude storm tracks (the typical paths of extratropical cyclones), a decrease in the frequency of weaker and medium-strength extratropical cyclones, and an increase in the frequency of strong cyclones (resulting in a net decrease in the overall cyclone frequency). An increase in the number of strong cyclones may yield more, and more intense,

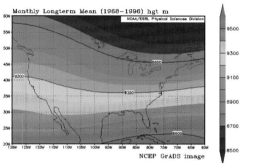
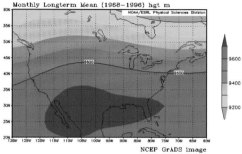

Figure 14.1. Mean January (left) and July (right) 300 hPa geopotential heights (1968–1996). The largest gradient in 300 hPa heights is located in the Midwestern USA in January, and is in southern Canada in July. Figure from the NCEP-NCAR Reanalysis data with images provided by the NOAA/ESRL Physical Sciences Division, Boulder, Colorado, from their web site at http://www.cdc.noaa.gov/.

high wind events (e.g., Pryor, Barthelmie, and Kjellström 2005). Such changes also will affect the frequency, intensity, and duration of heat waves, cold spells, drought, and floods (e.g., Liang, Wang, and Dudek 1996).

This chapter provides a brief introduction to basic North American atmospheric circulation climatology as it relates to the Midwestern USA. Our primary focus is to provide an overview of typical North American mid-to upper-level tropospheric flow patterns and their associated surface conditions. We discuss major modes of atmospheric circulation variability, particularly in reference to atmospheric and oceanic teleconnections and the implications of potential climate change scenarios for the future climate of our region. Subsequent chapters in this section will expound upon these themes, particularly in reference to Midwestern synoptic-scale systems and surface wind characteristics.

Climatology

UPPER AND MID-TROPOSPHERIC CIRCULATION

The seasonal variation in the strength and position of the polar jet stream significantly influences Midwestern USA atmospheric circulation regimes and regional climate. During the winter, the jet stream is stronger and more southerly; in summer, it is weaker and more northerly. This seasonality can be seen in the climatological average 300 hPa geopotential heights (figure 14.1); the jets are located in the areas with the strongest height gradients. The strongest synoptic-scale storms (cyclones) typically occur in winter and spring, when the equator-to-pole temperature gradient is strongest; cyclones are weaker in summer, when temperature gradient also is weak (figure 14.1) (e.g.,

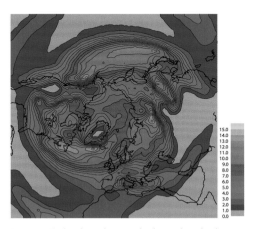

Figure 14.2. Northern hemisphere annual cyclone track number density (expressed as number per month per unit area, unit area is equivalent to a 5° spherical cap) calculated from the ECMWF forty-year reanalysis (ERA-40) for the period 1979–2003. Courtesy of Kevin Hodges, University of Reading.

Harman 1991). The frequency and persistence of different weather types are closely related to the paths (tracks) followed by cyclones and anticyclones across the United States (figure 14.2).

LOWER TROPOSPHERIC CIRCULATION AND NEAR-SURFACE CONDITIONS

As with other mid-latitude locations, Midwestern lower tropospheric conditions are characterized by seasonal changes in baroclinic zones that develop from the movement and modification of air masses. Regions of low-level baroclinic instability promote surface cyclogenesis, especially in conjunction with an upper-level longwave trough, and produce winds speeds that strengthen with increasing height (Rohli and Vega 2008). Baroclinic zones most frequently arise during the non-summer seasons along the polar front. North American topography encourages air masses to propagate from their source regions in Canada and the Gulf of Mexico into the central United States, funneling between the Western Cordillera and the Appalachians. These air masses are significantly altered before they reach the Midwestern USA. Sheridan (2003) finds continental polar air masses warm nearly 25°C on average between southern Canada and the southern Great Plains. Mid-latitude baroclinic zones also generate numerous transitional air mass days (i.e., more than one air mass type present during a twenty-four-hour period), a condition common with frontal passage. For the southern half of the Midwestern USA, transitional conditions may account for approximately 25% of all days, especially during the spring and early summer, when strong frontal passages are the most common (Schwartz 1991).

Climatological shifts in semi-permanent circulation features (subtropical highs, subpolar lows) and their accompanying air masses affect the mean sea level pressure (SLP) patterns across North America. Surface wind fields in the Midwestern USA are closely related to these shifts in SLP (figure 14.3). In the winter, the dominance of high pressure over the western United States brings into the Midwest the cold, dry air masses from Canada and the Arctic. During summer, the semi-permanent high pressure area over the southeastern United States and subtropical Atlantic Ocean brings warm, moist air masses originating in the Gulf of Mexico into the Midwest. These seasonal shifts are accompanied by changes in near-surface wind speeds. Mean monthly wind speeds are fastest in winter and spring and slowest in summer (Klink 1999), corresponding to the seasonality in the strength of hemispheric baroclinic zones and of upper-level height and SLP gradients. Largely following the pattern of SLP, wind directions in the northern part of the region (Minnesota, Wisconsin, Iowa, Michigan) are westerly-northwesterly in winter and shift to southerly-southwesterly in summer; in the southern part of the region (Missouri, Illinois, Indiana, Ohio), wind directions show a smaller seasonal shift, primarily from the west-southwest in winter and from the south-southwest in the summer (figure 14.4).

Inter-annual Variability and Change

Atmospheric and oceanic teleconnection patterns constitute a major mode of annual large-scale atmospheric circulation vari-

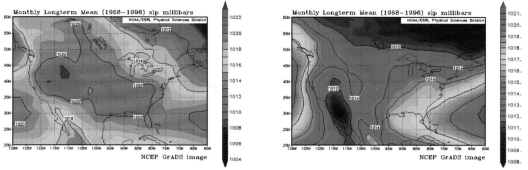

Figure 14.3. Mean January (left) and July (right) mean sea level pressure (SLP) (1968–1996) over the coterminous United States. Figure from the NCEP-NCAR Reanalysis data with images provided by the NOAA/ESRL Physical Sciences Division, Boulder, Colorado, from their web site at http://www.cdc.noaa.gov/.

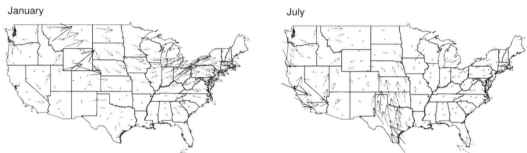

Figure 14.4. Mean January (left) and July (right) wind velocity (the wind vector) at 10 m above ground level (AGL). Speeds measured at heights other than 10 m AGL were adjusted to 10 m using the power law wind profile with an exponent of $1/7$ (see chapter 15, this volume, for a discussion of wind speed height adjustment). Means were computed for the period 1961–1990. Figure reproduced with permission of the Royal Meteorological Society from Klink (1999), first published by John Wiley and Sons, Ltd.

ability across the North American continent. Teleconnections are hemispheric-scale statistical relationships that result from the spatial interdependencies of atmosphere and ocean dynamics (Leathers, Yarnal, and Palecki 1991). On the order of weeks to years, these global associations are recognized by a transient wave disturbance, forming oscillations in the upper-level geopotential height and/ or sea-level pressure fields (Esbensen 1984). Locations in the vicinity of these "centers of action" (the climatological positions of lower and higher pressure areas associated with particular teleconnection patterns) shift from characteristic meteorological conditions to distinct patterns of climatic anomalies that occur over a range of time periods. This sec-

tion briefly describes the Pacific and Atlantic teleconnections most likely to influence Midwestern USA circulation changes and related climate regimes; more detailed information can be found in the references cited herein.

PACIFIC TELECONNECTIONS

The El Niño–Southern Oscillation (ENSO) is probably the most widely known teleconnection pattern. ENSO is an inter-annual periodic abnormal warming or cooling of the eastern equatorial Pacific sea surface temperatures (SSTs) that is regularly accompanied by a shift in the typical surface pressure field between the Australian and Peruvian coasts (e.g., Diaz, Hoerling, and Eischeid

2001). The positive (warm) phase of ENSO (i.e., El Niño) transports energy and moisture from the tropics to North America, thus displacing the cold California current and destabilizing the overlying atmosphere along the U.S. Pacific coast (Rohli and Vega 2008; Glantz 2001). Consequently, the amplitude and structure of the jet streams across North America alter in response to this low-latitude forcing, often with the polar jet stream bifurcating into a northern and a southern branch starting in the eastern North Pacific (figure 14.5a). The poleward displacement of the northern polar jet branch typically limits the penetration of Canadian arctic air masses into the North American interior, while the southern branch steers mid-latitude low pressure systems equatorward along the U.S. Gulf Coast. For the Midwestern USA, this signifies warmer than normal winters with fewer polar outbreaks and a reduction in winter storm activity (Wang and Ting 2000). Smith and O'Brien (2001) report decreased snowfall in the northern Great Lakes region during the ENSO warm phase, most likely as a result of higher surface temperatures and diminished jet stream dynamics. It is plausible that low-level wind speeds in the Midwest would decrease as the polar jet shifts northward, and Klink (2007) presents some preliminary evidence to support this hypothesis. Global climate model simulations suggest a possible future shift to a more El Niño–like pattern, but suggest that the ENSO teleconnection impact on the North American mid-latitudes may weaken (Meehl et al. 2006; 2007).

During the negative (cold) phase of ENSO (i.e., La Niña), the Walker circulation of the equatorial Pacific is reinforced, thus magnifying "normal" (or La Nada)

conditions. Cold event atmospheric circulation impacts for North America typically show the jet stream entering the Pacific Northwest and continuing along the U.S.-Canadian border (figure 14.5b). In comparison to non-ENSO years (figure 14.5c), the La Niña–induced upper tropospheric winds flow in a single, continuous jet stream across the continent. For the central United States, mean surface wind speeds significantly increase (decrease) during the cold (warm) phase months from November to March, most likely attributable to increased (decreased) cyclonic activity in the region (Enloe, O'Brien, and Smith 2004; Higgins, Leetmaa, and Kousky 2002). Midwestern and Ohio River Valley snowfall amounts associated with both ENSO phases tend to be much lower than when compared to neutral winters and may be related to the availability of moisture from the Gulf of Mexico (Smith and O'Brien 2001); however, some studies (e.g., Kunkel and Angel 1999) show a significant snowfall increase during La Niña events for the upper Midwest–Great Lakes region. The North American extratropical climate response to ENSO phases is frequently nonlinear, demonstrating asymmetry between its warm and cold phases (e.g., Rogers and Coleman 2003; Montroy, Richman, and Lamb 1998; Hoerling, Kumar, and Zhong 1997). Wang, Chang, and Wang (2007) note that for the boreal summer (the ENSO development period) the 500-hPa geopotential height response over North America is appreciably stronger for La Niña compared to El Niño events: during La Niña, anomalously high pressure occurs over Hudson Bay and into the central United States, but for El Niño events, the opposite (anomalously low pressure) does not occur.

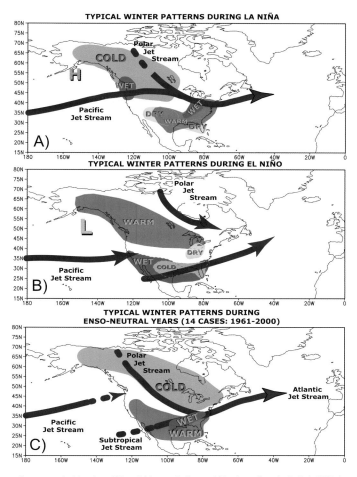

Figure 14.5. Mean jet stream flow position for (a) eighteen El Niño, (b) fourteen La Niña, and (c) eighteen Neutral or La Nada ENSO phases based on January–March composite anomalies for the 1950–1999 period. Reproduced with permission of the American Meteorological Society from Higgins et al. (2002).

ENSO-like conditions can persist past the typical periodicity of two to seven years, often favoring a dominant phase regime on periods ranging from twenty to thirty years where the primary oceanic signature extends to the northern Pacific (Bridgman and Oliver 2006). This long-term persistence of an ENSO-like pattern is known as the Pacific Decadal Oscillation (PDO) or the North Pacific Oscillation (NPO) (Mantua and Hare 2002, Verdon and Franks 2006). Positive (negative) PDO values indicate warmer (colder) than normal SSTs along the North American Pacific Northwest coast.

North American climate anomalies are most pronounced when the PDO and ENSO are in similar modes or "in-phase" (i.e., warm [cold] ENSO events are occurring during warm [cold] PDO periods) (Gershunov and Barnett 1998). Independently, the PDO mostly influences the strength of the westerlies around the Aleutian Low, becoming stronger (weaker) during warm (cold) PDO phases. Chen and Yoon (2002) describe a noticeable trend toward increased blocking activity during the latter half of the twentieth century as the PDO switched from a cold (about 1947–1976) to a warm period

(1977–late 1990s). Increased blocking would be expected during a warm PDO: a deeper Aleutian Low accentuates the amplitude of the Pacific Northwest ridge, leading to a typical "omega" block pattern. As such, the PDO can partially impact the jet stream configuration across North America on a decadal time scale, influencing the hemispheric and regional circulation patterns downstream. Synoptic-scale circulation regimes, such as those characterized by surface low situated over the central United States and upper-level ridge along the U.S. east coast, become less (more) common during PDO warm (cold) periods (Coleman and Rogers 2007).

ENSO is also linked to notable shifts in the location and magnitude of pressure systems in the mid-latitudes. Thermal forcing from the equatorial Pacific can change the trough-ridge configuration across North America (Hoerling and Ting 1994; Horel and Wallace 1981). Two regionally important teleconnections are partly produced in this manner, the Pacific North American (PNA) and Tropical Northern Hemisphere (TNH) patterns. Warm (cold) ENSO phases frequently prompt positive (negative) PNA and negative (positive) TNH modes (Rohli and Vega 2008; Robertson and Ghil 1999).

The PNA pattern is a broad feature of atmospheric low-frequency variability most prominent during the boreal winter and early spring in the mid-lower troposphere (Leathers, Yarnal, and Palecki 1991; Esbensen 1984). The PNA is distinguished by a Rossby wave configuration such that the pressure (or height) anomalies of the northern Pacific and southeastern United States are the reverse of those in the western United States

(Bridgman and Oliver 2006). Expressed as an index, the PNA characterizes the amplitude of the typical ridge-trough flow pattern across North America. The index is derived from the normalized 500-hPa or 700-hPa seasonal height anomalies for a set of locations (e.g., Wallace and Gutzler 1981) or using statistical pattern analysis methods, such as empirical orthogonal functions (EOFs) (e.g., Barnston and Livezey 1987). A positive (negative) PNA index typically indicates a meridional (zonal) circulation flow regime, thus intensifying (weakening) the baroclinic zone and increasing (decreasing) 500-hPa vorticity over the eastern United States (Walsh, Richman, and Allen 1982).

The extratropical PNA centers of action (i.e., the Aleutians, northwest Canada, the southeastern United States) are evident in the mean winter 500-hPa height differences between the negative and positive PNA phases (figure 14.6). PNA positive (negative) winters are linked to decreased (increased) precipitation and streamflow discharge for the Ohio River Valley region, an area extending from southeastern Missouri to Ohio (Rogers and Coleman 2004; Coleman and Rogers 2003; Leathers, Yarnal, and Palecki 1991). For the Great Lakes area, Isard, Angel, and Van Dyke (2000) and Angel and Isard (1998) find meridional (zonal) flow during PNA positive (negative) periods is correlated with higher (lower) cyclone frequency, with most extratropical storms originating from northwestern Canada (western and southwestern United States). In the eastern United States, Sheridan (2003) notes PNA positive periods in this sector are dominated by dry, polar air masses, and days influenced by moist, tropical air masses and transitional

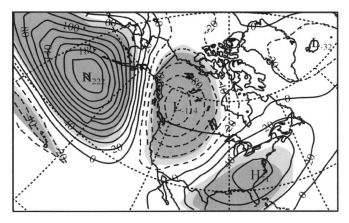

Figure 14.6. Pacific–North American pattern centers of action as shown by the mean winter (December–February) 500-hPa height differences in the NCEP-NCAR reanalysis between the lower and upper PNA index quintiles (1947–1998). Statistical significance from the student t-test is shown for the 95% (light gray) and 99% (dark gray) confidence levels. Reproduced with permission of the American Meteorological Society from Coleman and Rogers (2003).

situations (e.g., passing frontal systems) are significantly diminished, particularly for the upper Midwestern USA.

Reminiscent of the PNA structure, the Tropical Northern Hemisphere (TNH) pattern appears as a significant mode of inter-annual variability, most prominent from December to February at the 500-hPa and 700-hPa levels (Mo and Livezey 1986). The TNH constitutes a tripole structure with two action centers of similar pressure oscillations located along coastal British Columbia and the Southeast U.S. Gulf Coast and another of opposing sign situated over the Hudson Bay. The positive (negative) phase is connected to a stronger (weaker) Hudson Bay Low and positive (negative) height anomalies over the other two action centers. The pattern characterizes the northern trough-ridge position over North America, thereby identifying variability in the strength and position of the polar front jet stream and controlling the southward transport of continental polar air into the central United States (Rohli and Vega 2008; Climate Prediction Center

2005b). Hence, for the Midwestern USA, the TNH positive phase would be associated with below-average surface temperature and stable atmospheric conditions.

ATLANTIC TELECONNECTIONS

In comparison to the Pacific, the influence of Atlantic-based teleconnections is weaker for the Midwest because the main development sector is downstream of the region. One Atlantic teleconnection that does effect atmospheric circulation and climate in the eastern half of North America is the North Atlantic Oscillation (NAO). The NAO is often considered a regional inter-annual manifestation of the hemisphere-spanning Arctic Oscillation (AO) (also known as the Northern Annular Mode [NAM]) that describes the strength and equatorward extent of the polar vortex on a decadal time scale (Bridgman and Oliver 2006; Thompson and Wallace 2000). These two modes are nearly inseparable during the winter season, when the patterns are most defined (Rogers and

McHugh 2002). The North Atlantic Oscillation (NAO) itself refers to a north-south seesaw in SLP between two centers of action, one located near the Azores in the subtropical Atlantic Ocean and the other near Iceland in the subpolar North Atlantic Ocean. The positive (negative) NAO phase is associated with a stronger (weaker) than normal Icelandic Low and Azores (Bermuda) High that in turn produces a stronger (weaker) pressure gradient across the North Atlantic basin. This alters the strength and position of the polar jet stream over eastern North America and Western Europe, modulating the typical patterns of heat and moisture transport in these regions (Climate Prediction Center 2005a).

A positive (negative) phase of the NAO/AO is associated with a stronger, faster moving (weaker, slower moving) polar jet stream and more (less) frequent synoptic-scale cyclone passages (Rauthe and Paeth 2004). During the positive phase, the stronger westerlies inhibit the southward advection of Canadian arctic air masses into the Midwestern USA, thereby producing milder and wetter winters than when the NAO/AO is in the negative phase (Bridgman and Oliver 2006). For the eastern half of the United States, Sheridan (2003) finds that NAO-positive conditions reduce the frequency of polar air mass types during the winter and increase the likelihood of dry tropical air mass occurrence during the spring.

In addition to inter-annual pressure oscillations, the North Atlantic is accompanied by a cycling of warm and cold ocean water SST over time periods ranging from fifteen to thirty years, known as the Atlantic Multidecadal Oscillation (AMO) (Rohli and Vega 2008). The AMO is characterized by the av-erage of North Atlantic sea surface temperature anomalies and was in a positive/warm (negative/cold) phase from 1930 to 1959 (1965–1994) (Enfield, Mestas-Nuñez, and Trimble 2001). Warm periods are generally linked to an increase in Midwestern drought frequency, particularly when combined with a PDO negative state (McCabe, Palecki, and Betancourt 2004). Rogers and Coleman (2003) report statistically significant decreases in winter streamflow over the upper Mississippi Valley between the AMO warm and cold phases; however, the signal in the lower Mississippi and Ohio river basins weakens and modifies in accordance with Pacific conditions (i.e., PNA and ENSO). These anomalies may be related to shifts in the pressure field and availability of moisture from the Gulf of Mexico. For the boreal summer, Sutton and Hodson (2005) find that, over the southern United States, mean SLP is about 0.6 hPa higher during the cold (negative) phase of the AMO as compared to the warm (positive) phase. Such a pressure shift has important regional implications for long-term drought prediction because higher pressures inhibit precipitation processes.

Concluding Remarks

Midwestern climate is largely a product of long-term and short-term fluctuations in the Rossby wave configuration across North America. The strength and position of the polar and subtropical jet streams—which are strongly connected to large-scale temperature patterns—influence the development and propagation of synoptic-scale pressure systems and result in seasonal and annual variability in local meteorological conditions. Model simulations of the climate over the next century indeed suggest that the northern hemisphere is likely to experience a poleward shift in the mean position of storm tracks (Meehl et al. 2007) with accompanying changes in the frequency and intensity of extratropical cyclones (Christensen et al. 2007). A shift in the frequency, location, and/or intensity of atmospheric circulation patterns, such as those associated with a particular Pacific or Atlantic teleconnection pattern, may alter the typical climate state and increase the likelihood of anomalous weather events (e.g., heat waves, droughts). Many future climate scenarios do indicate a trend toward more frequent positive phases of the NAO/AO circulation modes; there is less consistency between model simulations of future changes in the frequency and intensity of ENSO events (Meehl et al. 2007). Other work (e.g., Liang, Wang, and Dudek 1996) suggests that the future climate of North America may be characterized by a higher frequency of positive PNA patterns; the accompanying meridional circulation pattern would increase the occurrence of polar outbreaks into the Midwest. Factors such as land cover change also may alter the nature and frequency of circulation patterns and accompanying climate characteristics: urbanization has well-known effects on surface wind speeds and on the temperature and humidity characteristics of the air over cities (e.g., Oke 1987), and regional land cover changes may also affect the characteristics of the air masses that typically occur over a region (e.g., Chase et al. 2000; Kalnay et al. 2006; Mahmood et al. 2006).

The chapters that follow describe some of the ongoing work that investigates patterns and trends in wind speed and atmospheric circulation across the Midwestern USA.

Chapters 15 and 16 examine near-surface wind speed trends utilizing model and/or station data, while chapters 17 and 18 illustrate methodologies for the identification of general Midwestern synoptic-scale circulation flow regimes, describing their pattern variability and associations with teleconnection pattern phase; however, the former focuses more on the ability of two GCMs in simulating the effects of multiple teleconnection patterns (PNA and NAO) on Midwestern atmospheric circulation. Each of these lines of research contributes to more accurate representations of the long-term regional climate variability. Investigation of near-surface wind speed trends and the connections with regional- and global-scale circulation regimes will improve our ability to relate model-predicted changes in the large-scale circulation to more regional and local flow patterns. Intercomparisons of observational and model-assimilated data (e.g., reanalysis products) help us to better interpret modeled changes in the weather and climate of the Midwest, both for individual stations and for the region as a whole. Understanding the possible alterations in the frequency and/or characteristics of Midwestern synoptic-scale circulation patterns will aid in forecasting the likelihood of adverse climate conditions. Together with investigations of other climate variables (such as temperature and precipitation), analyses of atmospheric circulation regimes and their related wind fluctuations will provide a more multifaceted assessment of climate variability and change in the Midwestern USA.

ACKNOWLEDGMENTS

We are indebted to R. J. Barthelmie for her thoughtful comments on this chapter.

REFERENCES

Angel, J. R., and S. A. Isard. 1998. "The Frequency and Intensity of Great Lake Cyclones." *Journal of Climate* 11: 61–71.

Barnston, A. G., and R. E. Livezey. 1987. "Classification, Seasonality and Persistence of Low-Frequency Atmospheric Circulation Patterns." *Monthly Weather Review* 115: 1083–1126.

Bridgman, H. A., and J. T. Oliver. 2006. *Global Climate System: Patterns, Processes, and Teleconnections.* New York: Cambridge University Press.

Changnon, S. A. 1985. "Climate Fluctuations and Impacts: The Illinois Case." *Bulletin of the American Meteorological Society* 66: 142–151.

Chase, T. N., et al. 2000. "Simulated Impacts of Historical Land Cover Changes on Global Climate in Northern Winter." *Climate Dynamics* 16: 93–105.

Chen, T. C., and J. H. Yoon. 2002. "Interdecadal Variation of the North Pacific Wintertime Blocking." *Monthly Weather Review* 130: 3136–3143.

Christensen, J. H., et al. 2007. "Regional Climate Projections." In *Climate Change 2007: The Physical Science Basis.* Contribution of Working Group I to the Fourth Assessment Report of the Intergovernmental Panel on Climate Change, 847–940. [Ed. S. Solomon, et al.] New York: Cambridge University Press.

Climate Prediction Center. 2005a. *North Atlantic Oscillation (NAO).* Available at: http://www.cpc.ncep.noaa.gov/data/teledoc/nao.shtml. Last modified on 23 May 2005. Accessed on 20 January 2008.

———. 2005b. *Tropical/Northern Hemisphere (TNH).* Available at: http://www.cpc.ncep.noaa.gov/data/teledoc/tnh.shtml. Last modified on 23 May 2005. Accessed on 17 February 2008.

———. 2006. *Storm Tracks.* Available at: http://www.cpc.noaa.gov/products/precip/CWlink/stormtracks/strack_NH.shtml. Last modified on 1 May 2006. Accessed on 3 February 2008.

Coleman, J. S. M. 2008. "Atmospheric Circulation Patterns Associated with Mortality Changes during the Transitional Seasons." *Professional Geographer* 60: 190–206.

Coleman, J. S. M., and J. C. Rogers. 2003. "Ohio River Valley Winter Moisture Conditions Associated with the Pacific–North American Teleconnection Pattern." *Journal of Climate* 16: 969–981.

———. 2007. "The Synoptic Climatology of the Central United States and Linkages to Pacific Teleconnection Indices." *Journal of Climate* 20: 3485–3497.

Diaz, H. F., M. P. Hoerling, and J. K. Eischeid. 2001. "ENSO Variability, Teleconnections, and Climate Change." *International Journal of Climatology* 21: 1845–1862.

Enloe, J., J. J. O'Brien, and S. R. Smith. 2004. "ENSO Impacts on Peak Wind Gusts in the United States." *Journal of Climate* 17: 1728–1738.

Enfield, D. B., A. M. Mestas-Nuñez, and P. J. Trimble. 2001. "The Atlantic Multidecadal Oscillation and Its Relationship to Rainfall and River Flows in the Continental U.S." *Geophysical Research Letters* 28: 2077–2080.

Esbensen, S. K. 1984. "A Comparison of Intermonthly and Interannual Teleconnections in the 700mb Height Field during the Northern Hemisphere Winter." *Monthly Weather Review* 112: 2016–2032.

Gershunov, A., and T. P. Barnett. 1998. "Interdecadal Modulation of ENSO Teleconnections." *Bulletin of the American Meteorological Society* 79: 2715–2726.

Glantz, M. 2001. *Currents of Change: Impacts of El Niño and La Niña on Climate and Society.* 2nd ed. London: Cambridge University Press.

Harman, J. R. 1991. *Synoptic Climatology of the Westerlies: Process and Patterns.* Washington, D.C.: Association of American Geographers Resource Publications in Geography.

Harrington, J., and J. E. Oliver. 2000. "Understanding and Portraying the Global Atmospheric Circulation." *Journal of Geography* 99: 23–31.

Higgins, R. W., A. Leetmaa, and V. E. Kousky. 2002. "Relationships between Climate Variability and Winter Temperature Extremes in the United States." *Journal of Climate* 15: 1555–1572.

Hoerling, M. P., A. Kumar, and M. Zhong. 1997.

"El Niño, La Niña, and the Nonlinearity of Their Teleconnections." *Journal of Climate* 10: 1769–1786.

Hoerling, M. P., and M. Ting. 1994. "Organization of Extratropical Transients during El Niño." *Journal of Climate* 7: 745–766.

Horel, J. D., and J. M. Wallace. 1981. "Planetary-Scale Atmospheric Phenomena Associated with the Southern Oscillation." *Monthly Weather Review* 109: 813–829.

IPCC. 2007. *Climate Change 2007: The Physical Science Basis.* Contribution of Working Group I to the Fourth Assessment Report of the Intergovernmental Panel on Climate Change. [Ed. S. Solomon, et al.] New York: Cambridge University Press.

Isard, S. A., J. R. Angel, and G. T. Van Dyke. 2000. "Zones of Origin for Great Lakes Cyclones in North America, 1899–1996." *Monthly Weather Review* 128: 474–485.

Kalnay, E., et al. 2006. "Estimation of the Impact of Land-Surface Forcings on Temperature Trends in the Eastern United States." *Journal of Geophysical Research* 111, doi:10.1029/2005JD006555.

Kalnay, E., et al. 1996. "The NCEP/NCAR 40 Reanalysis Project." *Bulletin of the American Meteorological Society* 77: 437–471.

Klink, K. 1999. "Climatological Mean and Inter-annual Variance of United States Surface Wind Speed, Direction, and Velocity." *International Journal of Climatology* 19: 471–488.

———. 2007. "Atmospheric Circulation Effects on Wind Speed Variability at Turbine Height." *Journal of Applied Meteorology and Climatology* 46: 445–456.

Kunkel, K. E., and J. R. Angel. 1999. "The Relationship of ENSO to Snowfall and Related Cyclone Activity in the Contiguous United States." *Journal of Geophysical Research* 104: 19425–19434.

Leathers, D. J., B. Yarnal, and M. A. Palecki. 1991. "The Pacific/North American Teleconnection Pattern and United States Climate, Part 1: Regional Temperature and Precipitation Associations." *Journal of Climate* 4: 517–528.

Liang, X. Z., W. C. Wang, and M. P. Dudek. 1996. "Northern Hemispheric Interannual Teleconnection Patterns and Their Changes Due to the Greenhouse Effect." *Journal of Climate* 9: 465–475.

Mahmood, R., et al. 2006. "Impacts of Irrigation on 20th Century Temperature in the Northern Great Plains." *Global and Planetary Change* 54: 1–18.

Mantua, N. J., and S. R. Hare. 2002. "The Pacific Decadal Oscillation." *Journal of Oceanography* 58: 35–44.

McCabe, G. J., M. A. Palecki, and J. L. Betancourt. 2004. "Pacific and Atlantic Ocean Influences on Multidecadal Drought Frequency in the United States." *Proceedings of the National Academy of Sciences* 101: 4136–4141.

Meehl, G. A., et al. 2006. "Climate Change Projections for the Twenty-First Century and Climate Change Commitment in the CCSM3." *Journal of Climate* 19: 2597–2616.

Meehl, G. A., et al. 2007. "Global Climate Projections." In *Climate Change 2007: The Physical Science Basis.* Contribution of Working Group I to the Fourth Assessment Report of the Intergovernmental Panel on Climate Change, 747–845. [Ed. S. Solomon, et al.] New York: Cambridge University Press.

Mo, K. C., and R. E. Livezey. 1986. "Tropical-Extratropical Geopotential Height Teleconnections during the Northern Hemisphere Winter." *Monthly Weather Review* 114: 2488–2515.

Montroy, D. L., M. B. Richman, and P. J. Lamb. 1998. "Observed Nonlinearities of Monthly Teleconnections between Tropical Pacific Sea Surface Temperature Anomalies and Central and Eastern North American Precipitation." *Journal of Climate* 11: 1812–1835.

Oke, T. R. 1987. *Boundary Layer Climates.* 2nd ed. New York: Routledge.

Patten, J. M., S. R. Smith, and J. J. O'Brien. 2003. "Impacts of ENSO on Snowfall Frequencies in the United States." *Weather and Forecasting* 18: 965–980.

Pryor, S. C., R. J. Barthelmie, and E. Kjellström. 2005. "Analyses of the Potential Climate Change Impact on Wind Energy Resources in Northern Europe Using Output from a Regional Climate Model." *Climate Dynamics* 25: 815–835.

Rauthe, M., and H. Paeth. 2004. "Relative Importance of Northern Hemisphere Circulation Modes in Predicting Regional Climate Change." *Journal of Climate* 17: 4180–4189.

Robertson, A.W., and M. Ghil. 1999. "Large-Scale Weather Regimes and Local Climate over the Western United States." *Journal of Climate* 12: 1796–1813.

Robeson, S. M. 2002. "Increasing Growing-Season Length in Illinois during the 20th Century." *Climatic Change* 52: 219–238.

Rogers, J. C., and J. S. M. Coleman. 2003. "Interactions between the Atlantic Multidecadal Oscillation, El Niño/La Niña, and the PNA in Winter Mississippi Valley Stream Flow." *Geophysical Research Letters* 30, doi:10.1029/2003GL017216.

———. 2004. "Ohio Winter Precipitation and Stream Flow Associations to Pacific Atmospheric and Oceanic Teleconnection Patterns." *Ohio Journal of Science* 104: 51–59.

Rogers, J. C., and M. L. McHugh. 2002. "On the Separability of the North Atlantic Oscillation and Arctic Oscillation." *Climate Dynamics* 19: 599–608.

Rohli, R. V., and A. J. Vega. 2008. *Climatology.* Subury, Mass.: Jones and Bartlett Publishers.

Schwartz, M. D. 1991. "An Integrated Approach to Air Mass Classification in the North Central United States." *Professional Geographer* 43: 77–91.

Sheridan, S. C. 2003. "North American Weather-Type Frequency and Teleconnection Indices." *International Journal of Climatology* 23: 27–45.

Smith, S. R., and J. J. O'Brien. 2001. "Regional Snowfall Distributions Associated with ENSO: Implications for Seasonal Forecasting."

Bulletin of the American Meteorological Society 82: 1179–1191.

Southworth J., et al. 2000. "Consequences of Future Climate Change and Changing Climate Variability on Maize Yields in the Midwestern United States." *Agriculture, Ecosystems, and Environment* 82: 139–158.

Sutton, R. T., and L. R. Hodson. 2005. "Atlantic Forcing of North American and European Summer Climate." *Science* 309: 115–118.

Thompson, D. W. J., and J. M. Wallace. 2000. "Annular Modes in the Extratropical Circulation, Part 1: Month-to-Month Variability." *Journal of Climate* 13: 1000–1016.

Verdon, D. C., and S. W. Franks. 2006. "Long-Term Behaviour of ENSO: Interactions with the PDO over the Past 400 Years Inferred from Paleoclimate Records." *Geophysical Research Letters* 33, doi: 10.1029/2005GL025052.

Wallace, J. M., and D. S. Gutzler. 1981. "Teleconnections in the Geopotential Height Field during the Northern Hemisphere Winter." *Monthly Weather Review* 109: 784–812.

Walsh, J. E., M. B. Richman, and D. A. Allen. 1982. "Spatial Coherence of Monthly Precipitation in the United States." *Monthly Weather Review* 110: 272–286.

Wang, H., and M. Ting. 2000. "Covariabilities of U.S. Winter Precipitation and Pacific Sea Surface Temperatures." *Journal of Climate* 13: 3711–3719.

Wang, Z., C. P. Chang, and B. Wang. 2007. "Impacts of El Niño and La Niña on the U.S. Climate during Northern Summer." *Journal of Climate* 20: 2165–2177.

15. Historical Trends in Near-Surface Wind Speeds

S. C. PRYOR AND R. J. BARTHELMIE

Introduction

MOTIVATION AND OBJECTIVE

The USA is investing in renewable energy resources to meet national and international goals related to reducing greenhouse gas/pollutant emissions and to improve the diversity of energy portfolios and security of supply. During 2005–2007 over 10,000 MW of wind energy developments came online, increasing installed capacity to 17,000 MW (AWEA 2008). A further 4,000 MW of capacity was installed in the first three quarters of 2008. While much of the installed capacity is in California and Texas, some of this development has been focused on the Midwestern states of Iowa, Minnesota, and Illinois (see chapter 1 of this volume), and in 2008 the first wind farm was installed in Indiana. Much of Iowa and Minnesota are defined as exhibiting wind power class 3 and 4 in the NREL wind energy atlas, equating to wind power densities at 50 m of 300–400 W m^{-2} and 400–500 W m^{-2}, respectively (Elliott et al. 1986).

Although many factors dictate the deployment and success of wind farm developments (Barthelmie 2007), identification of optimal sites for development of wind farms relies in part on detailed knowledge regarding the local wind climates and hence likely power production. In the context of wind energy applications, it is necessary to estimate the power output over the twenty-to-thirty-year lifetime of the wind farm for economic feasibility (Pryor, Barthelmie, and Schoof 2005). In this context the questions that arise are: "what is the current inter-annual variability in likely power production and will non-stationarities in the global climate system cause that variability or magnitude of a *normal wind year* to evolve on timescales of relevance to wind energy developments?" (Pryor, Barthelmie, and Kjellström 2005; Pryor, Schoof, and Barthelmie 2005; Pryor, Schoof, and Barthelmie 2006). Because energy density is proportional to the cube of the wind speed (see equation 15.1), comparatively small changes in the wind speed at turbine hub-height have large consequences for power production and hence for the overall economics of wind projects. This places unprecedented demands for both accuracy and precision on wind speed data and forecasts; wind farm developers expect to quantify the mean wind speed at hub-height to within ±5% of the true value, and mean annual power production to better than ±10%.

Herein we focus on quantifying historical variability and trends using in situ observations and reanalysis-derived wind speeds.

ENERGY DENSITY AND WIND SPEED DISTRIBUTIONS

The energy density (E) is the total energy in the wind that is potentially available to be harnessed using wind turbines:

$$E = \frac{1}{2}\rho U^3 \qquad (15.1)$$

ρ is density and U is wind speed

Hence the cumulative energy density over a time interval (i.e., a year) is dominated by the upper percentiles of the wind speed probability distribution. The probability distribution of wind speeds is typically represented using the two-parameter Weibull probability density function:

$$p(U) \equiv \frac{k}{A}\left(\frac{U}{A}\right)^{k-1} \quad \exp\left[-\left(\frac{U}{A}\right)^k\right] \quad (15.2)$$

Where k is a dimensionless shape parameter (a measure of the peakedness of the distribution), A is the scale parameter (a measure of the central tendency), U is the time series of wind speed observations, and p(U) is the probability density function.

The Weibull distribution parameters can be derived from, or used to derive, other descriptors of the probability distribution (Pryor et al. 2004) as follows:

$$\overline{U} = A\Gamma\left(1+\frac{1}{k}\right) \qquad (15.3)$$

$$U_{X*100} = A(-1.*\ln(1-X))^{1/k} \quad (15.4)$$

where $\overline{U}$ is the mean wind speed, $U_{50\%}$ is the median wind speed, U_{x*100} is the $(X*100)^{th}$ percentile wind speed, and Γ is the incomplete gamma function.

The energy density is related to the Weibull A and k via:

$$E = \frac{1}{2}\rho A^3 \Gamma\left(1+\frac{3}{k}\right) \qquad (15.5)$$

Data and Methods

DATA

Wind speed time series from four sources are analyzed here:

1. Near-surface wind speeds from land-based sites across the contiguous United States extracted from the data archive held by the National Climatic Data Center (NCDC) (http://www.ncdc.noaa.gov/oa/mpp/freedata.html, DS3505 surface data, global hourly). Data were obtained for all land-based sites that have records from 1973 to 2005, along with comprehensive station histories. All valid data that passed all NCDC data quality control procedures were selected for sites where the station histories (including anemometer height) were available and indicated the station had not moved more than 5 km over the study period. The wind speed data were reported in whole knots (the ASOS anemometers are rated to ±2 knots from 0–125 knots, http://www.nws.noaa.gov/asos/pdfs/aum-toc.pdf). The result is that the data are not continuous, which can generate difficulties in use of linear trend tools. Hence log-nor-

mally distributed random numbers were added to the time series to "recover" some of the missing variability. Additionally, not all wind speed data were collected at 10 m above the ground, and several anemometer heights changed during the study period, so the data were corrected to a common measurement height of 10 m based on the recorded anemometer heights and the power law wind profile and an exponent (α) of $1/7$ (Manwell, McGowan, and Rogers 2002):

$$Uz_1 = Uz_2 \left(\frac{z_1}{z_2} \right)^{\alpha} \quad (15.6)$$

Where Uz_x is the wind speed at height z_x.

Application of this approximation assumes near-neutral stability and a flat relatively smooth surface around the sites. However, the correction to 10 m is relatively small for most measurement heights encountered in the data records (figure 15.1).

Although the data records are nominally hourly, at least in the initial part of the study period at many sites the data were actually

archived at three-hourly intervals or only during daylight hours. Accordingly here we focus on observations from 0000 UTC and 1200 UTC, and analyze data from these two observation times separately to examine if there is any "time of observation" bias in observed temporal trends. We consider only sites where over half of the possible observations are present in every year of record (1973–2005) and more than half of valid observations are available in each climatological season of each year. The data selection criteria resulted in time series from 193 stations being available for analysis. All stations are airports (190) or military installations (3).

Time series from five-year running means of the annual percentiles from these sites were used to determine the year in which the largest change in the percentiles was observed, to assess whether the trends derived herein are dominated by data inhomogeneities associated with the introduction of new measurement technologies or protocols, such as the Automated Surface Observing System (ASOS) which commenced in the

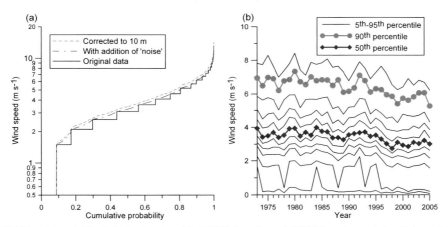

Figure 15.1. (a) Cumulative probability distribution of wind speeds for 0000 UTC observations at site 724320 (in southwestern Indiana) in the raw data, with log-normally distributed white noise added, and then adjusted to a nominal measurement height of 10 m. (b) Annual percentiles for 0000 UTC observations from site 724320 (5th, 10th, 20th . . . 90th, 95th percentile, where the 50th and 90th percentiles are shown in the blue and red, respectively). Despite considerable inter-annual variability, data from this station exhibit a significant downward trend in both the 50th percentile (of approximately 0.7%/year) and the 90th percentile (of approximately 0.6%/year) wind speed.

early 1990s. This does not appear to be the case; trimming the data sets (to remove years at the end or beginning of the time series) did not reverse temporal trends, though the number of significant trends does decline if the latter portion of the time series is removed from consideration (Barthelmie and Pryor 2007). This finding illustrates that important homogeneity issues with wind speeds did occur with the introduction of the ASOS. We specifically chose here to focus on percentile-based approaches to avoid biases in mean wind speed deriving from known changes in extreme/high wind speeds with the introduction of ASOS; however, the frequencies with which calm winds were reported were also "significantly affected after the ASOS implementation" (McKee et al. 2000; Groisman 2001). Specifically, an analysis of twelve stations showed ASOS-derived wind speeds were an average of 0.2 ms^{-1} lower than with the prior observing system, with a range of −0.65 ms^{-1} to 0.15 ms^{-1} (McKee et al. 2000), though the higher wind speeds were higher from the ASOS instrumentation. Evidence of this effect is evident in figure 15.1b, which shows the 5th and 10th percentile dropped to nearly zero after the introduction of the ASOS system at this site.

2. Near-surface (10 m) U (west-east) and V (south-north) components of the flow were extracted from the four-times-daily output of the NCEP-NCAR reanalysis data archive (http://www.cdc.noaa.gov/cdc/data.ncep.reanalysis.html) for 1948–2006, inclusive (Kalnay et al. 1996). This data set is referred to as NCEP Reanalysis 1 and has a spatial resolution ~2.5×2.5°.

3. Near-surface (10 m) U (west-east) and V (south-north) components of the flow

were extracted from the four-times-daily output of the NCEP-DoE reanalysis data (http://www.cdc.noaa.gov/cdc/data.ncep.reanalysis.html) for 1979–2006 (the data record starts in 1979), inclusive (Kanamitsu et al. 2002). This data set is referred to as NCEP Reanalysis 2, and it has a resolution ~1.9×1.9°. NCEP Reanalysis 2 was undertaken in response to several issues and errors in the NCEP/NCAR Reanalysis 1 model and products and is generally seen as a superior reanalysis product.

4. Near-surface (10 m) U (west-east) and V (south-north) components of the flow were extracted from the four-times-daily output of the ERA-40 reanalysis data set (Uppala et al. 2005). This data set is referred to as ERA-40 reanalysis, and has a spatial resolution ~2.5×2.5°. Data were extracted for 1973–2001 (the reanalysis product ends in the middle of 2002), inclusive.

As with the in situ observations, reanalysis data from 0000 UTC and 1200 UTC are analyzed separately.

Reanalysis projects such as these draw data from a range of sources, which are quality controlled and assimilated with a consistent data simulation system (models). These reanalysis products are thus a hybrid of the observations that are assimilated and "background" information used to provide complete representations of the atmosphere that are derived from a short-range forecast initiated from the most recent previous analysis. The data and forecasts are integrated using error statistics, and the reanalysis products thus comprise four-dimensional, homogenized, and systematic data sets. The degree of dependence on (and to some degree, association with) observations depends on the density and relative accuracy of the observations and with the dynamics and physics of the forecast model, and is variable with geophysical parameter (Kistler et al. 2001; Kanamitsu et al. 2002; Uppala et al. 2005).

The reanalysis output and observational station wind speed time series are analyzed separately herein, and the results are inter-compared because while the reanalysis procedures integrate the surface observations, they also assimilate data from other sources, and the reanalysis products are also dependent on the formulation of the numerical model used.

The reanalysis products have been extensively evaluated and inter-compared by both the groups that derive them and independent researchers (recent examples include; Kanamitsu et al. 2002; Wang, Swail, and Zwiers 2006; Zhao and Fu 2006; Dell'Aquila et al. 2007; Song and Zhang 2007). A key issue in the use of wind components from reanalysis products is that there is tremendous evolution in the number and nature of assimilated data sources over the period of record considered here; 1973–2005 (see for example figure 1 in Uppala et al. [2005]). Wind speeds are inherently difficult to model due to the high spatial variability, influence of local obstacles, and relative scarcity of observations. As just one example, previous analysis of ERA-15 near-surface wind speeds indicated a negative bias over oceanic surfaces. These "deficiencies were subsequently corrected by changes to the assimilation system. Unrepresentative island wind observations were no longer used, and ship winds were applied at the anemometer height where known and otherwise as a more representative height than 10 m. . . . Surface winds in the later years of ERA-40 also benefit directly from the assimilation of

scatterometer and SSM/I data, and indirectly from interaction with the ocean-wave model that in turn benefits from the assimilation of altimeter measurements of ocean-wave height" (Uppala et al. 2005).

PRIOR ANALYSIS OF WIND SPEED TEMPORAL TRENDS

Trends in near-surface wind speeds have relevance beyond the wind energy industry, and may also impact; the insurance industry (Changnon, Fosse, and Lecomte 1999), construction (Ambrose and Vergun 1997), actions to prevent or minimize coastal erosion (Bijl 1997, Viles and Goudie 2003), and forest and infrastructure protection (Jungo, Goytette, and Beniston 2002). However, wind speed time series have been subject to far fewer analyses than temperature and precipitation records, in part because of the data homogeneity issues articulated above. Nevertheless, in our previous research over Europe we have shown:

• Based on the NCEP Reanalysis 1 data set, annual mean wind speeds over the Baltic significantly increased over the period 1953–1999, with the majority of the increase being associated with increases in the upper quartile of the wind speed distribution and extreme winds (Pryor and Barthelmie 2003). More recent analyses (by both the authors and others) have suggested near-surface wind speeds have subsequently declined relative to the high values observed in the early 1990s (Barthelmie and Pryor 2006).

• Wind speed projections for the twenty-first century indicate no evidence of substantial evolution relative to the end of the twentieth century, particularly relative to current variability manifest in downscaling from different coupled Atmosphere-Ocean General Circulation Models (AOGCMs) (Pryor, Barthelmie, and Kjellström 2005; Pryor, Schoof, and Barthelmie 2006).

• Generally mean 10-m wind speeds are lower in the ERA-40 reanalysis products than in the NCEP Reanalysis 1 (Pryor, Barthelmie, and Schoof 2006).

The latter observation is equally true of much of the North American domain analyzed here. Herein we focus on comparisons of temporal trends in three reanalysis products and situ observations over the contiguous United States, but with a focus on the Midwestern USA.

METHODOLOGY

Wind speed time series exhibit variability on multiple temporal scales. Here we focus on the annual time series and analyze percentiles of the wind speed distribution computed at the annual time scale for trends using linear regression and boot-strapping techniques to determine whether trends are robust to the stochastic effects in the time series. In brief, this involves bootstrap resampling of the residuals from the linear regression analysis of annual Xth percentile wind speed on year. These residuals are randomly selected using a bootstrapping technique and added onto the linear fit line from the trend analysis, and the trend is re-estimated (Kiktev et al. 2003). This procedure is repeated 1,000 times to generate 1,000 plausible trends for each station. If a zero trend falls within the middle 900 values in an ordered sequence of the distribution of 1,000 realizations, the trend is not significant at the 90% confidence level.

We also examine the wind speed time series using a wind indexing technique in order to quantify both the inter-annual variability in wind indices and temporal trends therein. Wind energy indices are simple mechanisms to assess inter- and intra-annual variability of wind energy (Pryor, Barthelmie, and Schoof 2006) and to provide a context for shorter-term, site-specific time series measured by wind energy developers to assess the feasibility of wind farm installations. The index as used here is:

$$\text{Index} = \overline{\sum_{j=1}^{n} \frac{U_j^3}{U_{i\ldots k}^3}} *100 \qquad (15.7)$$

Where j=1, n indicates the time series from the period of interest, i . . . k indicates the normalization period.

Naturally the index for any given year is determined in part by the normalization period used. Wind indices are computed here for the Midwestern USA using the 1948–2006 data from the NCEP Reanalysis 1 data set to provide a context for the 1973–2005 study period. Wind indices are also computed for the Midwest using the other two shorter reanalysis time series. In each case a normalization period of 1992 to 2001 is used, and data from all grid-cells within the Midwest are merged into a single time series to generate a single wind index for the entire region.

Objectives

The objectives of this research are fourfold:

1. To quantify historical (1973–2005) trends in near-surface wind speeds across the contiguous United States based on in

situ observations and to assess their statistical significance.

2. To compare trends in time-series of near-surface wind speeds from individual stations with those derived from gridded reanalysis data.

3. To evaluate whether the temporal trends exhibit substantial bias with hour of observation.

4. To assess the impact of any temporal trends on the magnitude of the wind energy resources and the feasibility of wind energy developments.

Results

TEMPORAL TRENDS OVER THE CONTIGUOUS UNITED STATES

An example of the time series subjected to trend analysis from a single site is given in figure 15.1b. As shown, there is considerable inter-annual variability in the magnitude of the percentile values, but at this site, as at the majority of those studied (figures 15.2 and 15.3), both the 50th and 90th percentile observed wind speeds exhibit statistically significant declines over the period 1973–2005. When results are generated for all 193 stations, 150 exhibit declines in the 0000 UTC 50th percentile values, 33 stations exhibit no trend, and only 10 stations exhibit increases. One hundred forty-six stations exhibit declines in the 90th percentile wind speeds, 36 stations exhibit no trend, and 11 stations exhibit increases (figure 15.2). Similar results are found for the 1200 UTC observations (figure 15.3). These results thus indicate a prevalence of trends toward decreasing wind speeds over the period of study, 1973–2005, and they are in accord with prior work in

and around Minnesota that indicated a decline in the upper percentiles of annual mean daily wind speeds from about 1960 to 1995 (Klink 2002). There is no evidence for substantially different trend signs or magnitude with hour of observation (cf. figures 15.2 and 15.3), but the magnitude of the temporal trends in the annual percentile values is generally slightly larger for the 50th percentile than the 90th percentile values when the change is expressed as a percentage change.

The same trend methodology was also applied to 10-m wind speeds derived from the three reanalysis products (see figures 15.2 and 15.3). One example for 0000 UTC from the NCEP Reanalysis 1 data set for the grid cell containing station 724320 and the time series from that site are depicted in figure 15.4. As described above, data from the station indicate significant downward trends in the 50th and 90th percentile values, while the NCEP Reanalysis 1 gridded data exhibit no trend. Also worthy of note is that the reanalysis product exhibited higher values for all percentiles than the station observations. Figures 15.2 and 15.3 show trend results for all reanalysis data sets and grid cells for 0000 UTC and 1200 UTC, respectively. As shown:

- The magnitudes of the trends in the observed wind speed records are substantial—up to 1%/year. Similarly large trends are observed in the reanalysis data sets, but over a smaller fraction of the study region.
- In contrast to the in situ observations, the NCEP Reanalysis 1 data set generally indicates a tendency toward increased values of the 50th and 90th percentile annual wind speeds over the course of 1973–2005, particularly in the central United States and Midwest.

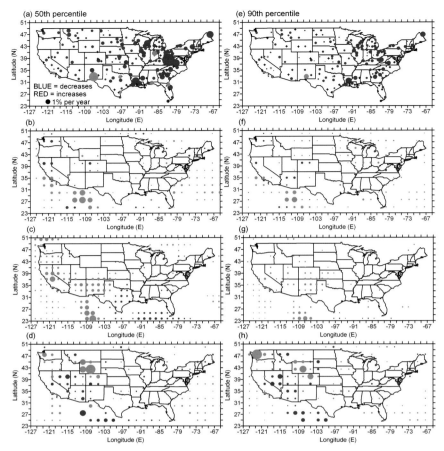

Figure 15.2. Results of the trend analysis applied to data from 0000 UTC. The individual frames show results from: (a) 50th percentile wind speed, observations (1973–2005), (e) 90th percentile wind speed, observations (1973–2005), (b) 50th percentile, NCEP Reanalysis 1 dataset (1973–2005), (f) 90th percentile, NCEP Reanalysis 1 dataset (1973–2005), (c) 50th percentile, NCEP Reanalysis 2 dataset (1979–2005), (g) 90th percentile, NCEP Reanalysis 2 dataset (1979–2005), (d) 50th percentile, ERA-40 dataset (1973–2001), (h) 90th percentile, ERA-40 dataset (1973–2001). In each frame the size of the dot scales linearly with the magnitude of the trend, and the color of the dot indicates the sign of the trend. Where the station time series did not indicate a statistically significant trend, a + symbol is shown. Where time series from a reanalysis grid cell did not exhibit a trend no symbol is shown.

•Also in contrast to the station observations, the trends in the NCEP Reanalysis 1 data set are frequently of larger magnitude in the 90th percentile values.

• There are differences in both sign and magnitude in trends in the NCEP Reanalysis 1 and 2 data sets spatial across the United States. While NCEP Reanalysis 1 exhibit data most spatially consistent increasing trends over the Midwest, largest positive trends in NCEP Reanalysis 2 data are evident over the western United States, and NCEP Reanalysis 2 gener-

ally does not indicate significant trends over the Midwest.

• Comparable trends in the 50th and 90th percentile wind speeds from ERA-40 are almost evenly divided between increasing, decreasing, and no-change over the contiguous United States, and as in the observational data set, some of the percentage trends are slightly larger in the 50th percentile values.

Recall that the time windows used in the trend analysis differ by data set with the sta-

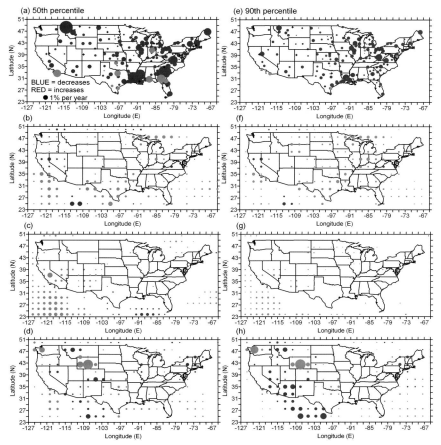

Figure 15.3. Results of the trend analysis applied to data from 1200 UTC. The individual frames show results from: (a) 50th percentile wind speed, observations (1973–2005), (e) 90th percentile wind speed, observations (1973–2005), (b) 50th percentile, NCEP Reanalysis 1 dataset (1973–2005), (f) 90th percentile, NCEP Reanalysis 1 dataset (1973–2005), (c) 50th percentile, NCEP Reanalysis 2 dataset (1979–2005), (g) 90th percentile, NCEP Reanalysis 2 dataset (1979–2005), (d) 50th percentile, ERA-40 dataset (1973–2001), (h) 90th percentile, ERA-40 dataset (1973–2001). In each frame the size of the dot scales linearly with the magnitude of the trend, and the color of the dot indicates the sign of the trend. Where the station time series did not indicate a statistically significant trend, a + symbol is shown. Where time series from a reanalysis grid cell did not exhibit a trend no symbol is shown.

tion data and NCEP Reanalysis 1 focusing on 1973–2005, while those from the ERA-40 data set are for 1973–2001, and those for NCEP Reanalysis 2 are for 1979–2005. However, truncation of the NCEP Reanalysis 1 data to match the NCEP Reanalysis 2 or ERA-40 temporal windows does not lead to improved agreement in trend sign or magnitude.

As mentioned above, a major discrepancy between the station data and the three reanalysis data sets is observed over the Mid-

west (figure 15.5). The overwhelming majority of station observations exhibit negative trends in both the 50th and 90th percentiles, while the NCEP Reanalysis 1 data exhibit a dominance of positive trends, particularly in the 90th percentile value. The NCEP Reanalysis 2 data exhibit fewer significant trends, and the number of grids showing significant trends is approximately the number expected by random chance. The ERA-40 data set exhibits few significant trends, almost evenly distributed between increases and declines.

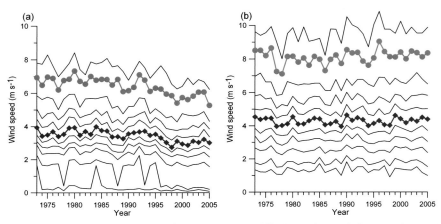

Figure 15.4. Annual percentiles for 0000 UTC observations from: (a) station 724320 and (b) NCEP Reanalysis 1 grid cell containing that station. Individual lines show the 5th, 10th, . . . 90th, 95th percentile, where the 50th and 90th percentiles are shown in blue and red.

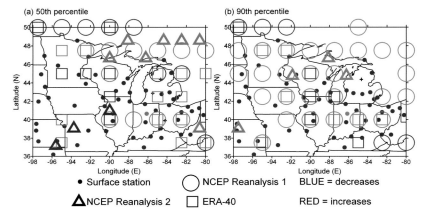

Figure 15.5. Results from the trend analysis over the Midwestern USA applied to annual percentiles for the 0000 UTC observations from the four datasets. The dots depict the station observations, the outer grid squares the NCEP Reanalysis 1, the inner grid squares the ERA-40 data, and the triangles the NCEP Reanalysis 2 data. Where no grid square is shown from the reanalysis data, the grid-cell showed "no trend." Frame (a) shows results for the 50th percentile wind speed, while frame (b) shows results for the 90th percentile wind speed.

SYNTHESIS OF TEMPORAL TRENDS

Based on the analysis presented herein we conclude there are substantial differences between trends derived from carefully quality controlled observational wind speed data and the reanalysis products, and indeed between wind speeds from reanalysis data sets. The source of the discrepancy between the four data sets is currently undetermined, but as in our prior research across the European continent (Pryor, Barthelmie, and Schoof 2006) there are quantitative differences in mean wind speeds and trends in wind speed percentiles between reanalysis data sets. While these differences can not be fully explained, they must be acknowledged, and their presence strongly advocates for use of all four in assessing flow climates. Further, as we propose in our prior work, such differences/discrepancies are "physically consistent with previous analyses of cyclone climatologies and might reasonably be invoked to provide a range of conditions (confidence bounds) for comparison with AOGCM simulations" (Pryor, Barthelmie, and Schoof 2006). Interestingly, results from a recent study over the Netherlands also resolved: "The results for moderate wind events (that occur on average 10 times per year) and strong wind events (that occur on average twice a year) indicate a decrease in storminess over the Netherlands between 5 and 10%/decade. This result is inconsistent with National Centers for Environmental Prediction–National Center for Atmospheric Research or European Centre for Medium-Range Weather Forecasts reanalysis data, which suggest increased storminess during the same 41 year period" (Smits, Tank, and Konnen 2005).

IMPLICATIONS FOR THE WIND ENERGY INDUSTRY

For trends in wind speeds to have relevance for the power output from wind turbines they must occur above the cut-in for wind turbines (i.e., the minimum wind speed at which a wind turbine will generate electrical power). The power curve for a sample 2 MW wind turbine is shown in figure 15.6, and illustrates the electrical power produced by a turbine as a function of prevailing wind speed at the hub-height (i.e., center of the three turbine blades).

Based on this power curve, we "corrected" the wind speeds from a nominal height of 10 m above ground level to 70 m AGL using the power law approximation, and then divided the observations and re-analysis data into four classes: wind speeds below 4 ms^{-1}, wind speeds above 4 ms^{-1}, wind speeds above 15 ms^{-1}, and wind speeds above 25 ms^{-1}. It is important to emphasize that use of the power law to vertically extrapolate wind speeds leads to some uncertainty in the 70-m wind speeds presented here. In the

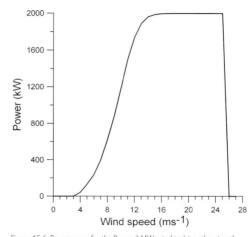

Figure 15.6. Power curve for the Bonus 2 MW wind turbine, showing the electrical output from the turbine as a function of the prevailing wind speed at hub-height (70 m). Below approximately 4 ms^{-1} (the "cut-in") no power is produced, between approximately 4 and 15 ms^{-1} there is a sharp increase in electrical power produced with increasing wind speed, and at wind speeds of between 15 and 25 ms^{-1} the power produced is a constant "rated" (i.e., 2 MW) value. At wind speeds above 25 ms^{-1} the turbine shuts down for safety reasons, and no power is produced.

absence of detailed site characterization, the power law remains the industry standard for such calculations. Questions have been raised regarding the appropriate exponent to use. We use $1/7$ here since it is commonly applied and represents a conservative estimate. Recent research based on the Department of Energy tall towers project suggests the shear parameter (α in equation 15.6) may be above 0.143 ($1/7$). Observed values of the shear parameter computed using data from 50 to 70, 80, and 110 m over the central United States range from 0.138 to 0.254 (Schwartz and Elliott 2006).

Since wind speeds above 25 ms^{-1} were observed only relatively infrequently, only the frequency of observations in the three lower classes were subject to trend analysis. The results indicate an increase in the number of "calms," or zero-production wind speeds, in observations from almost all of the in situ sites (figure 15.7), with a commensurate decline in the frequency with which power-producing

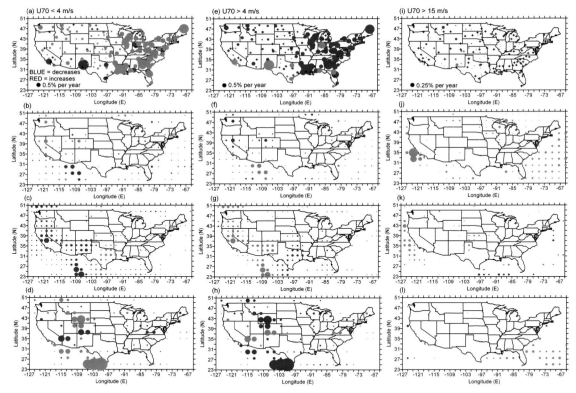

Figure 15.7. Results of trend analysis applied to data from 0000 UTC where the wind speeds from the four data sources were extrapolated to 70 m AGL, and analyzed in terms of the change in frequency with which wind speeds below 4 ms^{-1} were observed in (a) the observational time series, (b) the NCEP Reanalysis 1, (c) the NCEP Reanalysis 2, and (d) the ERA-40 reanalysis; wind speeds above 4 ms^{-1} were observed in (e) the observational time series, (f) the NCEP Reanalysis 1, (g) the NCEP Reanalysis 2, and (h) the ERA-40 reanalysis, and wind speeds above 15 ms^{-1} were observed in (i) observations, (j) the NCEP Reanalysis 1, (k) the NCEP Reanalysis 2, and (l) the ERA-40 reanalysis. In each frame the size of the dot scales linearly with the magnitude of the trend, and the color of the dot indicates the sign of the trend. Where the observational data did not indicate a statistically significant trend, a + symbol is shown, and where the reanalysis data do not indicate a significant trend no symbol is shown.

classes were observed. These trends are in accord with other analyses presented herein and known instrument biases associated with ASOS implementation. These trends are not, however, replicated in the reanalysis data.

In contrast to the observations that show larger magnitude trends in the eastern United States (of the relative frequency of above and below 4 ms^{-1} wind speeds), the majority of the NCEP Reanalysis 2 and ERA-40 grid cells show largest trend magnitudes in the western United States. In the observed data sets the frequency with which wind speeds of above 15 ms^{-1} were observed at 70 m was largely unchanged over the period of study. This is also

true of the reanalysis data sets, though all reanalysis data sets exhibit a greater frequency of increasing than of decreasing trends.

To provide a further context for the temporal trends presented herein, 10-m wind speed data from the NCEP Reanalysis 1 output were conditionally sampled to compute annual wind indices for the Midwestern USA over 1948–2006 using a normalization period of 1992–2001. Data for 1973–2001 from the ERA-40 data set and NCEP Reanalysis 2 output for 1979–2005 were also used in similar calculations (figure 15.8). As shown, the period of record used in the trend analysis naturally has a critical impact on the trend results. According the NCEP Reanalysis 1 data set the early 1970s exhibit low wind speeds and energy low wind indices relative to the 1990s, leading to positive trends for 1973–2005. However, as shown, the trend is positive when the entire NCEP Reanalysis 1 record is used, in part because the data exhibit maxima in the late 1960s and mid-1980s to date, and minima in the late 1940s and early 1950s and in the early to middle 1970s. These features are not observed in the observational records (see for example figure 15.1) or in the shorter time series from the other reanalysis data. The ERA-40 reanalysis wind indices exhibit small (statistically insignificant) declines over 1973–2001, while wind indices from NCEP Reanalysis 2 exhibit small (statistically insignificant) increases over 1979–2005.

Concluding Remarks

Returning to our original objectives:

1. To quantify historical (1973–2005) trends in near-surface wind speeds across the contiguous United States based on in

situ observations and to assess their statistical significance.

2. To compare trends in time-series of near-surface wind speeds from individual stations with those derived from gridded reanalysis data.

3. To evaluate whether the temporal trends exhibit substantial bias with hour of observation.

4. To assess the impact of any temporal trends on the magnitude of the wind energy resources and the feasibility of wind energy developments.

Results presented herein indicate:

Objective 1 and 2: As in our prior research across the European continent there are quantitative differences in mean wind speeds and trends in wind speed percentiles between reanalysis data sets (Pryor, Barthelmie, and Schoof 2006). Based on the analysis presented herein we would further note there are substantial differences between trends derived from carefully quality controlled observational data and the reanalysis products. While these differences cannot be fully explained they must be

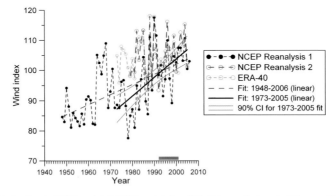

Figure 15.8. Annual wind indices from the NCEP Reanalysis 1, NCEP Reanalysis 2, and ERA-40 datasets for the entire Midwestern USA (98–80°W, 36–50°N) computed using a normalization period of 1992–2001. Results are shown for each year 1948–2006 for the NCEP Reanalysis 1, 1979–2005 for NCEP Reanalysis 2, and 1973–2001 for ERA-40. For the NCEP values trends are also shown for the entire data record and for 1973–2005, along with 90% confidence intervals (CI). The pink bar on the x-axis indicates the normalization period.

acknowledged, and their presence strongly advocates for use of multiple data sets in analyses of wind speed climates.

Objective 3: There is no strong evidence of substantial bias in temporal trends with the hour of the day in which observations were made.

Objective 4: It is difficult to quantify possible changes in the wind energy climate of the contiguous United States given the large discrepancies in trend analyses conducted using four independent data sets. It should be acknowledged that temporal trends in wind speeds and/or energy density need not be linear with time and that the analyses presented herein disregard periodic variations. Herein we focused on linear trends because of the inter-annual variability inherent in wind speeds and energy density and because we are seeking to identify a possible long-term trend associated with global climate change. The inter-annual variability in wind indices across the Midwestern USA implies the energy density in individual years varies by over ±20% from the 1992–2001 mean, which can be used as a context for the 6%/decade increase in WI computed from the NCEP Reanalysis 1 wind speeds.

Given the importance of the wind energy industry to meeting federal and state mandates for increased use of renewable energy supplies (see chapter 1 of this volume), further research on wind climate variability and evolution is required, as is a detailed analysis focused on reconciling the discrepancies illuminated herein.

ACKNOWLEDGMENTS
Financial support from the NSF Geography and Regional Science program (grants #0618364 and 0647868) and the Office of the Dean of the College of Arts and Science of Indiana University is gratefully acknowledged. NCEP Reanalysis 1 and NCEP Reanalysis 2 data were provided by the NOAA/OAR/ESRL PSD, Boulder, Colorado, from their web site at http://www.cdc.noaa.gov/. ERA-40 reanalysis data were provided by ECMWF from their web site at http://www.ecmwf.org/research/era/Data_Services/index.html. Near-surface observed wind speeds were provided by the National Climatic Data Center (NCDC) via their web site at http://www.ncdc.noaa.gov/oa/mpp/freedata.html.

REFERENCES
Ambrose, J., and D. Vergun. 1997. *Simplified Building Design for Wind and Earthquake Forces.* 3rd ed. New York: John Wiley and Sons.

AWEA. 2008. *AWEA Wind Energy Fact Sheets: 2007 a Record Year.* Accessed at: http://www.awea.org/pubs/factsheets/2008_Market_Update.pdf on 25 June 2008.

Barthelmie, R. J. 2007. *Wind Energy: Status and Trends Geography Compass* 1: 275–301.

Barthelmie, R. J., and S. C. Pryor. 2006. "Challenges in Predicting Power Output from Offshore Wind Farms." *Journal of Energy Engineering* 132: 91–103.

———. 2007. "Wind Speed Trends over the Contiguous USA." *Papers of the Applied Geography Conferences* 30: 344–353.

Bijl, W. 1997. "Impact of a Wind Climate Change on the Surge in the Southern North Sea." *Climate Research* 8: 45–59.

Changnon, S. A., E. R. Fosse, and E. L. Lecomte. 1999. "Interactions between the Atmospheric Sciences and Insurers in the United States." *Climatic Change* 42: 51–67.

Dell'Aquila, A., et al. 2007. "Southern Hemisphere Midlatitude Atmospheric Variability of the NCEP-NCAR and ECMWF Reanalyses." *Journal of Geophysical Research* 112: 1–11.

Elliott, D. L., et al. 1986. *Wind Energy Resource Atlas of the United States.* Washington, D.C.: Solar Technical Information Program, U.S. Department of Energy.

Groisman, P. Y. 2001. Data documentation for data set TD-6421: "Enhanced Hourly Wind Station Data for the Contiguous United

States." Asheville, N.C.: National Climatic Data Center.

Jungo, P., S. Goyette, and M. Beniston. 2002. "Daily Wind Gust Speed Probabilities over Switzerland according to Three Types of Synoptic Circulation." *International Journal of Climatology* 22: 485–499.

Kalnay, E., et al. 1996. "The NCEP/NCAR 40 Reanalysis Project." *Bulletin of the American Meteorological Society* 77: 437–471.

Kanamitsu, M., et al. 2002. "NCEP-DOE AMIP-II Reanalysis (R-2)." *Bulletin of the American Meteorological Society* 83: 1631–1643.

Kiktev, D., et al. 2003. "Comparison of Modeled and Observed Trends in Indices of Daily Climate Extremes." *Journal of Climate* 16: 3560–3571.

Kistler, R., et al. 2001. "The NCEP-NCAR 50 Year Reanalysis: Monthly Mean CD-ROM and Documentation." *Bulletin of the American Meteorological Society* 82: 247–267.

Klink, K. 2002. "Trends and Interannual Variability of Wind Speed Distributions in Minnesota." *Journal of Climate* 15: 3311–3317.

Manwell, J. F., J. McGowan, and A. L. Rogers. 2002. *Wind Energy Explained: Theory, Design and Application.* New York: John Wiley and Sons.

McKee, T. B., et al. 2000. "Climate Data Continuity with ASOS: Report for Period April 1996 through June 2000." Colorado Climate Center. Available at: http://ccc.atmos.colostate.edu/centerpublications.php.

Pryor, S. C., and R. J. Barthelmie. 2003. "Long Term Trends in Near Surface Flow over the Baltic." *International Journal of Climatology* 23: 271–289.

Pryor, S. C., R. J. Barthelmie, and E. Kjellström. 2005. "Analyses of the Potential Climate Change Impact on Wind Energy Resources in Northern Europe Using Output from a Regional Climate Model." *Climate Dynamics* 25: 815–835.

Pryor, S. C., R. J. Barthelmie, and J. T. Schoof. 2005. "The Impact of Non-stationarities in the Climate System on the Definition of 'a Normal Wind Year': A Case Study from the Baltic." *International Journal of Climatology* 25: 735–752.

———. 2006. "Inter-annual Variability of Wind Indices across Europe." *Wind Energy* 9: 27–38.

Pryor, S. C., J. T. Schoof, and R. J. Barthelmie. 2005. "Climate Change Impacts on Wind Speeds and Wind Energy Density in Northern Europe: Results from Empirical Downscaling of Multiple AOGCMs." *Climate Research* 29: 183–198.

———. 2006. "Winds of Change? Projections of Near-Surface Winds under Climate Change Scenarios." *Geophysical Research Letters* 33, doi:10.1029/2006GL026000.

Pryor, S. C., et al. 2004. "Can Satellite Sampling of Offshore Wind Speeds Realistically Represent Wind Speed Distributions? Part 2: Quantifying Uncertainties Associated with Sampling Strategy and Distribution Fitting Methods." *Journal of Applied Meteorology* 43: 739–750.

Schwartz, M., and D. Elliott. 2006. "Wind Shear Characteristics at Central Plains Tall Towers." Conference paper NREL/CP-500-40019. Available at: http://www.osti.gov/bridge.

Smits, A., A. Tank, and G. P. Konnen. 2005. "Trends in Storminess over the Netherlands, 1962–2002." *International Journal of Climatology* 25: 1331–1344.

Song, H., and M. H. Zhang. 2007. "Changes of the Boreal Winter Hadley Circulation in the NCEP-NCAR and ECMWF Reanalyses: A Comparative Study." *Journal of Climate* 20: 5191–5200.

Uppala, S. M., et al. 2005. "The ERA-40 Re-analysis." *Quarterly Journal of the Royal Meteorological Society* 131: 2961–3012.

Viles, H. A., and A. S. Goudie. 2003. "Interannual, Decadal and Multidecadal Scale Climatic Variability and Geomorphology." *Earth-Science Reviews* 61: 105–131.

Wang, X. L. L., V. R. Swail, and F. W. Zwiers. 2006. "Climatology and Changes of Extratropical Cyclone Activity: Comparison of ERA-40 with NCEP-NCAR Reanalysis for 1958–2001." *Journal of Climate* 19: 3145–3166.

Zhao, T. B., and C. B. Fu. 2006. "Comparison of Products from ERA-40, NCEP-2, and CRU with Station Data for Summer Precipitation over China." *Advances in Atmospheric Sciences* 23: 593–604.

16. Variability of Wind Speed Regimes in Minnesota

K. KLINK

Introduction

The State of Minnesota has long recognized the potential for significant wind energy development in the state and region. The Energy Office of the Minnesota Department of Commerce (DoC) has been monitoring wind speeds since the mid-1980s (Minnesota Department of Commerce 2002), originally at 30–50 m above ground level (AGL) and more recently at 70–90 m AGL—mirroring the increasing height of utility-scale wind turbines. State legislation has mandated that, by 2015, at least 10% of the electricity supplied by Minnesota utilities must come from renewable sources; this minimum rises to 25% by 2025 (Minnesota Statute 216B.1691). Revenues from this renewable-energy mandate benefit not only utilities but also citizens and communities in rural areas of the state, particularly via "community wind" projects (Minnesota Statute 216C.41).

In this work I use the tall-tower wind speed data set compiled by the Minnesota DoC to develop a first look at the climatology of wind speeds at turbine heights, and to examine patterns of wind speed variability that occur at these heights. I also discuss the potential impacts of large-scale climate variability on Minnesota's wind resources.

Wind Speed Data

The Minnesota DoC has instrumented a number of radio towers around the state with wind speed and wind direction monitoring equipment. Wind speeds at each site are measured using two standard three-cup anemometers (manufactured by NRG Systems), one each on the northwest and southeast sides of the tower. Anemometers typically are mounted around 30, 50, and 70 m AGL, with wind direction monitored around 30 and 70 m AGL. Dataloggers record the mean hourly values of wind speed and wind direction. Hourly wind speeds are recorded as the higher of the two measured wind speeds under the assumption that the lower speed is affected by tower shading. Minnesota DoC personnel examine data from each tower every one to two weeks, and anomalous data are flagged for quality checks and for possible sensor malfunction (J. Sheehy, personal communication). Wind speed measurements also are checked by the DoC for consistency between the paired anemometers at each location on the tower. Wind instruments

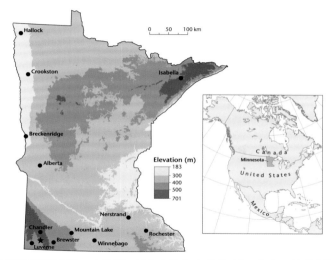

Figure 16.1. Location of tall-tower wind sites in Minnesota. St. Killian is indicated by the star. Station information is provided in Table 16.1. Map modified with permission of the American Meteorological Society from Klink (2007).

typically are not re-calibrated once deployed, but these quality checks allow DoC personnel to determine when it is necessary to repair, replace, or re-calibrate the instruments (R. Artig, personal communication). In this chapter I discuss wind speed patterns at the sites shown in figure 16.1. Table 16.1 provides additional information about the tower sites, and topographic maps and additional statis-

tics can be found in Minnesota Department of Commerce (2002).

Results

CLIMATOLOGY OF WIND SPEEDS AT TURBINE HEIGHT

Wind speeds in Minnesota have a distinct

Table 16.1. Tower characteristics and mean wind speed at the sites shown in figure 16.1. The period of record (POR) is 1995–2003 except where noted.

Station (abbreviation)	Location	Elevation above MSL (m)	Mean annual wind speed at 70 m AGL (ms⁻¹)
Isabella (POR 2002–2005)	47.62°N 91.37°W	594	6.04*
Hallock	48.76°N 96.93°W	250	6.91
Crookston	47.76°N 96.67°W	264	6.70
Breckenridge	46.26°N 96.53°W	292	7.07
Alberta	45.58°N 96.05°W	338	7.03
Chandler	43.89°N 95.93°W	555	8.49
Luverne	43.71°N 96.07°W	472	7.19
St. Killian (star symbol) (POR 1999–2005)	43.78°N 95.87°W	543	7.08†
Brewster	43.73°N 95.36°W	427	7.46
Mountain Lake	44.04°N 94.85°W	366	7.32
Winnebago	43.70°N 94.03°W	343	7.36
Nerstrand	44.35°N 93.04°W	365	6.99
Rochester	43.97°N 92.42°W	364	6.36

*Measured at 75 m AGL above a canopy of trees approximately 25 m tall
†Measured at 90 m AGL

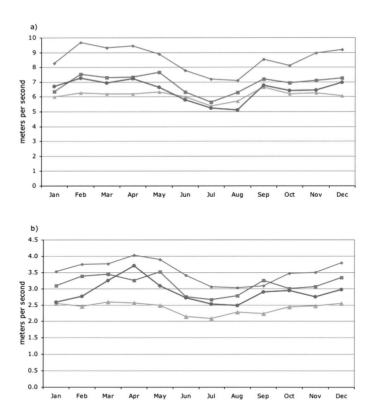

Figure 16.2. a) Mean monthly wind speed and b) standard deviation of monthly wind speed (based on hourly data) at a subset of the tall-tower sites. Monthly values are based on data from 2002 to 2004. Redrawn from Wichser and Klink (2008).

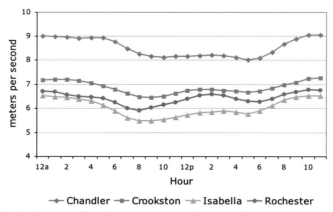

Figure 16.3. Mean diurnal wind speed at a subset of the tall-tower sites, based on data from 2002 to 2004.

annual cycle, with the highest speeds in spring and autumn, and the lowest speeds during summer (figure 16.2a). This seasonality reflects the climatological variation in mid-latitude temperature and pressure gradients. The variability of wind speeds within a month (computed from hourly wind speeds for that month, figure 16.2b) follows a similar seasonal cycle: months with higher mean speeds tend also to have higher hour-to-hour variability because wind speed is zero-bounded and the stronger and more frequent cyclone activity in the spring and autumn (e.g., Angel and Isard 1998, Key and Chan 1999, McCabe, Clark, and Serreze 2001) leads to higher mean speeds and to higher variability

Mean hourly wind speeds at 70 m AGL have a clear diurnal cycle, with the fastest winds occurring at night and slower winds during the day (figure 16.3). This diurnal cycle is opposite that for near-surface (e.g., 10 m AGL) wind speeds, which typically are fastest during the day. The diurnal cycles at 10 m and 70 m AGL are connected: during the day, surface heating leads to convection and turbulence within the near-surface boundary layer, thus mixing the faster winds above the ground with the slower winds near the ground. This mixing produces the typical increase in near-surface speeds during the day as well as the reduced speeds at somewhat higher elevations. At night, as the surface becomes decoupled from the overlying atmosphere (due to reduced heating and turbulent mixing), near-surface wind speeds are less well mixed with speeds at higher elevations, so that near-surface winds remain low and winds above the ground remain high (Oke 1987).

INTER-ANNUAL VARIABILITY IN MEAN MONTHLY WIND SPEED

Klink (2007) showed that, at monthly time scales, there were clear correlations between above- and below-normal mean monthly wind speeds across a range of sites in Minnesota (figure 16.4). Mean monthly wind speeds from 1995 to 2003 were shown to be directly related to variations in the north-south pressure gradient over Minnesota, as would be expected based on climatological theory. Wind speed anomalies also showed some covariability with the Arctic Oscillation (AO) and the El Niño–Southern Oscillation (ENSO) indices. The positive (warm) phase of the AO enhances the mid-latitude westerlies, and this was manifest as a positive correlation between the AO index and monthly wind speed anomalies at a majority of the Minnesota tower sites. El Niño conditions generally cause a northward shift in the polar jet stream, particularly in the winter. At these Minnesota sites, the strong El Niño in 1997–98 corresponded with a significant decrease in mean monthly wind speeds, most notably in the winter and spring. The jet's typical wintertime position is near the latitude of Minnesota, so a northward shift results in both weaker wind speeds and reduced cyclone activity during these meteorologically active times of the year.

The strength of the correlation between wind speed and these large-scale circulation indices differs depending on the monthly wind speed characteristic that is chosen. (Because the AO, pressure gradient, and Niño 3.4 index [lag 5 months, based on Diaz, Hoerling, and Eischeid 2001, Enloe,

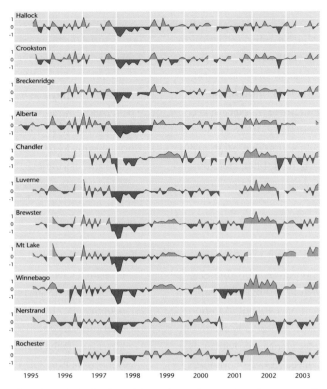

Figure 16.4. Mean monthly wind speed anomalies (ms^{-1}) at eleven of the sites shown in figure 16.1. Anomalies are computed from monthly means for the 1995–2003 base period. Reproduced from Klink (2007).

O'Brien, and Smith 2004, and Klink 2007] are nearly uncorrelated over this period [table 16.2], I interpret these indices as representing nearly independent sources of information.) Lower wind speeds, such as the 10th and 25th percentile values (based on the distribution of hourly speeds within a month), are more highly correlated with the AO than are higher wind speeds, such as the 75th and 90th percentile values (fig-ure 16.5a). In contrast, correlations with the Niño 3.4 index (lagged 5 months) show no obvious variation across the monthly wind speed distributions (figure 16.5b), and higher wind speeds are more strongly correlated with the north-south pressure gradient than are lower speeds (figure 16.5c). The similar correlation between the Niño 3.4 index and each of the wind speed percentiles suggests that El Niño events produce a more or less

Table 16.2. Pearson correlation coefficients for the circulation indices used to evaluate relationships with wind speed anomalies shown in figure 16.5. Correlations are calculated using monthly values of the indices and wind speed anomalies for the period 1995–2003.

	North-south pressure gradient[*]	Niño 3.4 index (lag 5 months)
Niño 3.4 index (lag 5 months)[†]	−0.0990	
AO	0.2606	−0.1221

[*]The north-south pressure gradient is represented here by the maximum difference in 500 hPa geopotential heights from 25°N to 75°N, along 95°W, which roughly bisects the state of Minnesota. Geopotential heights are taken from the NCEP-NCAR Reanalysis dataset available via the NOAA-CIRES Climate Diagnostics Center (http://www.cdc.noaa.gov/cgi-bin/DataMenus.pl?stat=mon.ltm&dataset=NCEP).
[†]Based on Diaz et al. 2001, Enloe et al. 2004, and Klink 2007.

uniform shift of the wind speed distribution toward lower values. A strong pressure gradient likely corresponds not only to faster winds, but to increased cyclonic activity, which may explain the stronger correlation between the pressure gradient and wind speeds at the 75th/90th percentiles (the upper tail of the distribution) as compared to the 10th/25th percentiles. It is unclear why low wind speeds would be more highly correlated to the AO than are higher speeds. Perhaps, in the absence of a strong pressure gradient, the positive phase of the AO (stronger westerlies) yields a slightly more narrow wind speed distribution: low wind speeds are slightly faster, but the frequency of high speeds is not appreciably changed. These results need further investigation to

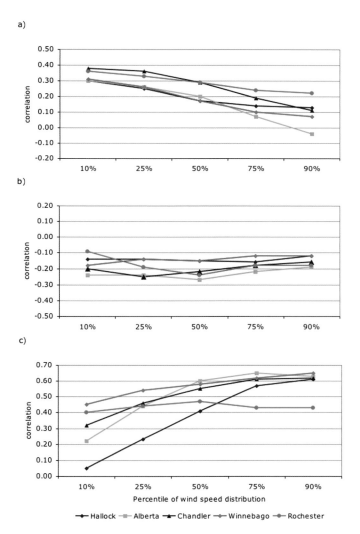

Figure 16.5. Pearson correlation for various percentiles of the wind speed distribution at selected locations in figure 16.1: a) correlation with the AO (Climate Prediction Center 2002); b) correlation with the Niño 3.4 index (lag 5 months) (Climate Prediction Center 2007); c) correlation with the north-south pressure gradient (see footnote to table 16.2 for description of pressure gradient calculation).

determine whether they are artifacts of this particular time period, or whether they are robust features of the relationship between turbine-level wind speeds and the large-scale atmospheric circulation.

SHORT-TERM VARIABILITY IN WIND SPEED AND POWER

Wind power density, P, in Wm^{-2} (Gipe 2004) is computed as:

$$P = \frac{1}{2}\left(\rho C_{PR} C_T\right)v^3 \quad (16.1)$$

where ρ is air density (kg m^{-3}), and v is wind speed (m s^{-1}). Pressure and temperature correction terms (C_{PR} and C_T respectively) are applied to account for deviations from standard atmospheric density (1.225 kg m^{-3}) due to differences from standard sea level pressure (SLP) (1013.25 hPa) and temperature (288.15 K). The correction factors are computed as

$$C_{PR} = \frac{PR_{obs}}{1013.25} \quad (16.2)$$

$$C_T = \frac{288.15}{T_{obs}} \quad (16.3)$$

where PR_{obs} is SLP (hPa) and T_{obs} is air temperature (K).

Temperature data are available at the tower site, but SLP was estimated from site-plus-tower elevation. Since the pressure correction is much smaller than the temperature correction, estimating C_{PR} based on elevation introduces a minimum amount of error in the power calculation.

Wind is a variable energy source. Geographic dispersion of wind turbines across large transmission areas helps to smooth out this variability (Milligan and Artig 1998, Nanahara et al. 2004), but utilities have concerns about whether short-term variation in wind power output may degrade operations or cause portions of the grid to go off-line (EnerNex Corporation 2006). Ten-minute wind speeds measured at 90 m AGL at St. Killian in southwestern Minnesota (figure 16.6) provide an initial look at the scale of wind power variability at short intervals. I focus on March in this preliminary analysis, because this is the month with the highest mean speeds, and therefore variability also is likely to be highest (Klink 1999). Available wind power then was estimated using the cut-in (3.5 ms^{-1}), cut-out (20 ms^{-1}), and rated speeds (14 ms^{-1}) for the General Electric (GE) 1.5 MW sl turbine (GE Energy 2005). In brief the method used is:

1. The wind power density was computed using equation (16.1).
2. If wind speeds were below the cut-in speed of the turbine (3.5 ms^{-1}), the wind power density was set to zero.
3. If wind speeds were above the cut-in speed of the turbine, but below the rated speed (14 ms^{-1}), the wind power density was used as computed.
4. If wind speeds were between the rated speed and the cut-out speed (25 ms^{-1}), the wind power density was set to the rated output.
5. If wind speeds were above the cut-out speed, the wind power density was set to zero.

There is a variation of ±10% in average March wind speed and of 15% or more in the monthly standard deviation and Weibull k parameter over the seven years of record (table 16.3). The *k* parameter is a commonly used statistical parameter that represents the peakedness of the wind speed distribution (see chapter 15 in this volume), and it was computed for these sites using the mean and standard deviation of monthly wind speed distributions, following Rohatgi and Nelson (1994). Despite the inter-annual variability in short-term wind speed, ten-minute variations in wind power are relatively small, with about 50% of the variability in the range of ±25 Wm^{-2} (except for 2002 and 2003) and with

Table 16.3. March wind speed characteristics at 90 m AGL at St. Killian, computed from ten-minute data

	Mean ($^{-1}$)	std dev (ms^{-1})	Weibull k (dimensionless)	% missing data
1999	8.47	4.27	2.08	0.0
2000	7.75	3.77	2.17	5.5
2001	6.85	3.51	2.04	4.6
2002	8.52	3.07	3.03	18.3
2003	8.90	3.29	2.94	0.2
2004	8.06	4.27	1.97	9.1
2005	7.47	3.66	2.15	2.8

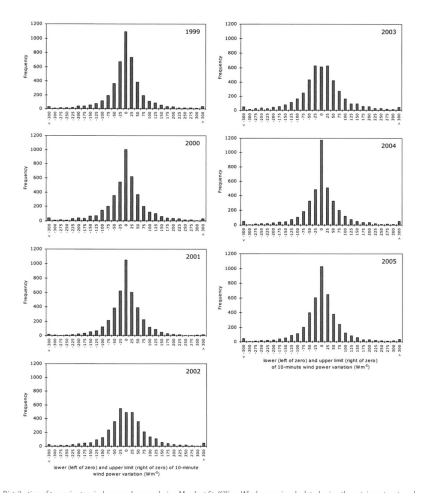

Figure 16.6. Distribution of ten-minute wind power changes during March at St. Killian. Wind power is calculated using the cut-in, cut-out, and rated speeds of the GE 1.5 MW sl turbine. Graphs courtesy of Brian Cacchiotti.

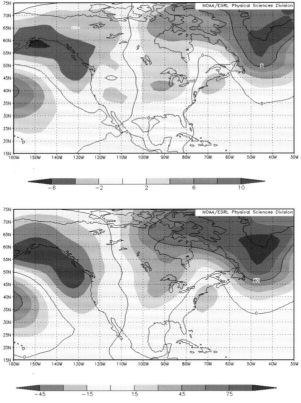

Figure 16.7. Sea level pressure anomalies (top) and 850 hPa height anomalies (bottom) for March 1999. Figure from the NCEP-NCAR Reanalysis data with images provided by the NOAA/ESRL Physical Sciences Division, Boulder, Colorado, from their website at http://www.cdc.noaa.gov/.

80% of the variability, or more, within ±100 Wm^{-2} (including 2002 and 2003) (figure 16.6). The distribution of ten-minute wind power variation in March 2002 and 2003 is noticeably wider than for the other months in this record, although the relatively high amount of missing data for March 2002 suggests that those data should be interpreted with caution. The small number of ten-minute variations that are above ±300 Wm^{-2} may be caused by instrument malfunction (e.g., icing on the anemometer); metadata for the tower site, however, are inconclusive on this point.

I use March 1999 and March 2003 (the months with the fewest missing observations) to represent the two main wind power "variability regimes" that occur in this seven-year period. Anomalies of sea level pressure (SLP) and 850 hPa heights for March 1999 (figure 16.7) and March 2003 (figure 16.8) show that, in 1999, there was a weaker than normal ridge-trough pattern at both the surface and aloft, with a weaker than normal north-south pressure gradient at 850 hPa. In contrast, the SLP map for March 2003 shows a stronger than normal eastern trough and an enhanced north-south pressure gradient at 850 hPa. Based on these and additional maps (not shown) for these two months, I infer that cyclones were stronger and more frequent in 2003 than in 1999, and that stronger and more frequent cyclones produced the broader distribution of ten-minute wind power variation in 2003 as compared to 1999 (cf. figure

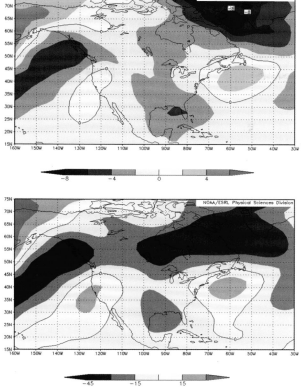

Figure 16.8. Sea level pressure anomalies (top) and 850 hPa height anomalies (bottom) for March 2003. Figure from the NCEP-NCAR Reanalysis data with images provided by the NOAA/ESRL Physical Sciences Division, Boulder, Colorado, from their website at http://www.cdc.noaa.gov/.

16.6). This interpretation is consistent with the discussion of inter-annual variability in mean monthly wind speed described above, but additional analysis is needed in order to refine and validate these inferences.

Concluding Remarks

I have used a unique tall-tower wind speed data set compiled by the State of Minnesota to examine patterns of wind speed and wind power variability at typical wind turbine hub heights. Analysis of nine years of data (1995–2003) at eleven sites across Minnesota, measured at 70 m AGL, showed that speeds are strongly and directly related to the strength of the north-south pressure gradient across the state, as would be expected from climatological theory. Inter-annual variability in mean monthly wind speeds also show a clear relationship with other hemispheric-scale circulation patterns: faster speeds at the towers are coincident with the positive phase of the Arctic Oscillation (which enhances mid- and high-latitude wind speeds aloft), and slower speeds are coincident with El Niño conditions (which results in a northward shift in the mid-latitude jet stream and thus weaker winds across Minnesota).

A preliminary look at very short term variability in wind power production used ten-minute wind speed data for March (a high-wind-speed month) from 1999 to 2005 at one site in southwestern Minnesota, along with the wind speed characteristics of the GE 1.5 MW sl turbine. Over these seven years, 80% or more of the ten-minute changes in wind power fell within ±100 Wm^{-2}, demonstrating that although wind is a variable power source, the degree of variability is small enough in most circumstances that it should not be a significant operating concern. The number of ten-minute power variations that exceed ±100 Wm^{-2} appears to increase when the north-south pressure gradient is stronger than normal, which I infer to be related to increased cyclone frequency and strength.

Although the analyses described in this chapter are based on fewer than ten years of data, the results so far suggest that future changes in hemispheric (Trenberth et al. 2007) and continental-scale (Christensen et al. 2007) pressure and circulation patterns will have important impacts on wind speed, and thus wind power production, in the Midwestern USA (see also chapter 15 in this volume). Enhanced high-latitude warming would reduce the north-south pressure gradient, thus may reduce mid-latitude wind speeds, as would an increase in the frequency and intensity of El Niño events (although there is uncertainty about the response of El Niño under future climate change; see Trenberth et al. 2007). Local modifications to the surface environment—such as urbanization and changes in vegetation cover (e.g., Balling and Cerveny 1987, Chase et al. 2000, Pielke 2005)—may increase or decrease local wind speeds depending on the type of surface change. As the Midwestern USA increases its investment in wind energy (see chapter 1 of this volume), it will be important to continue to examine the relationships be-

tween wind speed, wind power, and climate variability.

ACKNOWLEDGMENTS

I thank Brian Cacchiotti, Kenzie Johnson, and Corinne Wichser for their collaboration on several aspects of the work described here, and Rebecca Barthelmie for valuable comments on this chapter. Special thanks go to Sara Pryor and Indiana University for inaugurating and hosting the Midwest Climate Workshop.

REFERENCES

Angel, J. R., and S. A. Isard. 1998. "The Frequency and Intensity of Great Lake Cyclones." *Journal of Climate* 11: 61–71.

Balling, R.C., Jr., and R. S. Cerveny. 1987. "Long-Term Associations between Wind Speeds and the Urban Heat Island of Phoenix, Arizona." *Journal of Climate and Applied Meteorology* 26: 712–716.

Chase, T. N., et al. 2000. "Simulated Impacts of Historical Land Cover Changes on Global Climate in Northern Winter." *Climate Dynamics* 16: 93–105.

Christensen, J. H., et al. 2007. "Regional Climate Projections." In *Climate Change 2007: The Physical Science Basis.* Contribution of Working Group I to the Fourth Assessment Report of the Intergovernmental Panel of Climate Change, 847–940. [Ed. S. Solomon, et al.] New York: Cambridge University Press.

Climate Prediction Center. 2002. Monitoring and Data: *Daily Arctic Oscillation Index.* Available at: http://www.cpc.ncep.noaa.gov/products/precip/CWlink/daily_ao_index/ao_index .html. Last modified on 24 October 2002. Last accessed on 8 January 2008.

Climate Prediction Center. 2007. Monitoring and Data: *Current Monthly Atmospheric and Sea Surface Temperatures Index Values.* Available at: http://www.cpc.ncep.noaa.gov/data/indices. Last modified 27 March 2007. Last accessed on 8 January 2008.

Diaz, H. F., M. P. Hoerling, and J. K. Eischeid. 2001. "ENSO Variability, Teleconnections, and

Climate Change." *International Journal of Climatology* 21: 1845–1862.

Earth System Research Laboratory. n.d. *Visualize NCEP Data.* Available at: http://www.cdc.noaa.gov/cgi-bin/DataMenus.pl?stat=mon.ltm&dataset=NCEP. Last accessed on 8 January 2008.

EnerNex Corporation. 2006. Final Report: *2006 Minnesota Wind Integration Study.* Vol. 1. Prepared for the Minnesota Public Utilities Commission by EnerNex Corporation in collaboration with the Midwest Independent System Operator. Available at: http://www.state.mn.us/portal/mn/jsp/content.do?id=-536881350&subchannel=-536881511&sc2=null&sc3=null&contentid=536904447&contenttype=EDITORIAL&programid=536902421&agency=Commerce.

Enloe, J., J. J. O'Brien, and S. R. Smith. 2004. "ENSO Impacts on Peak Wind Gusts in the United States." *Journal of Climate* 17: 1728–1738.

GE Energy. 2005. *1.5MW Series Wind Turbine.* Product brochure dated September 2005. Available at: http://www.gepower.com/prod_serv/products/wind_turbines/en/downloads/ge_15_brochure.pdf.

Gipe, P. 2004. *Wind Power.* White River Junction, Vt.: Chelsea Green Publishing.

Key, J. R., and A. C. K. Chan. 1999. "Multidecadal Global and Regional Trends in 1000 mb and 500 mb Cyclone Frequencies." *Geophysical Research Letters* 26: 2053–2056.

Klink, K. 1999. "Climatological Mean and Interannual Variance of United States Surface Wind Speed, Direction, and Velocity." *International Journal of Climatology* 19: 471–488.

———. 2007. "Atmospheric Circulation Effects on Wind Speed Variability at Turbine Height." *Journal of Applied Meteorology and Climatology* 45: 445–456.

McCabe, G. J., M. P. Clark, and M. C. Serreze. 2001. "Trends in Northern Hemisphere Surface Cyclone Frequency and Intensity." *Journal of Climate* 14: 2763–2768.

Milligan, M., and R. Artig. 1998. *Reliability Benefits of Dispersed Wind Resource Development.* National Renewable Energy Laboratory Report NREL/CP-500-24314. Available at: http://www.nrel.gov/docs/legosti/fy98/24314.pdf.

Minnesota Department of Commerce. 2002. *Wind Resource Analysis Program 2002.* St. Paul, Minn.: Minnesota Department of Commerce. Available at: http://www.state.mn.us/mn/externalDocs/Commerce/WRAP_Report_110702040352_WRAP2002.pdf.

Nanahara, T., et al. 2004. "Smoothing Effects of Distributed Wind Turbines. Part 2. Coherence among Power Output of Distant Wind Turbines." *Wind Energy* 7: 75–85.

Oke, T. R. 1987. *Boundary Layer Climates.* 2nd ed. Routledge.

Pielke, R. A., Sr. 2005. "Land Use and Climate Change." *Science* 310: 1625–1626, doi:10.1126/science.1120529.

Rohatgi, J. S., and V. Nelson. 1994. *Wind Characteristics: An Analysis for the Generation of Wind Power.* Canyon, Tex.: Alternative Energy Institute, West Texas A&M University.

Trenberth, K. E., et al. 2007. "Observations: Surface and Atmospheric Climate Change." In *Climate Change 2007: The Physical Science Basis.* Contribution of Working Group I to the Fourth Assessment Report of the Intergovernmental Panel of Climate Change, 235–336. [Ed. S. Solomon, et al.] New York: Cambridge University Press.

Wichser, C., and K. Klink. 2008. "Low Wind Speed Turbines and Wind Power Potential in Minnesota, USA." *Renewable Energy* 33: 1749–1758.

17. Teleconnections and Circulation Patterns in the Midwestern USA

J. T. SCHOOF AND S. C. PRYOR

Introduction

The climate of the Midwestern USA is characterized by high day-to-day variability associated with both the behavior of the polar jet stream and the relative intensity of several semi-permanent pressure systems, including the subtropical (Bermuda) high (often located within or to the east/southeast of the region) and the subpolar (Hudson Bay) low (often located within or to the north/northeast of the region) (see chapter 14 of this volume). Variations in the intensity of these pressure systems are manifest in larger modes of variability, specifically the North Atlantic Oscillation (NAO) and Pacific–North American (PNA) patterns, and cause variations in the tracking and intensity of synoptic scale phenomena in the study region (Leathers, Yarnal, and Palecki 1991; Hurrell 1995). These synoptic scale variations are in turn manifest in surface parameters. For example, the NAO accounts for 31% of the variance in northern hemisphere winter mean surface air temperatures (Hurrell 1996). The PNA is closely linked with both thermal and hydrologic regimes within the study area across a range of timescales (e.g., Coleman and Rogers 2003), including daily

(Sheridan 2003), and accounts for as much as 40% of the variance in surface air temperature during the winter months (Leathers, Yarnal, and Palecki 1991). Recently, there has been renewed interest in the joint behavior of the NAO and PNA as it relates to the Arctic Oscillation (Ambaum, Hoskins, and Stephenson 2001), energy usage in the Midwestern USA (Changnon et al. 2002), and Northern Hemisphere atmospheric blocking (Croci-Maspoli, Schwierz, and Davies 2007).

Given the importance of these teleconnections in dictating conditions at the regional scale here we:

1. Investigate links between the teleconnection indices and conditions at the synoptic scale (as manifest in map pattern frequency) over the Midwestern USA as manifest in observational records.

2. Evaluate the ability of coupled Atmosphere-Ocean General Circulation Models (AOGCMs) to reproduce:

a. The teleconnection patterns and indices.

b. Synoptic map type frequencies.

c. Scale linkages between the map types and teleconnection indices.

Because the NAO is primarily a cold-season

mode (Barnston and Livezy 1987), we focus here on wintertime conditions.

Data

To investigate the observed teleconnections and circulation patterns in the study region, the NCEP/NCAR reanalysis data (Kistler et al. 2001), hereafter referred to as NNR, were used. Note, this data set is referred to as the NCEP Reanalysis 1 data set in chapter 15. These data are available on a 2.5°×2.5° grid as shown in figure 17.1. The PNA index and circulation patterns were derived from daily 500-mb geopotential height fields, and the NAO index was derived from daily sea-level pressure data as described in Schoof and Pryor (2006), and briefly outlined below.

The AOGCM output is derived from two transient simulations covering the period 1990–2099 (HadCM3) and 1990–2100 (CGCM2). HadCM3 is a Cartesian AOGCM with horizontal resolution of approximately 2.5° latitude by 3.75° longitude (Gordon et al. 2000; Pope et al. 2000). CGCM2 is a spectral AOGCM with horizontal resolution

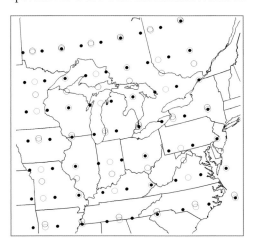

Figure 17.1. Map of the study region showing NCEP-NCAR Reanalysis (·), HadCM3 (o) (red) and CGCM2 (o) (blue) grid points.

of approximately 3.75°×3.75° as described by Flato et al. (2000) and McFarlane et al. (1992). Unlike HadCM3, CGCM2 employs flux adjustments as first described in Hansen et al. (1984). The model resolutions are as depicted in figure 17.1, but for development of the map pattern classification, both AOGCMs were interpolated to the 2.5°×2.5° NNR grid (figure 17.1) using an inverse distance based interpolation algorithm. For calculation of the teleconnection indices, the NNR data and the AOGCMs were interpolated to specific point locations (see the following section). The AOGCM simulations used here were conducted using the SRES A2 emissions scenario (IPCC 2000). Although newer model simulations are available as part of the archive for the IPCC Fourth Assessment Report, the data required for derivation of the PNA teleconnection index and circulation characteristics (geopotential heights) are not routinely archived.

Methodology

THE NAO AND PNA INDICES

Several NAO definitions exist, ranging from point-based pressure differences to weighted pressure extremes and empirical orthogonal functions (EOFs) of sea level pressure fields (see Hurrell et al. 2003). Here, we use the simplest metric: the difference in standardized SLP between gridpoints near the NAO "centers of action," at Ponta Delgada, Azores (37.7°N, 25.7°W) and Stykkisholmur, Iceland (65.1°N, 22.7°W).

The PNA index likewise has multiple definitions. Here, we use the equation of Wallace and Gutzler (1981):

$$PNA = \frac{1}{4}\Big[Z(20^\circ N, 160^\circ W) - Z(45^\circ N, 165^\circ W) + Z(55^\circ N, 115^\circ W) - Z(30^\circ N, 85^\circ W)\Big]$$

where Z are standardized 500 hPa geopotential height values.

In each case calculation of the teleconnection indices is computed independently for each data source. For each day and each data source, the daily teleconnection indices are considered either positive (the value of the teleconnection index was more than one standard deviation greater than the seasonal mean of the teleconnection index), negative (the value of the teleconnection index was more than one standard deviation less than the seasonal mean of the teleconnection index), or neutral (the value of the teleconnection index was within one standard deviation of the seasonal mean of the teleconnection index).

In earlier research (Schoof and Pryor 2006) the authors evaluated the performance of the AOGCM teleconnection indices with respect to (1) probability distributions (using a Kolmogorov-Smirnov test) and (2) temporal evolution (lag 1–10 autocorrelation) to those derived from NNR. We found that the NNR and AOGCM-derived teleconnection indices exhibited good agreement during the period of overlap (1990–2001), although both AOGCMs, and particularly CGCM2, overestimated the autocorrelation of the NAO index and underestimated the autocorrelation of the PNA index. Further, the differences between the observed and AOGCM-derived teleconnection probability distributions were smaller than those associated with their historical evolution or their projected evolution in the transient AOGCM simulations.

Here we focus on analysis of the spatial signatures of the teleconnection indices. Composites showing the mean sea level pressure (MSLP) fields during positive and negative phases of the NAO as derived from the NNR and AOGCM output, according to the definitions of positive and negative phase events above, for 1990–2001 are given in figure 17.2. As shown, the AOGCMs generally reproduce the spatial patterns of MSLP during positive and negative phases of the NAO. However, the Icelandic low is more intense in the NNR data during the positive phase of the NAO than is depicted by either AOGCM. Equally the gradient in MSLP fields during the negative phase appears to be underestimated in the CGCM2 output. Figure 17.3 shows similar results for the PNA, and again indicates relatively good reproduction of the spatial fields by both AOGCM. However, again, the spatial gradients (of 500 hPa height) are weaker in the CGCM2 output for positive phase PNA than in either HadCM3 or the NNR data.

MAP PATTERN CLASSIFICATION

Synoptic classifications have been derived using a range of methodologies, input parameters, and data sets. Here we base our analysis on a correlation-based (Kirchhofer) typing derived using 500 hPa fields from the NNR data set for 1990–2001 (Schoof and Pryor 2006). This classification derives fifteen map types for the region (figure 17.4), which account for 91.4% of all days and 94.7% of all winter (DJF) days. The three most commonly observed winter map patterns represent the "average" meridional

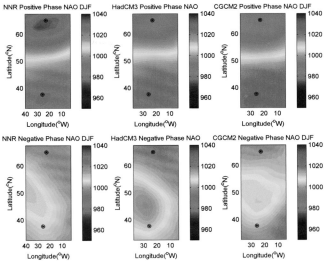

Figure 17.2. Composites of sea level pressure (mb) for positive and negative NAO phases from the reanalysis data (NNR), HadCM3, and CGCM2 output for 1990–2001.

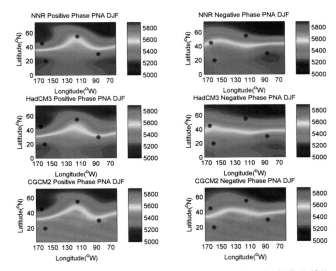

Figure 17.3. Composites of 500-mb height (gpm) for positive and negative PNA phases from the reanalysis data (NNR), HadCM3, and CGCM2 output for 1990–2001.

(map pattern 1) and zonal (map patterns 8 and 11) conditions over the study area. Weak north-south height gradients and a zonal pattern characterize the most prevalent winter pattern (map pattern 11), which accounts for 23.2% of the observations. The second most prevalent pattern (map pattern 1) accounts for 18.0% of the observations and is characterized by a strong southwest-northeast height gradient and a trough located north/northeast of the study area. With the exception of patterns 7 and 8 (10.4% and 12.5%, respectively) each of the remaining patterns accounts for less than 10% of the observations, although each represents a unique and meaningful synoptic scale circulation regime. Map pattern 15 does not occur during the DJF season and therefore is not considered in subsequent discussions.

Due to the short period of overlap between the NNR and AOGCM output, a bootstrap resampling method (Efron 1982) was used to generate a "climatology" comprising one thousand random samples of NNR data drawn from this twelve-year period, which were presented in the form of confidence intervals for the means and standard deviations of map pattern frequencies. A targeted Kirchhofer analysis of the HadCM3 and CGCM2 simulations for 1990–2001 was then performed, where each of the seeds derived from the NNR data were used to map the AOGCM output. This procedure led to classification of 86.7% of winter days from HadCM3 output and 93.6% of simulated winter days from the CGCM2 output. Both AOGCM classifications reproduce the entire range of map patterns found in the NNR data (figure 17.5).

The largest discrepancy between the observed and HadCM3 map pattern frequencies is an overestimation of the occurrence of the most common map pattern (1) and an underestimation of the occurrence of the second and third most common map patterns (8 and 11) (figure 17.5). HadCM3 also slightly overestimates the frequency of map type 2. Based on the descriptions of the map types, these results suggest that the representation of the synoptic scale climate of the study region within HadCM3 may be biased toward meridional conditions. The map pattern frequencies from the targeted analysis of CGCM2 show better agreement with the observed classification. Unlike HadCM3, CGCM2 produces the most common map pattern too seldom (figure 17.5). The largest difference between observed and CGCM2

is associated with the partitioning of days between two similar map patterns (1 and 3). Both types are associated with low geopotential height anomalies to the northeast of the study region, although map pattern 1 exhibits much stronger height gradients (figure 17.4).

LINKS BETWEEN TELECONNECTIONS AND MAP PATTERNS

An important diagnostic of AOGCM simulations is the strength and nature of the relationship between the hemispheric (or sub-hemispheric) and synoptic- or regional-scale circulation. To examine this, for each day, the "state" of the teleconnection index was determined: positive (the value of the teleconnection index was more than one standard deviation greater than the seasonal mean of the teleconnection index), negative (the value of the teleconnection index was more than one standard deviation less than the seasonal mean of the teleconnection index), or neutral (the value of the teleconnection index was within one standard deviation of the seasonal mean of the teleconnection index). Then the strength of the link between the teleconnections and map patterns is determined by comparing the frequency of map-pattern occurrences during positive and negative teleconnection phases using the two-sample difference of proportions test (see Ott 1993, Sheridan 2003).

As an example, the frequency of map pattern 1 is shown in table 17.1 for the positive, neutral, and negative phases of the NAO and PNA as derived from the NNR, HadCM3, and CGCM2 output. The table suggests that this pattern occurs more commonly under the positive phase of the PNA than the negative phase, although the bias with phase of

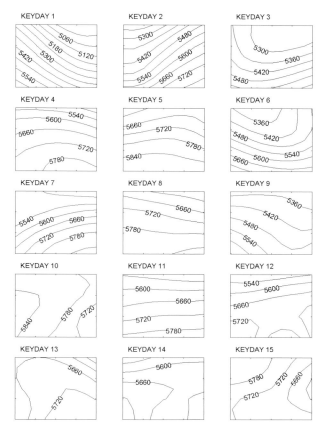

Figure 17.4. Key day maps for each of the fifteen Kirchhofer map patterns associated with 1990–2001 NCEP-NCAR Reanalysis 500-mb geopotential height fields (gpm). The area shown is as depicted in figure 17.1.

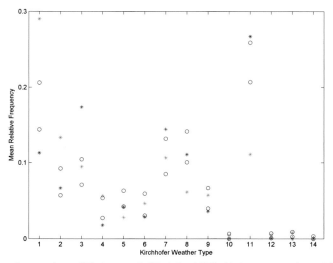

Figure 17.5. Mean relative frequency of winter (DJF) 1990–2001 HadCM3 (red) and CGCM2 (blue) map-patterns relative to NCEP-NCAR Reanalysis-derived bootstrap confidence intervals (o).

Table 17.1. The frequency of map pattern 1 under positive, neutral, and negative phases of the NAO and PNA in the NNR (1954–2001), HadCM3 (1990–2099), and CGCM2 (1990–2100) datasets

	NNR		HadCM3		CGCM2	
	NAO	PNA	NAO	PNA	NAO	PNA
Positive	119	246	467	514	152	372
Neutral	563	540	1604	1685	747	671
Negative	147	43	340	212	190	46

the PNA is more pronounced in the NNR and CGCM2 output than in the HadCM3 simulations. In accord with a priori expectations, the influence of the NAO on the frequency with which this map type is observed appears to be much weaker in the NNR data and both AOGCMs. The stronger influence of the PNA on map type frequencies is also indicated by results of the two-sample difference of proportions test (table 17.2). The results generally indicate that links between observed map patterns and the NAO index are not statistically significant in the NNR, while the PNA index is significantly linked to the frequency of four map types in the NNR data set, five map types in the HadCM3 output, and six map types in

the CGCM2 simulations. None of the map patterns is linked to both the positive and negative phase of the same teleconnection index, providing further evidence that the associations are physically derived rather than the product of stochastic effects.

Here we attempt to gain additional insight into these scale links by considering the joint influence of the teleconnection indices. This is achieved by conducting a two-sample difference of proportions test for each map pattern and teleconnection index under each phase of the other teleconnection index (table 17.3). For example, map pattern 1 occurs more frequently under the positive phase of the PNA. What is less clear is whether or not the positive phase frequencies are also sensi-

Table 17.2. Results of the two-sample difference of proportions tests performed on observed (NCEP/NCAR reanalysis), HadCM3, and CGCM2 map patterns and teleconnection indices. The table provides the p-value of the test. Underlined entries are significantly (at $\alpha=0.05$) linked to the negative phase of the teleconnection, and bold entries are significantly linked to the positive phase of the teleconnection.

Map pattern	NAO			PNA		
	OBS	HadCM3	CGCM2	OBS	HadCM3	CGCM2
1	0.24	<u>0.03</u>	0.21	**0.01**	**0.00**	**0.00**
2	0.41	**0.04**	<u>0.01</u>	0.13	<u>0.00</u>	0.07
3	0.14	0.26	<u>0.02</u>	0.08	0.65	**0.00**
4	0.22	0.08	0.68	0.08	0.07	<u>0.02</u>
5	0.57	0.14	0.14	<u>0.01</u>	<u>0.00</u>	<u>0.00</u>
6	0.23	0.05	0.11	0.66	0.57	0.59
7	0.20	**0.04**	0.15	<u>0.04</u>	<u>0.00</u>	<u>0.03</u>
8	0.27	0.53	**0.01**	0.56	0.54	0.67
9	0.50	0.32	0.35	0.50	0.50	0.61
11	0.14	**0.00**	0.15	<u>0.00</u>	<u>0.02</u>	<u>0.00</u>

Table 17.3. Results of the conditional two-sample difference of proportions tests performed on NNR (a), HadCM3 (b), and CGCM2 (c) winter (DJF) map patterns and teleconnection indices. The table provides the p-value of the test. Underlined (bold) entries are significantly (at α=0.05) linked to the negative (positive) phase of the teleconnection specified in the top row of the table when jointly considered with the phase of the teleconnection given in the second row.

a) NNR

Map Pattern	NAO			PNA		
	−PNA	Neutral PNA	+PNA	−NAO	Neutral NAO	+NAO
1	0.05	0.48	0.53	**0.00**	**0.00**	**0.00**
2	0.09	0.89	0.06	0.07	0.12	0.36
3	<u>0.00</u>	0.15	0.87	0.36	**0.02**	**0.01**
4	0.75	0.24	0.28	1.00	0.22	0.56
5	0.73	1.00	0.08	0.08	<u>0.02</u>	0.12
6	0.42	0.57	0.07	0.63	0.27	0.34
7	0.68	0.24	0.51	<u>0.02</u>	<u>0.04</u>	<u>0.02</u>
8	0.59	0.27	0.27	0.10	0.91	1.00
9	0.67	0.94	1.00	0.50	0.72	0.29
11	**0.01**	0.69	0.09	<u>0.01</u>	<u>0.00</u>	<u>0.00</u>

b) HadCM3

Map Pattern	NAO			PNA		
	−PNA	Neutral PNA	+PNA	−NAO	Neutral NAO	+NAO
1	0.59	0.10	<u>0.00</u>	**0.00**	**0.00**	**0.00**
2	**0.01**	0.06	0.29	<u>0.03</u>	<u>0.00</u>	<u>0.00</u>
3	**0.05**	0.76	<u>0.00</u>	**0.00**	0.59	<u>0.03</u>
4	0.36	0.17	**0.00**	<u>0.02</u>	0.14	0.31
5	0.35	0.20	1.00	0.12	<u>0.02</u>	0.07
6	0.07	0.18	0.11	0.73	0.45	0.29
7	**0.01**	0.19	**0.00**	<u>0.00</u>	<u>0.01</u>	<u>0.00</u>
8	0.11	0.97	0.53	1.00	0.75	0.08
9	0.85	0.81	<u>0.01</u>	0.72	0.18	<u>0.05</u>
11	**0.00**	**0.00**	**0.03**	<u>0.00</u>	<u>0.00</u>	<u>0.00</u>

c) CGCM2

Map Pattern	NAO			PNA		
	−PNA	Neutral PNA	+PNA	−NAO	Neutral NAO	+NAO
1	**0.00**	0.66	0.15	**0.00**	**0.00**	**0.00**
2	0.08	<u>0.02</u>	0.10	<u>0.04</u>	<u>0.04</u>	<u>0.03</u>
3	<u>0.00</u>	0.10	<u>0.01</u>	**0.00**	**0.00**	**0.00**
4	0.71	0.61	1.00	0.62	0.66	0.47
5	0.63	0.25	**0.00**	<u>0.00</u>	<u>0.03</u>	<u>0.04</u>
6	<u>0.00</u>	0.54	<u>0.01</u>	0.50	0.54	0.37
7	<u>0.01</u>	0.56	0.32	<u>0.00</u>	<u>0.00</u>	<u>0.00</u>
8	**0.00**	**0.01**	**0.00**	0.12	0.15	0.10
9	0.64	0.86	0.28	**0.03**	0.14	0.12
11	0.74	0.18	0.28	<u>0.00</u>	<u>0.00</u>	<u>0.00</u>

tive to the NAO phase. Due to sample size considerations, this test is performed only for map patterns with more than one hundred occurrences. This removes map patterns 10, 12, 13, and 14 from the analysis.

The analyses presented in table 17.3 sug-gest that there is considerable interdependence between the occurrence of specific regional map patterns and the joint probability of various phases of the NAO and PNA. Evaluation of this dependence on both teleconnections can reveal physical linkages but may also illustrate reasons for the noted differences between the NNR and AOGCM-simulated map-pattern frequencies. When the relationship between the NAO and regional map patterns is considered separately for each PNA phase, statistical significance that is not present in table 17.2 emerges for some map patterns (table 17.3). Similarly, the analysis provides additional insight into the dependence of several map patterns on the PNA index. Here, we illustrate the diagnostic power of this analysis using examples drawn from map types 1, 3, 5, 7 and 11:

• The first and second most prevalent map patterns correspond to strong zonal (pattern 11) and meridional (pattern 1) circulation patterns, respectively, and are therefore expected to be closely linked to the PNA teleconnection since positive PNA conditions are associated with a considerable meridional component to the large-scale circulation in the region, while negative PNA conditions correspond to zonal flow. As indicated in table 17.3, significant links between the negative PNA index and pattern 11 and between the positive PNA index and pattern 1 are not sensitive to NAO phase. The strong dependence on PNA phase and lack of co-dependence on NAO as manifest in the NNR data are correctly reproduced by the two AOGCMs.

• As shown in table 17.3, HadCM3's map pattern 1 is also significantly more likely to occur under a combination of positive phase PNA and negative phase NAO, while CGCM2's map pattern 1 is significantly more likely under the positive NAO phase if PNA is also in its positive phase. These differences in the co-dependence of the map patterns on the NAO and PNA may be responsible for the underestimation of map pattern 1 frequency by CGCM2

and overestimation by HadCM3 (figure 17.5) given this co-dependence is not observed in the NNR data.

• HadCM3 underestimates the frequency of map pattern 11 (figure 17.5) although the link between the map pattern and the negative (i.e., zonal) PNA mode is properly simulated by the model. In the NNR data, a significant link emerges between map pattern 11 and the positive phase of the NAO, but only when the PNA is in its negative phase. In HadCM3, the link extends to all PNA phases, suggesting that the link between the NAO and map pattern in HadCM3 is too strong relative to observations (table 17.3). In CGCM2, the PNA-dependent link between the NAO and map pattern 11 is not significant (table 17.3), suggesting that model does not adequately capture the joint dependence of map pattern 11 on the NAO and PNA.

• Map pattern 3 is significantly more likely under the positive phase of the PNA (table 17.2) in the NCEP/NCAR reanalysis data. Closer examination shows that this link is not significant when the NAO is in its negative phase (table 17.2). However, map pattern 3 is also shown to be more likely during the negative NAO phase if the PNA is also in its negative phase. Neither of the AOGCMs accurately portrays these observed linkages, and indeed the relationship between map pattern 3 and the NAO within HadCM3 changes sign as the PNA phase changes. Specifically, if the PNA is negative, the pattern is more likely to occur under the positive NAO phase, but if the PNA is positive, the pattern is more likely to occur under the negative phase of the NAO. This suggests that there may be more than one hemispheric circulation pattern that is conducive to occurrence of map pattern 3 in HadCM3. In CGCM2, map pattern 3 is strongly linked the positive PNA phase regardless of NAO phases and also exhibits links to the negative NAO phase for both positive and negative PNA phases.

• Map pattern 5 is more likely to occur during the negative PNA phase (Schoof and Pryor 2006; table 17.2), but the link is only significant during the neutral NAO phase (table 17.3). HadCM3 reproduces this link, but still produces map pattern 5 too infrequently. In CGCM2, the negative (zonal) PNA phase is significantly linked to map pattern 5 during all NAO phases and is also significantly linked to the positive NAO phase when the positive PNA phase is occurring.

• Like map patterns 1 and 11, map pattern 7 is most likely to occur when the PNA is negative, regardless of NAO phase. Both AOGCMs are able to reproduce this link, but HadCM3 also has a significant link between the positive NAO and map pattern 7 when the PNA is not neutral, while in CGCM2 map pattern 7 is more likely under the negative NAO when the PNA is negative.

Concluding Remarks

The North Atlantic Oscillation (NAO) and Pacific–North American (PNA) teleconnections play important roles in dictating conditions at the synoptic scale over the Midwestern USA. Two AOGCMs exhibit a relatively high degree of skill in reproducing the indices at the daily scale during the winter months in terms of both the spatial patterns and temporal structure. However, these AOGCMs exhibit lower skill in correctly simulating the frequency with which certain synoptic scale phenomena (as described using map type frequencies) are observed. Herein we have demonstrated that a key diagnostic in understanding these failures

is the scale linkages between the teleconnection indices and the synoptic scale, and particularly the co-dependence of synoptic map-type frequencies with the phase of both the NAO and PNA. Naturally, the results presented above do not explain all of the discrepancies between the observed (NCEP/NCAR) and AOGCM-simulated (HadCM3 and CGCM2) circulation patterns. However, they do provide additional insight into the results of Schoof and Pryor (2006) regarding the links between hemispheric and sub-hemispheric scale circulation and regional map patterns. For some of the links reported here, the physical mechanism is obvious. For example, positive NAO and negative PNA indices coincide to provide strong zonal flow in the eastern part of the study region, resulting in more occurrences of map pattern 11. The causal mechanisms for many other links remain elusive and will be the focus of additional research. Nevertheless, the research presented here highlights the need for additional research into the multi-scale evaluation of AOGCMs, and offers one potential mechanism for such analyses. Similar efforts should be undertaken with larger suites of models (e.g., those prepared for IPCC AR4 or CMIP3) to further assess model reliability at a range of scales.

ACKNOWLEDGMENTS
Financial support from the NSF Geography and Regional Science program (grants #0618364 and 0647868) is gratefully acknowledged.

REFERENCES
Ambaum, M. H., B. J. Hoskins, and D. B. Stephenson. 2001. "Arctic Oscillation or North Atlantic Oscillation?" *Journal of Climate* 14: 3495–3507.
Barnston, A. G., and R. E. Livezey. 1987.

"Classification, Seasonality, and Persistence of Low-Frequency Atmospheric Circulation Patterns." *Monthly Weather Review* 115: 1083–1126.

Changnon, D., et al. 2002. "Efforts to Improve Predictions of Urban Winter Heating Anomalies Using Various Climate Indices." *Meteorological Applications* 9: 105–111.

Coleman, J. S. M., and J. C. Rogers. 2003. "Ohio River Valley Winter Moisture Conditions Associated with the Pacific-North American Teleconnection Pattern." *Journal of Climate* 16: 969–981.

Croci-Maspoli, M., C. Schwierz, and H. C. Davies. 2007. "A Multi-Faceted Climatology of Atmospheric Blocking and Its Recent Linear Trend." *Journal of Climate* 20: 633–649.

Efron, B. 1982. *The Jackknife, the Bootstrap, and Other Resampling Plans.* CBMS-NSF Regional Conference Series in Applied Mathematics. Philadelphia: Society for Industrial and Applied Mathematics.

Flato, G., et al. 2000. "The Canadian Centre for Climate Modelling and Analysis Global Coupled Model and Its Climate." *Climate Dynamics* 16: 451–467.

Gordon, C., et al. 2000. "The Simulation of SST, Sea Ice Extents and Ocean Heat Transports in a Version of the Hadley Centre Coupled Model without Flux Adjustments." *Climate Dynamics* 16: 147–168.

Hansen, J., et al. 1984. "Climate Sensitivity Analysis of Feedback Mechanisms." In *Climate Processes and Climate Sensitivity,* ed. James E. Hansen and Taro Takahashi, 130–163. Geophysical Monographs no. 29. Washington, D.C.: American Geophysical Union.

Hurrell, J. W. 1995. "Decadal Trends in the North Atlantic Oscillation and Relationships to Regional Temperature and Precipitation." *Science* 269: 676–679.

———. 1996. "Influence of Variations in Extratropical Wintertime Teleconnections on Northern Hemisphere Temperatures." *Geophysical Research Letters* 23: 665–668.

Hurrell, J. W., et al. 2003. "An Overview of the North Atlantic Oscillation." In *The North Atlantic Oscillation: Climatic Significance and Environmental Impact,* 1–36. AGU Geophysical Monograph 134. Washington D.C.: American Geophysical Union.

IPCC. 2000. "Special Report on Emissions Scenarios." Cambridge: Cambridge University Press.

Kirchhofer, W. 1974. "Classification of European 500 mb Patterns." *Schwiezerische Meteorologische Anstalt, Institut Suisse de Meteorologie, Zurich* 43: 1–16.

Kistler, R., et al. 2001. "The NCEP-NCAR 50-Year Reanalysis: Monthly Means CD-ROM and Documentation." *Bulletin of the American Meteorological Society* 82: 247–267.

Leathers, D. J., B. Yarnal, and M. A. Palecki. 1991. "The Pacific North-American Teleconnection Pattern and United States Climate, part 1: Regional Temperature and Precipitation Associations." *Journal of Climate* 4: 517–528.

McFarlane, N., et al. 1992. "The Canadian Centre Second Generation General Circulation Model and Its Equilibrium Climate." *Journal of Climate* 5: 1013–1044.

Ott, R. L. 1993. *An Introduction to Statistical Methods and Data Analysis.* Belmont, Calif.: Duxbury Press.

Pope, V. D., et al. 2000. "The Impact of New Physical Parameterizations in the Hadley Centre Climate Model: HadAM3." *Climate Dynamics* 16: 123–146.

Schoof, J. T., and S. C. Pryor. 2006. "An Evaluation of Two GCMs: Simulation of North American Teleconnection Indices and Synoptic Phenomena." *International Journal of Climatology* 26: 267–282.

Sheridan, S. C. 2003. "North American Weather-Type Frequency and Teleconnection Indices." *International Journal of Climatology* 23: 27–45.

Wallace, J. M., and D. S. Gutzler. 1981. "Teleconnections in the Geopotential Height Field during the Northern Hemisphere Winter." *Monthly Weather Review* 109: 784–812.

Yin, Z.-Y. 1994. "Moisture Conditions in the South-Eastern USA and Teleconnection Patterns." *International Journal of Climatology* 14: 947–967.

18. Regional Synoptic Classification:
A Midwestern USA Example

J. S. M. COLEMAN AND J. C. ROGERS

An Overview of Atmospheric Circulation Classification Schemes

Identification and analysis of synoptic- and larger-scale circulation patterns and their impact on regional and local climates is a major research theme in the field of synoptic climatology (Barry and Carleton 2001). Weather and circulation types, or weather regimes, represent the principal modes of atmospheric flow from which other weather elements (e.g., temperature, precipitation, humidity) might be analyzed at the regional or local scale. The purpose or applicability of the classification procedure often dictates the methodology employed, spatial extent studied, and variables included (e.g., Davis and Walker 1992; Sheridan 2002).

Techniques by which synoptic classifications are created can be manual or automated, or hybrid schemes that utilize aspects of both (Yarnal 1993). Manual methods rely upon the skill of the researcher to determine map pattern classification and meteorologically significant modes of variability. The strength in manual (and hybrid modes) lies in their ability to link synoptic-scale atmospheric circulation conditions with local conditions more readily than automated

methods since the researcher supervises the classification for optimal performance. Perhaps the most recognized regional manual classification systems are the Lamb (1972) and Muller (1977) weather types created for the United Kingdom and the U.S. Gulf Coast, respectively, which are constructed using the classic mid-latitude cyclone model as a reference frame.

Automated methods derive classes usually from more statistical related approaches and are commonly subdivided into correlation-based techniques (see chapter 17 of this volume), use of self-organizing maps, and eigenvector techniques. Correlation-based map-pattern classification examines the similarity (or dissimilarity) between two locations for a particular variable. The selection of the gridpoint density and the minimum correlation threshold by the examiner determines the resolution and type of circulation features retained as distinct map pattern classes (Yarnal 1993). A drawback of correlation-based approaches, however, is their inability to extract within-type variability (i.e., days with very different atmospheric characteristics grouped as the same type) (e.g., Brinkmann 1999; 2000; 2002).

Self-organizing maps (SOMs) are a relatively new automated classification procedure to synoptic climatology that extracts a user-specified number of map patterns from an input data matrix. The SOM technique allows for assessment of a high-dimensional data space through a set of reference vectors (or cluster centroids) that can be projected onto a two-dimensional grid representing an atmospheric circulation feature (Gutiérrez et al. 2005). Hewitson and Crane (2002) note the SOM places similar patterns adjacent to one another and very dissimilar patterns far apart in the SOM space. The constructed SOMs can be used to describe the range of atmospheric states over time, including major climate modes and anomalies that can be used in seasonal forecasts (Gutiérrez et al. 2005; Hewitson and Crane 2002; Cavazos 2000). Yet, standard automated clustering algorithms (e.g., k-means) may produce superior results compared to SOMs, particularly for smaller, regional classifications (Balakrishnan et al. 1994)

Eigenvector-based techniques, particularly principal component analysis (PCA), have been used extensively in synoptic climatology to detect major modes of atmospheric variability. These techniques extract the eigenvalues and eigenvectors from a dispersion matrix in order to isolate the variation in a multidimensional dataset (Yarnal 1993) and maximize the association between the component and as many locations and times as possible (Rohli and Vega 2008). PCA does not distinguish variables as dependent and independent, but rather treats the dataset as a single unit of interrelated elements, and reduces the overall size of the original dataset through a linear transformation series of orthogonal components, which

are independent in nature. Each successive component explains a lesser amount of the overall amount of variance present in the original dataset. A rotation method, such as Varimax, is often then executed on the retained components to aid interpretation of the spatial pattern of the loadings (e.g., Barnston and Livezey 1987).

PCA has often been used in conjunction with other statistical techniques such as cluster analysis (CA). The primary objective of CA is to establish relatively uniform groups of cases (e.g., weather types or regimes) from the input data (which may be PCA-based), such that within-group variance is minimized while between-group variance is maximized. CA techniques are often classified as nonhierarchical and hierarchical (Cheng and Wallace 1993). Nonhierarchical procedures include a partitioning method often known as the k-means clustering method (Michelangeli, Vautard, and Legras 1995; Straus and Molteni 2004), in which a randomly pre-selected number of "seed points" are chosen and the clusters are built around them as each individual element is compared to the seed points using a pre-chosen similarity measure. Hierarchical clustering methods find clusters of observations within a dataset by either partitioning clusters from a larger cluster set (a divisive method) or joining smaller clusters into larger grouping (a agglomerative method). For synoptic climate classification, hierarchical-agglomerative cluster algorithms (e.g., average linkage) alone or combined with nonhierarchical methods have proven to be the most successful in procuring synoptically meaningful homogenous categories for large multivariate datasets (e.g., Kalkstein,

Tan, and Skindlov 1987; Kalkstein 1991; Davis and Walker 1992; Schoof and Pryor 2003; Coleman and Rogers 2007).

Hybrid classification schemes are a compromise between the automated statistical techniques and manual classification methods for synoptic categorization, attempting to capture the most constructive attributes of both (Schwartz 1991; Sheridan 2002). Whether categorized as manual or automated, each methodology introduces an element of subjectivity to a greater or lesser degree. Manual and hybrid methods are usually the most difficult to reproduce since decisions on the "ideal" physical characteristics and number of weather types are based on graphical weather maps, data interpretation, and researcher expertise. Manually based map classes do not always have a precise, unbiased justification for initial categorization. Hence, these weather types are not easily replicated by another set of individuals given the same criteria that often include indefinable, qualitative terminology.

Automated techniques introduce subjectivity via the selection of the statistical classification technique and input variables. While these choices can significantly impact the outcome, an automated method has the most potential for replication. Automated methods are usually less labor intensive and time consuming than many manual methods. The utility of automated methods is evident in its broad-spectrum appeal in synoptic-scale atmospheric circulation studies and applied climatology research. This chapter describes the development of an automated PCA-CA-based weather classification scheme for the central United States. The ensuing synoptic types are discussed in

light of known atmospheric circulation features and are evaluated against local surface conditions at stations throughout the Midwestern USA.

Regional Synoptic Classification Schemes: An Example for the Central United States

The approach applied here is based on the work conducted by Coleman and Rogers (2007). To capture the atmospheric flow characteristics of the lower-mid troposphere, the approach presented utilizes a broad range of dynamic and thermodynamic air mass information from a multiple-point daily surface and upper-air dataset. This information is then used in a multi-step classification procedure whereby the advancement into each step is dependent on the outcome of the preceding step (El-Kadi and Smithson 1992). In the first stage, the procedure employs S-mode PCA, a data reduction method for analyzing multiple data points or stations over time. For the second stage, the results of the PCA are used in conjunction with CA, a data-categorization technique to create the final groupings. The resulting synoptic types are then compared to recognized atmospheric circulation features, teleconnection patterns, and regional climate variability in the Midwestern USA.

METHODOLOGY

To capture a regional climatology for the Midwestern USA, the study area expands outside the Midwestern core, extending in latitude from the Gulf of Mexico to southern Canada (25°–55°N) and in longitude

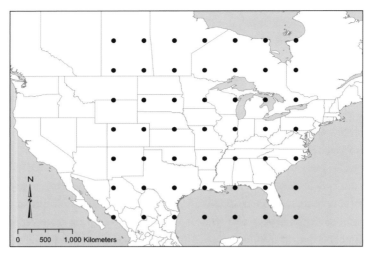

Figure 18.1. Study region in the central United States bounded by latitudes 25–55° N and longitudes 77.5–107.5°W from which NCEP-NCAR Reanalysis data grid points at five-degree latitude and longitude intervals (shown as circles) are taken.

from the Rocky and Appalachian mountain ranges (77.5°–107.5°W; figure 18.1). These boundaries contain atmospheric circulation characteristics that are critical to the development and propagation of synoptic scale systems (e.g., Colorado Lows) and associated air mass fluctuations (e.g., maritime Tropical, continental Polar). Yet, the spatial domain diminishes the impact of both peripheral circulation features (e.g., tropical storm systems, winter nor'easters) not vital to generalizing the synoptic climatology of the Midwestern USA as well as measurement or modeling errors associated with high terrain areas.

To classify atmospheric phenomena at the synoptic scale, continuous data records from locations well distributed throughout the study area are required. The National Centers for Environmental Prediction (NCEP)–National Center for Atmospheric Research (NCAR) Reanalysis was first developed in the early 1990s to accommodate climate research by assimilating disparate weather information from several sources into a spatially uniform gridded format (Kalnay et al. 1996; Kistler et al. 2001). The NCEP-NCAR Reanalysis and other reanalysis products (see chapter 15) have attributes that are advantageous for multi-pressure-level synoptic-scale analysis, such as complete spatial and temporal resolution, consistent error agreement across data points, and spatial continuity, particularly in remote regions. Analysis of synoptic-scale features have shown good conformity between the Reanalysis data and rawinsondes over North America, producing similar short-range forecast errors (Schoof and Pryor 2003; Kistler et al. 2001). Numerous research papers have incorporated Reanalysis data since its inception (e.g. Gulev, Zolina, and Grigoriev 2001; Finley and Raphael 2007; Perry, Konrad, and Schmidlin 2007).

NCEP-NCAR Reanalysis data for the period 1948–2004 acquired from the National Oceanic and Atmospheric Administration–Cooperative Institute for Research in the Environmental Sciences (NOAA-CIRES) Climate Diagnostics Center are used herein.

This is the same dataset as used in chapter 17 and is referred to as NCEP Reanalysis 1 in chapter 15. Although the relatively large grid spacing of the Reanalysis dataset (2.5°×2.5° latitude and longitude grid) precludes detailed boundary layer analyses at the local level, the resulting synoptic types allow for interpretation of general tropospheric characteristics for locations between gridpoints. Individual locations may have considerably different weather characteristics based on juxtaposition to prevailing synoptic features. Yet, the advantage lies in that *all locations* within the region can be compared according to a single classification system, improving its interregional utility. This varies considerably from other site-specific classification schemes (e.g., Stone 1989; Kalkstein 1991; McGregor, Walters, and Wordley 1999; Kassomenos, Gryparis, and Katsouyanni 2007) that make multi-location comparisons, regardless of proximity, difficult, if not impossible (Kalkstein and Greene 1997).

Daily mean variables from the NCEP-NCAR Reanalysis dataset employed in the classification are: surface temperature and relative humidity; temperature and geopotential height at the 925, 850, 700 and 500 hPa levels and specific humidity at these lowest four levels; pressure reduced to mean sea level pressure (SLP); and the 1200 UTC zonal (u) and meridional (v) wind components. The data were reduced to a 5°×5° latitude and longitude grid due to computational considerations; thus the study region contains forty-nine data points for each pressure level across its boundaries (figure 18.1). The selected unstandardized NCEP-NCAR Reanalysis data matrix (25 variables per day over 20,820 days [from 1948–2004] at 49 gridpoints) is subjected to a correla-tion-based S-mode PCA, which reduces the dimensionality of the dataset such that the first few components retain most of the variance of the original data matrix. In S-mode PCA, individual gridpoints are considered the variables and the temporal observations the cases; thus detecting those gridpoints that vary similarly over time (Serrano et al. 1999). The seasonal cycle was kept in the data analysis so as to help identify and demonstrate the overlap in some of the patterns between, for example, autumn and winter or winter and spring, producing a readily observed continuum of the occurrences of certain synoptic types from one month (or season) to the next.

Nine principal components were retained based on guidelines (Joliffe 1986) suggesting that the retained components comprise 70–90% of the cumulative percentage of variance (these nine components account for 70.1% of the variance) and that extraction communalities (the squared multiple correlation for a variable using the factors as predictors) should be 0.5 or higher to effectively represent the variability of the individual data points. With the exception of a few gridpoints for the u and v wind components, the extraction communalities are 0.6 or higher, thus indicating the nine components embody the primary attributes of the data matrix. Nearly half of the retained cumulative variance (34.3%) is explained by the first component that loads chiefly on the temperature and specific humidity variables at all five pressure levels while the remaining components are largely related to pressure and wind variability.

Based on the nine retained principal components, the PCA also produces component scores for each day. The scores are

Table 18.1. Number of days from 1948 to 2004 assigned to one of ten clusters or synoptic types (in total and by season) and the associated name abbreviations. Seasons are identified as: Winter: December, January, and February (DJF); Spring: March, April, and May (MAM); Summer: June, July, and August, (JJA); and Autumn: September, October, and November (SON).

Synoptic Type Name	Total	DJF	MAM	JJA	SON
Eastern Zonal (EZ)	1866	1099	310	0	457
Moderate Zonal (MZ)	2120	465	1233	60	362
Lake Effect Favorable (LEF)	1581	394	573	53	561
East Coast Ridge (ECR)	1954	638	941	32	343
Southeast Low (SEL)	1393	683	412	5	293
East Coast Trough (ECT)	2120	1017	544	1	558
Polar Intrusion (PI)	1417	846	304	0	267
Central Convergence Zone (CCZ)	3461	0	416	2439	606
Warm Central Ridge (WCR)	2531	0	257	1556	718
Warm Central Trough (WCT)	2377	3	254	1098	1022

orthogonal and signify the representation of the original data matrix onto the principal component axes. The daily unrotated PCA scores are used in a two-step CA that generates groups of cases with similar component scores. The two-step CA method is intended for optimal performance on very large datasets that may not necessarily be normally distributed. The procedure starts by assuming that first day's element forms the first cluster. Then in the second step, standard hierarchical clustering is applied as the algorithm proceeds sequentially and agglomeratively through the 20,820 data elements (the days) either by adding new clusters or by placing a daily data element into a preexisting cluster based on the Bayesian Information Criterion (BIC), a log-likelihood distance measurement. Smaller BIC values and relatively large change in the distance measures between sequential clusters indicate the most optimal solution.

The largest BIC change (after that occurring between cluster 1 and 2) occurs between clusters 10 and 11, indicating a ten-cluster so-lution. While other relatively large changes in BIC values occurred at other intervals, it was found that retaining too few clusters tended to minimize within-season synoptic variability, whereas retaining too many clusters tended to produce groups with disproportionately fewer days assigned to them. Cases or days assigned to the same cluster possess comparable component scores and thus represent an atmospheric circulation regime distinct from that of other cluster memberships.

The final cluster solution associates each day from 1948–2004 into one of ten predominant atmospheric circulation types capturing the regional and seasonal synoptic circulation characteristics of the central United States (see table 18.1). Distribution of days among the clusters is relatively uniform over the entire time period, with no exceedingly small or large clusters. The final CA solution is designed to show more typical synoptic situations rather than to identify exceptional weather events (e.g., heat waves) due to the small number of clusters.

An examination of the months of peak cluster occurrence reveals distinct seasonality across the groupings (see table 18.1), an expected result since no transformations were done on the input data matrix to remove the seasonal cycle. The Moderate Zonal (MZ) and East Coast Ridge (ECR) patterns are most prominent during the spring months of March, April, and May, whereas the warmer patterns of Central Convergence Zone (CCZ), Warm Central Ridge (WCR), and Warm Central Trough (WCT) are the governing synoptic types during the summer months of June, July, and August, and into early autumn. The remaining five synoptic types—Lake Effect Favorable (LEF), Polar Intrusion (PI), Eastern Zonal (EZ), East Coast Trough (ECT), and Southeastern Low (SEL)—are overall cold-season dominant, occurring primarily from November through March, but also having a significant presence during early spring and late autumn. Distinctive seasonal/annual tendencies toward one synoptic type over another have also been shown (Coleman and Rogers 2007).

For each of the ten synoptic types identified, the mean values of each variable (temperature, specific humidity, SLP or geopotential height, and wind) are plotted for each of the forty-nine gridpoints at four different pressure levels (surface, 850 hPa, 700 hPa, 500 hPa levels) (figures 18.2a–j; adapted from Coleman and Rogers 2007). Mean u and v wind values are converted to wind magnitude (in knots, where 1 knot equals approximately 0.51 m s^{-1}) and direction. These wind variables are shown as sta-

tion model wind barbs, indicating the direction from which the wind is coming and the wind speed in increments of 10 knots, with solid flags equaling 50 knots.

The resulting synoptic patterns are characterized by distinctive surface circulations, baroclinic vertical structure, and thermal advection. For example, the LEF and SEL patterns (figures 18.2c and e) imply areas of a mid-tropospheric divergence and positive vorticity advection, illustrating a circulation configuration favorable for surface low pressure system development and strengthening in the Great Lakes (LEF) and Southeast regions (SEL). In contrast, the warm-season dominant patterns (CCZ, WCR, and WCT) exhibit a more barotropic structure typical of the mid-latitude boreal summer, with a comparatively narrow mean temperature range (<15°C difference across the region), weak mean surface and mid-tropospheric winds, and a larger tropical moisture influx.

The time series of the annual frequencies of the ten atmospheric circulation types shows sizeable inter-annual variability (Coleman and Rogers 2007), including the order of synoptic pattern occurrence. Table 18.2 displays the number of times each synoptic pattern immediately follows another pattern (i.e., the succeeding day) during the 1949–2004 study period. The overall sequencing of the synoptic types reveals a propensity toward a persistence forecast model, whereby each atmospheric circulation regime tends to be followed immediately by itself. This is especially true for standing wave patterns, such as zonal (e.g., EZ, MZ) or meridional flow (e.g., ECT), that dominate the Midwest for extensive time periods during the cooler seasons. Exceptions are with the more transient synoptic patterns, such as the

LEF, which are just as likely to be followed by more zonal conditions.

An examination of a three-day sequence of synoptic types by season (not shown) also indicates persistence patterns are the most common, differing primarily by the seasonality of the circulation types itself. Sequential days of EZ, MZ, WCR, and WCT are the most likely three-day patterns to occur during winter, spring, summer, and autumn, respectively, since these are also the seasons with peak annual frequency for these patterns. The LEF and SEL are the least likely persistent patterns for a three-day cycle, owing again to the transient nature of these synoptic types. In comparison to summer and early autumn, winter and spring have more variability in the range of synoptic sequences. In particular, these seasons tend to have alternating combinations involving ECT-ECR and PI-ECT, coinciding with shifts or intensification of mid-tropospheric flow prevalent during this time of year across the central United States.

To examine the actual ambient conditions at specific locations, mean surface

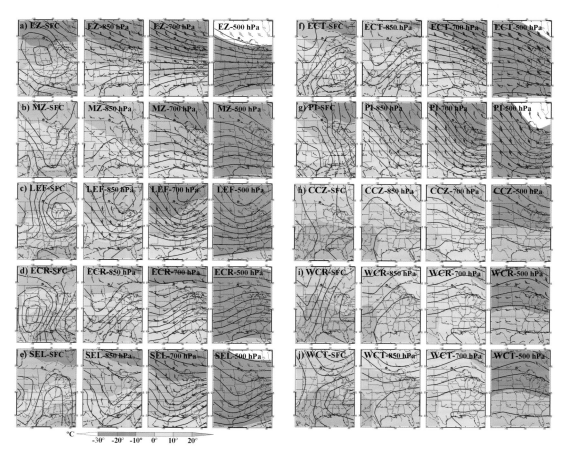

Figure 18.2a–j. Mean meteorological conditions for each of the ten synoptic circulation types at the surface (sfc), 850 hPa, 700 hPa, and 500 hPa (adapted from Coleman and Rogers 2007). The following variables are plotted: temperature in °C at 10°C intervals as shaded contours; specific humidity in grams per kilogram as dashed lines (except for 500 hPa level); mean SLP (hPa) as solid lines; and wind direction and speed in knots given as station model (or grid point) wind barbs (see text for description).

Table 18.2. Synoptic-type frequency associated with the day after (Day 2) the initial synoptic pattern occurrence for the 1948–2004 study period. Highlighted portion indicates when a pattern is followed by itself.

	EZ Day 2	MZ Day 2	LEF Day 2	ECR Day 2	SEL Day 2	ECT Day 2	PI Day 2	CCZ Day 2	WCR Day 2	WCR Day 2
EZ	967	75	363	187	88	117	18	8	0	43
MZ	57	1028	271	117	238	59	146	106	47	51
LEF	99	130	360	476	166	147	26	34	54	89
ECR	248	354	47	634	35	466	28	54	37	51
SEL	141	156	93	396	440	90	13	19	5	39
ECT	259	139	92	9	70	992	496	13	19	31
PI	36	54	175	7	303	152	665	5	0	20
CCZ	11	66	102	62	10	23	7	1896	362	922
WCR	18	57	21	29	18	26	6	1048	1061	247
WCT	30	61	56	37	25	48	12	278	946	884

weather variables for the 1975–2004 were acquired from the "Global Summary of the Day" database from the National Climatic Data Center (NCDC) at their website http://www.ncdc.noaa.gov/oa/ncdc.html. Meteorological variables include daily means of temperature, dew point temperature, sea level pressure, wind speed, and precipitation as well as maximum and minimum temperature. For the seven cool-season dominant synoptic types, the mean surface variables for winter are shown for four Midwestern cities (table 18.3). Based on the analysis of variance (ANOVA) F-test statistic, all weather variables among the synoptic types for each location are significantly different from one another ($p<0.001$).

The annual frequencies of the synoptic types reveal high inter-annual variability linked to known low-frequency atmospheric circulation modes, particularly stemming from the Pacific basin. Coleman and Rogers (2007) showed that the positive (meridional flow) and negative (zonal flow) phases of the Pacific–North American (PNA) teleconnec-

tion pattern are particularly linked to the PI and EZ winter synoptic types, a relationship that continues, albeit weaker, into spring. These oscillations in atmospheric circulation tend to produce significant hydrological changes within the Midwestern USA, particularly for the Ohio River Valley region centering on southern Indiana, Illinois, and Ohio (Rogers and Coleman 2004; Coleman and Rogers 2003). Other synoptic types' frequencies (e.g., SEL and MZ) are also significantly correlated to other Pacific teleconnection patterns, such as the Pacific Decadal Oscillation (PDO) and El Niño–Southern Oscillation (ENSO).

Concluding Remarks

The ten synoptic types produced for the central United States reveal the generalized atmospheric flow regimes, including strong baroclinic patterns, which have linkages to Pacific teleconnections and local near-surface parameters. Applications of this research could include investigations on topics

Table 18.3. Mean surface weather variables for each of the seven winter (DJF) synoptic types at four Midwestern locations: Chicago, Ill.; Columbus, Ohio; Minneapolis, Minn.; and St. Louis, Mo. Daily mean meteorological variables include temperature (T), dew point temperature (T_d), maximum temperature (T_{max}), minimum temperature (T_{min}), sea level pressure (SLP), wind speed (WND), and precipitation (P). Data are from 1975 to 2004.

Chicago	T (°C)	T_d (°C)	T_{max} (°C)	T_{min} (°C)	SLP (hPa)	WND (kts)	P (cm)
EZ	−5.4	−10.0	0.2	−10.0	1024.6	9.3	0.28
MZ	−4.1	−8.4	0.4	−8.3	1021.3	8.3	0.18
LEF	−2.8	−6.5	2.5	−7.5	1012.8	12.6	0.76
ECR	2.5	−0.5	6.4	−2.3	1014.5	9.9	0.74
SEL	−1.1	−4.4	3.0	−5.0	1014.6	9.9	0.84
ECT	−3.1	−7.9	2.6	−8.7	1021.8	9.3	0.05
PI	−5.5	−9.9	1.3	−9.9	1017.0	10.2	0.18
Columbus							
EZ	−1.0	−5.6	4.8	−5.6	1024.1	7.6	0.23
MZ	−2.6	−7.1	2.5	−6.5	1019.9	7.4	0.15
LEF	2.3	−2.0	7.7	−3.3	1013.8	10.8	0.48
ECR	4.0	−0.5	9.5	−1.8	1019.2	6.9	0.20
SEL	2.5	−1.3	7.4	−2.2	1014.6	7.9	0.48
ECT	−2.7	−8.0	3.3	−7.9	1024.8	7.2	0.05
PI	−3.7	−8.1	1.3	−7.9	1015.8	8.8	0.18
Minneapolis							
EZ	−11.8	−16.5	−5.6	−17.1	1025.5	8.1	0.05
MZ	−7.9	−12.2	−3.0	−12.6	1022.9	7.5	0.05
LEF	−9.1	−13.5	−3.3	−13.7	1016.5	12.0	0.20
ECR	−2.7	−5.9	1.3	−7.3	1013.2	9.0	0.16
SEL	−6.1	−9.9	−1.3	−10.6	1017.9	8.6	0.11
ECT	−5.2	−9.5	−0.3	−11.1	1018.4	8.3	0.03
PI	−8.3	−13.0	−2.6	−13.5	1019.5	9.1	0.05
St. Louis							
EZ	−1.3	−5.9	4.8	−5.9	1025.8	8.1	0.18
MZ	−1.3	−5.4	3.9	−5.7	1021.9	8.1	0.07
LEF	−0.4	−4.8	7.0	−4.3	1018.1	12.3	0.28
ECR	7.2	3.7	12.6	1.4	1013.0	9.6	0.47
SEL	1.3	−1.8	6.5	−2.8	1014.8	9.6	0.51
ECT	−2.0	−4.5	9.3	−4.7	1023.3	8.0	0.04
PI	−1.6	−6.8	4.7	−6.6	1020.5	9.9	0.05

such as agricultural productivity, pollution concentrations, water resource management, or long-term weather forecasting. These synoptic types, for instance, have been used to evaluate the impact of weather types on human mortality changes in the central United States (Coleman 2008). Identification of these circulation regimes also provides an environmental baseline from which to assess future changes in Midwestern USA atmospheric flow, including evaluation against coupled Atmosphere-Ocean General Circulation Models (AOGCMs) (e.g., Schoof and Pryor 2006, and chapter 17 of this volume).

REFERENCES

Balakrishnan, P. V., et al. 1994. "A Study of the Classification Capabilities of Neural Networks Using Unsupervised Learning: A Comparison with K-Means Clustering." *Psychometrika* 59: 509–525.

Barnston, A. G., and R. E. Livezey. 1987. "Classification, Seasonality and Persistence of Flow Frequency Atmospheric Circulation Patterns." *Monthly Weather Review* 115: 1083–1126.

Barry, R. G., and A. M. Carleton. 2001. *Synoptic and Dynamic Climatology.* London: Routledge.

Brinkmann, W. A. R. 1999. "Within-Type Variability of 700 hPa Winter Circulation Patterns over the Lake Superior Basin." *International Journal of Climatology* 19: 41–58.

———. 2000. "Modification of a Correlation-Based Circulation Pattern Classification to Reduce Within-Type Variability of Temperature and Precipitation." *International Journal of Climatology* 20: 839–852.

———. 2002. "Local versus Remote Grid Points in Climate Downscaling." *Climate Research* 21: 27–42.

Cavazos, T. 2000. "Using Self-Organizing Maps to Investigate Extreme Climate Events: An Application to Wintertime Precipitation in the Balkans." *Journal of Climate* 13: 1718–1732.

Cheng, X., and J. M. Wallace. 1993. "Cluster Analysis of Northern Hemisphere Wintertime 500-hPa Height Field: Spatial Patterns." *Journal of Atmospheric Science* 50: 2674–2696.

Coleman, J. S. M. 2008. "Atmospheric Circulation Patterns Associated with Mortality Changes in the Transitional Seasons." *Professional Geographer* 60: 190–206.

Coleman, J. S. M., and J. C. Rogers. 2003. "Ohio River Valley Winter Moisture Conditions Associated with the Pacific/North American Teleconnection Pattern." *Journal of Climate* 16: 969–981.

———. 2007. "A Synoptic Climatology of the Central United States and Associations with Pacific Teleconnection Pattern Frequency." *Journal of Climate* 20: 3485–3497.

Davis, R. E., and D. R. Walker. 1992. "An Upper-Air Synoptic Climatology of the Western United States." *Journal of Climate* 5: 1449–1467.

El-Kadi, A. A., and P. Smithson. 1992. "Atmospheric Classification and Synoptic Climatology." *Progress in Physical Geography* 16: 432–455.

Finley, J., and M. Raphael. 2007. "The Relationship between El Niño and the Duration and Frequency of Santa Ana Winds in Southern California." *Professional Geographer* 59: 184–192.

Gulev, S. K., O. Zolina, and S. Grigoriev. 2001. "Extratropical Cyclone Variability in the Northern Hemisphere Winter from the NCEP/NCAR Reanalysis Data." *Climate Dynamics* 17: 795–809.

Gutiérrez, J. M., et al. 2005. "Analysis and Downscaling Multi-Model Seasonal Forecasts in Peru Using Self-Organizing Maps." *Tellus* 57: 435–447.

Hewiston, B., and R. Crane. 2002. "Self-Organizing Maps: Applications to Synoptic Climatology." *Climate Research* 22: 13–26.

Joliffe, I. T. 1986. *Principal Component Analysis.* New York: Springer-Verlag.

Kalkstein, L. S. 1991. "A New Approach to Evaluate the Impact of Climate on Human Mortality." *Environmental Health Perspectives* 96: 145–150.

Kalkstein, L. S., and J. S. Greene. 1997. "An Evaluation of Climate/Mortality Relationship in Large U.S. Cities and the Possible Impacts of a Climate Change." *Environmental Health Perspectives* 105: 84–93.

Kalkstein, L. S., G. Tan, and J. A. Skindlov. 1987. "An Evaluation of Objective Clustering Procedures for Use in Synoptic Climatological Classification." *Journal of Climate and Applied Meteorology* 26: 717–730.

Kalnay, E., et al. 1996. "The NCEP/NCAR 40-Year Reanalysis Project." *Bulletin of the American Metrological Society* 77: 437–471.

Kassomenos, P. A., A. Gryparis, and K. Katsouyanni. 2007. "On the Association between Daily Mortality and Air Mass Types in Athens, Greece during Winter and Summer." *International Journal of Biometeorology* 51: 315–322.

Kistler, R., et al. 2001. "The NCEP/NCAR 50-Year

Reanalysis." *Bulletin of the American Meteoro-logical Society* 82: 247–267.

Lamb, H. 1972. "British Isles Weather Types and a Register of Daily Sequence of Circulation Patterns, 1861–1971." *Geophysical Memoir* 116: HMSO, London.

McGregor, G. R., S. Walters, and J. Wordley. 1999. "Daily Hospital Respiratory Admissions and Winter Air Mass Types, Birmingham, UK." *International Journal of Biometeorology* 43: 21–30.

Michelangeli, P.-A., R. Vautard, and B. Legras. 1995. "Weather Regimes: Recurrence and Quasi-Stationarity." *Journal of Atmospheric Science* 52: 1237–1256.

Muller, R. A. 1977. "A Synoptic Climatology for Environmental Baseline Analysis: New Orleans." *Journal of Climate* 16: 20–33.

National Center for Environmental Prediction (NCEP). "Daily Reanalysis Data from 1948–2004." Available at http://www.cdc.noaa.gov.

National Climatic Data Center (NCDC). "Global Surface Summary of the Day." Available at: http://www.ncdc.noaa.gov/oa/ncdc.html. Last accessed 1 December 2007.

Perry, L. B., C. E. Konrad, and T. W. Schmidlin. 2007. "Antecedent Air Trajectories Associated with Northwest Flow Snowfall in the Southern Appalachians." *Weather and Forecasting* 22: 334–352.

Rogers, J. C., and J. S. M. Coleman. 2004. "Ohio Winter Precipitation and Stream Flow Associations to Pacific Atmospheric and Oceanic Teleconnection Patterns." *Ohio Journal of Science* 104: 51–59.

Reusch, D. B., R. B. Alley, and B. C. Hewitson. 2007. "North Atlantic Climate Variability from a Self-Organizing Map Perspective."

Journal of Geophysical Research 112, doi:10.1029/2006JD007460.

Rohli, R. V., and A. J. Vega. 2008. *Climatology.* Sudbury, Mass.: Jones and Bartlett.

Schwartz, M. D. 1991. "An Integrated Approach to Air-Mass Classification in the North Central United States." *Professional Geographer* 43: 77–91.

Serrano, A., et al. 1999. "Monthly Modes of Variation of Precipitation over the Iberian Peninsula." *Journal of Climate* 12: 2894–2919.

Schoof, J. T., and S. C. Pryor. 2003. "Evaluation of the NCEP-NCAR Reanalysis in Terms of Synoptic-Scale Phenomena: A Case Study from the Midwestern USA." *International Journal of Climatology* 23: 1725–1741.

———. 2006. "An Evaluation of Two GCMs: Simulation of North American Teleconnection Indices and Synoptic Phenomena." *International Journal of Climatology* 26: 267–282.

Sheridan, S. C. 2002. "The Redevelopment of a Weather-Type Classification Scheme for North America." *International Journal of Climatology* 22: 51–68.

———. 2003. "North American Weather-Type Frequency and Teleconnection Indices." *International Journal of Climatology* 23: 27–45.

Stone, R. C. 1989. "Weather Types at Brisbane, Queensland: An Example of the Use of Principal Components and Cluster Analysis." *International Journal of Climatology* 9: 3–32.

Straus, D. M., and F. Molteni. 2004. "Circulation Regimes and SST Forcing: Results from Large GCM Ensembles." *Journal of Climate* 17: 1641–1656.

Yarnal, B. 1993. *Synoptic Climatology in Environmental Analysis: A Primer.* London: Belhaven Press.

19. Overview: *Climate Hazards*

D. A. R. KRISTOVICH AND K. E. KUNKEL

Importance of Hazard Responses to Climate Variability Impacts

Atmospheric hazards represent the deadliest hazards in the USA as a whole and specifically in the Midwestern USA (table 19.1, Riebsame et al., 1986). Besides the direct impacts and costs of extreme and severe weather, there are also secondary issues associated with factors such as loss of mobility. Although, as discussed herein, there are tremendous challenges to accurately quantifying these inherently rare events, these events represent one of the primary mechanisms by which weather and climate impact socioeconomic processes.

Some key impacts of climate change and variability are due to increases in the frequency and/or intensity of severe and extreme events (Parry et al. 2007), which have important impacts on the Midwestern USA and the rest of the country. For example, Pielke (2001) estimated that from 1955 to 1999, the U.S. damages due to hurricanes, floods, and tornadoes averaged over $200 million per week in 1997 dollars. Recently updated figures for 1955–2006 indicate total flood, hurricane, and tornado damages of $600 trillion in 2006 dollars. Because of the societal importance

of extreme or hazardous weather conditions, there is an increasing interest in making seasonal forecasts of these phenomena (Murphy et al. 2001). In the Midwestern USA, extreme events can affect a wide range of scales, from individual cities affected by severe thunderstorms or tornadoes (see chapter 22), to numerous communities within lake-effect snowbelts (see chapter 21), to large regions of the Midwest affected by synoptic-scale weather systems generating severe weather, heavy

Table 19.1. Presidential disaster declarations 1953–2006

Hazard	National	Region 5	Region 7
Flooding	689	91	68
Flood and tornadoes	272	68	32
Hurricanes and tropical storms	209	0	0
Snow	119	14	8
Severe ice storm	1	0	0
Levee/dam failure	3	0	0
Earthquake	23	0	0
Volcanic eruption	4	0	0
Fire	36	0	0
Landslide	2	1	0
Drought	10	0	0
Total	1657	206	133

Source: Data from http://www.peripresdecusa.org/mainframe.htm.
Note: Not all causes of disaster declarations are shown in the table, so the sum of the individual hazards will not equal the total.
Region 5 FEMA includes Indiana, Illinois, Michigan, Minnesota, Ohio, and Wisconsin.
Region 7 FEMA includes Iowa, Kansas, Missouri, and Nebraska.

rain, snow, and freezing rain (see chapter 20; e.g., Han et al. 2007).

Parry et al. (2007) pointed out that one of the most important impacts of climate variability and change is how it could be manifested by changes in the frequency, severity, or duration of short-term hazardous weather conditions. If changes in climate conditions gave rise to significant changes in extreme weather events, for example, the changes would have greater impacts on socioeconomic or ecological responses than shifts in mean conditions or decreases in the variations of these conditions (e.g., Meehl et al. 2000). Indeed, concern about changes in extreme events in response to climate variability and change has led to a great deal of research on past and future changes in heavy precipitation (e.g., Griffiths and Bradley 2007; Benestad 2006; Kunkel, Wescott, and Kristovich 2002; Kunkel, Andsager, and Easterling 1999; and several chapters in this volume), droughts (e.g., Sheffield and Wood 2008; Burke, Brown, and Christidis 2006), extreme temperatures (e.g., Griffiths and Bradley 2007; Lynn, Healy, and Druyan 2007; Portis et al. 2006; Colombo, Etkin, and Karney 1999), winds (e.g., van den Brink, Können, and Opsteegh 2004; Pryor and Barthelmie 2003), and other atmospheric and surface characteristics.

Approaches to Investigating Hazard Responses to Climate Variability

A formidable difficulty in assessing the influence of climate variability and change on hazardous weather is that many of the intense weather systems occur on small scales that are not well represented in coupled Atmosphere-Ocean General Circulation Models

(AOGCMs; Parry et al. 2007). For example, severe thunderstorms are mesoscale phenomena, often confined to only a few tens of kilometers size scale. Lake-effect snowstorms are additionally difficult because they are mesoscale systems generally confined to shallow near-surface convective boundary layers (e.g., Chang and Braham 1991). Such small-scale phenomena are often not well simulated in AOGCMs, but as demonstrated by table 19.1 are major reasons for presidential disaster declarations. Even intense rain and snowstorms associated with larger cyclonic weather systems often occur in narrow bands associated with mesoscale processes (e.g., Han et al. 2007, Market and Moore 1998, Blier and Wakimoto 1995, Rauber, Ramamurthy, and Tokey 1994). Types of hazardous weather that frequently occur over much larger regions, such as droughts and extreme temperatures, also may have mesoscale aspects such as interactions with the local land surface.

A number of approaches have been employed to examine the mesoscale responses to climate variability and change. A common approach is to isolate the synoptic-scale atmospheric conditions that are associated with severe weather, such as the approach taken for intense lake-effect snowstorms by Kunkel, Wescott, and Kristovich (2002). Quite a number of approaches have been applied to downscaling large-scale atmospheric patterns, which are thought to be better represented by AOGCMs, to scales appropriate for hazardous weather (e.g., to name a very few, Coulibaly, Dibike, and Anctil 2005; Rebora et al. 2006; Friederichs and Hense 2007). Regional climate modeling represents a relatively recent advance in numerical modeling approaches, whereby higher-resolution models are embedded within large-scale conditions predicted by AOGCMs or other large-scale datasets (e.g., Giorgi 1990). This approach has been applied to a wide range of atmospheric and hydrologic processes (e.g., to name a very few, Xue et al. 2007; Liang et al. 2007; Bielli and Laprise 2006; van den Hurk et al. 2005), but has generally not focused on the climate hazards described herein due in part to the resolution issues described above.

In addition to studies of variability and trends in hazardous weather conditions for predicted future climate scenarios, examination of how such hazardous weather responded to past climate variations has similar difficulties. The following chapters use several observational methods to examine several types of hazardous weather systems and their responses to climate variations and trends. Chapter 20 gives information on the frequency and trends in a range of hazardous weather conditions throughout the Midwestern USA, determined through compilation of these statistics at surface weather sites throughout the region. Chapter 22 describes how an urban area influences the climatology of convective severe storms through comparison of surface, sounding, and radar observations near to, and far from, the Twin Cities, Minnesota, area. Chapter 21 examines previous studies examining lake-effect snowstorms using a combination of approaches—examination of statistical information from a wide region of surface sites and comparison of sites within and outside of lake-effect snowbelt regions.

Summary of Findings

The following three chapters give information on a wide variety of hazardous weather

systems, their relationships to local surface characteristics, and trends in their occurrence and severity in the historical record.

Kunkel and Changnon (chapter 20) use surface observations throughout the Midwestern USA to examine spatial variations and long-term trends in the frequency of occurrence of several types of hazardous weather systems. The chapter presents a comprehensive synthesis of historical observations of convective systems (thunderstorms, hail storms, tornadoes), heavy rain and snow, and freezing rain and sleet. Where appropriate (thunderstorms, tornadoes, freezing rain and sleet), diurnal and annual variations are also summarized.

Blumenfeld (chapter 22) seeks to determine the impact of the Twin Cities, Minnesota, region on the occurrence of severe convective storms through a detailed examination of a number of observational datasets, noting observational biases and using radar observations. In general, the author notes that these observations do not support the postulate that large urban areas suppress severe storms.

Kristovich (chapter 21) reviews the state of knowledge about climatological impacts and trends of lake influences throughout the Great Lakes region. Most of the research in the scientific literature has focused on the portion of the year when the air tends to be colder than the underlying lake surface (cold-season, primarily fall and winter). The overall impacts of the Great Lakes on weather conditions at the surface and aloft are given. A summary of a number of studies reveals a general increasing trend in cold-season lake-effect precipitation during the twentieth century, but results of studies on expected future trends give somewhat conflicting results. The chapter also summarizes the much more limited research reported in the scientific literature on time periods when the air tends to be warmer than the underlying lake surface (warm-season, spring and summer). Overall influences of the lakes on surface weather conditions are examined, as well as recent evidence that the frequency of lake breezes may vary over multiple-year time periods.

Identification of Key Research Avenues That Should Be Pursued

The impacts of climate variability and change on hazardous weather phenomena, which often occur on size scales much smaller than are well simulated by AOGC-Ms, are a critical component of climate sciences. The following three chapters utilize several observational approaches to examination of the occurrence and severity of hazardous weather systems and possible relationships of these to climate variability and change. However, they also suggest several areas in which additional work is needed.

A common difficulty in observational climate studies is that variations and changes in the occurrence of hazardous weather can be carefully documented, but the physical processes giving rise to the variations and changes are often difficult to identify. Climatic and shorter-term (multi-year) variations have been identified for several types of hazardous weather, but longer-term variations and trends often remain unexplained. Careful analyses of changes that could influence long-term variations, such as ocean circulations and surface temperatures, land-use changes, and the like offer the opportunity to gain insight into the responsible mechanisms

and improve seasonal forecasts of hazardous weather (e.g., Murphy et al. 2001). These, in combination with ongoing numerical modeling studies, offer the opportunity to more fully understand the small-scale responses to large-scale climate variations.

There is an ongoing need to determine long-term variations and trends in societal and environmental impacts due to changing climate. Examples of this work can be found in chapter 20, Ebi et al. 2006, and elsewhere. However, continued work on a wider range of impacts on humans (finances, health, habits and decisions, etc.) and the environment (ecology, agriculture, hydrology, species changes in habit, etc.) are also needed.

Analyses of the frequency, intensity, or duration of large-scale conditions associated with hazardous weather phenomena give valuable insight into overall trends and variations in their occurrence. However, some hazardous weather systems, such as lake-effect snowstorms, are highly dependent on the organization of mesoscale systems. For example, if the overall cold-season wind direction changes by only a few tens of degrees, large changes in the locations and intensities of lake-effect snowstorms might result. The importance of the organization of mesoscale storm systems is not limited only to topographically forced hazardous weather types. For example, variations and trends in severe thunderstorm occurrence might be related to how often convective storms are organized into mesoscale convective systems in a given location and climate regime. Clearly, additional work is needed across a wide range of time and size scales to fully understand the influences of climate variability and change on hazardous weather in the Midwestern USA.

ACKNOWLEDGMENTS

The authors greatly appreciate review comments from Dr. James Angel, Dr. Nancy Westcott, and Prof. Sara Pryor. This synthesis chapter was funded by the Illinois State Water Survey. Any opinions, findings, and conclusions or recommendations expressed in this material are those of the authors and do not necessarily reflect the views of the Illinois State Water Survey.

REFERENCES

Benestad, R. E. 2006. "Can We Expect More Extreme Precipitation on the Monthly Time Scale?" *Journal of Climate* 19: 630–637.

Bielli, S., and R. Laprise. 2006. "A Methodology for the Regional-Scale-Decomposed Atmospheric Water Budget: Application to a Simulation of the Canadian Regional Climate Model Nested by NCEP-NCAR Reanalyses over North America." *Monthly Weather Review* 134: 854–873.

Blier, W., and R. M. Wakimoto. 1995. "Observations of the Early Evolution of an Explosive Oceanic Cyclone during ERICA IOP 5, Part 1: Synoptic Overview and Mesoscale Frontal Structure." *Monthly Weather Review* 123: 1288–1310.

Burke, E. J., S. J. Brown, and N. Christidis. 2006. "Modeling the Recent Evolution of Global Drought and Projections for the Twenty-First Century with the Hadley Centre Climate Model." *Journal of Hydrometeorology* 7: 1113–1125.

Chang, S. S., and R. R. Braham. 1991. "Observational Study of a Convective Internal Boundary-Layer over Lake Michigan." *Journal of the Atmospheric Sciences* 48: 2265–2279.

Colombo, A. F., D. Etkin, and B. W. Karney. 1999. "Climate Variability and the Frequency of Extreme Temperature Events for Nine Sites across Canada: Implications for Power Usage." *Journal of Climate* 12: 2490–2502.

Coulibaly, P., Y. B. Dibike, and F. Anctil. 2005. "Downscaling Precipitation and Temperature with Temporal Neural Networks." *Journal of Hydrometeorology* 6: 483–496.

Ebi, K. L., et al. 2006. "Climate Change and Human Health Impacts in the United States: An Update on the Results of the US National Assessment." *Environmental Health Perspectives* 114: 1318–1324.

Friederichs, P., and A. Hense. 2007. "Statistical Downscaling of Extreme Precipitation Events Using Censored Quantile Regression." *Monthly Weather Review* 135: 2365–2378.

Giorgi, F. 1990. "Simulation of Regional Climate Using a Limited Area Model Nested in a General Circulation Model." *Journal of Climate* 3: 941–963.

Griffiths, M. L., and R. S. Bradley. 2007. "Variations of Twentieth-Century Temperature and Precipitation Extreme Indicators in the Northeast United States." *Journal of Climate* 20: 5401–5417.

Han, M., et al. 2007. "Mesoscale Dynamics of the Trowal and Warm-Frontal Regions of Two Continental Winter Cyclones." *Monthly Weather Review* 135: 1647–1670.

Kunkel, K. E., K. Andsager, and D. R. Easterling. 1999. "Long-Term Trends in Extreme Precipitation Events over the Conterminous United States and Canada." *Journal of Climate* 12: 2515–2527.

Kunkel, K. E., N. E. Wescott, and D. A. R. Kristovich. 2002. "Assessment of Potential Effects of Climate Change on Heavy Lake-Effect Snowstorms near Lake Erie." *Journal of Great Lakes Research* 28: 521–536.

Liang, X. Z., et al. 2007. "Regional Climate Model Simulation of U.S.-Mexico Summer Precipitation Using the Optimal Ensemble of Two Cumulus Parameterizations." *Journal of Climate* 20: 5201–5207.

Lynn, B. H., R. Healy, and L. M. Druyan. 2007. "An Analysis of the Potential for Extreme Temperature Change Based on Observations and Model Simulations." *Journal of Climate* 20: 1539–1554.

Market, P. S., and J. T. Moore. 1998. "Mesoscale Evolution of a Continental Occluded Cyclone." *Monthly Weather Review* 126: 1793–1811.

Meehl, G. A., et al. 2000. "An Introduction to Trends in Extreme Weather and Climate Events: Observations, Socioeconomic

Impacts, Terrestrial Ecological Impacts, and Model Projections." *Bulletin of the American Meteorological Society* 81: 413–416.

Murphy, S. J., et al. 2001. "Seasonal Forecasting for Climate Hazards: Prospects and Responses." *Natural Hazards* 23: 171–196.

Parry, M. L., et al., eds. 2007. *Climate Change 2007: Impacts, Adaptation and Vulnerability.* Contribution of Working Group II to the Fourth Assessment Report of the Intergovernmental Panel on Climate Change. Cambridge: Cambridge University Press.

Pielke, R. 2001. *The Extreme Weather Sourcebook, 2001 Edition.* Boulder, Colo.: University Corporation for Atmospheric Research.

Portis, D. H., et al. 2006. "Low-Frequency Variability and Evolution of North American Cold Air Outbreaks." *Monthly Weather Review* 134: 579–597.

Pryor, S. C., and R. J. Barthelmie. 2003. "Long-Term Trends in Near-Surface Flow over the Baltic." *International Journal of Climatology* 23: 271–289.

Rauber, R. M., M. K. Ramamurthy, and A. Tokay. 1994. "Synoptic and Mesoscale Structure of a Severe Freezing Rain Event: The St. Valentine's Day Ice Storm." *Weather and Forecasting* 9: 183–208.

Rebora, N., et al. 2006. "RainFARM: Rainfall Downscaling by a Filtered Autoregressive Model." *Journal of Hydrometeorology* 7: 724–738.

Riebsame, W. E., et al. 1986. "The Social Burden of Weather and Climate Hazards." *Bulletin of the American Meteorological Society* 67: 1378–1388.

Sheffield, J., and E. F. Wood. 2008. "Global Trends and Variability in Soil Moisture and Drought Characteristics, 1950–2000, from Observation-Driven Simulations of the Terrestrial Hydrologic Cycle." *Journal of Climate* 21: 432–458.

Van den Brink, H. W., G. P. Können, and J. D. Opsteegh. 2004. "Statistics of Extreme Synoptic-Scale Wind Speeds in Ensemble Simulations of Current and Future Climate." *Journal of Climate* 17: 4564–4574.

Van den Hurk, B., et al. 2005. "Soil Control on Runoff Response to Climate Change in Regional Climate Model Simulations." *Journal of Climate* 18: 3536–3551.

Xue, Y., et al. 2007. "Assessment of Dynamic Downscaling of the Continental U.S. Regional Climate Using the Eta/SSiB Regional Climate Model." *Journal of Climate* 20: 4172–4193.

20. Severe Storms in the Midwestern USA

K. E. KUNKEL AND S. A. CHANGNON

Introduction

The Midwestern USA has a continental location far from the moderating effects of oceans; hence, wide extremes of both temperature and precipitation occur over days, weeks, months, and years. Summers are typically hot and humid, and winters are cold and often snowy. Severe storms occur in every season, and are an integral part of the Midwest climate. Winter can bring huge snowstorms, damaging ice storms, or both. Warmer months, typically March–October, have convective storms, including thunderstorms and lightning, flood-producing rainstorms, hail, and deadly tornadoes. All seasons experience damaging high winds.

This chapter presents analyses of severe storms that occur in the Midwestern USA based on data collected over the past fifty to one hundred years. Severe storms are defined here as precipitation-producing systems that cause human death or injury and damage to property, crops, and the environment. As described in chapter 1 of this volume, major business activities in the Midwest are highly climate sensitive. Hail, heavy rains, and high winds can decide the fate of the Midwest's crops. The Midwestern USA also serves as the nation's center for air and ground transportation, and weather extremes influence each form of transportation—commercial airlines, barges, trains, and trucks. Severe weather causes shipment delays, a major problem for manufacturers, and storms can indirectly cause profit losses for many businesses. Severe storms have impacts on human health and safety, including an average of 190 deaths due to tornadoes, lightning, winter storms, and floods each year, while property losses average $2.5 billion per year. Climate also shapes key environmental conditions. Record floods in 1993 breached many levees and renewed the growth of natural plants in floodplains. Various storm conditions, such as freezing rain and hail, are detrimental to growth of plants in natural, residential, and commercial landscapes.

Thunderstorms and Lightning

THUNDERSTORMS

Storm frequencies are higher in the south, where warmer temperatures, a longer warm season, and closer proximity to the Gulf of Mexico moisture source lead to more frequent conditions of atmospheric instability.

225

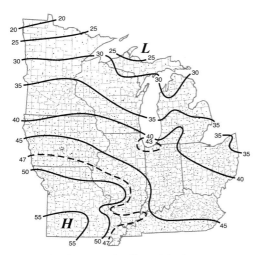

Figure 20.1. Annual average number of days with thunderstorms, 1901–1998. Dots indicate observing station locations. Figure reproduced with permission of the American Meteorological Society from Changnon (2001a).

Storm frequencies are also higher in regions where cyclonic activity is more frequent, and where topography, such as that of southern Missouri, accentuates uplift (figure 20.1). The average annual number of days with thunderstorms, based on data from 1901 to 1998, is fifty-five or more in southwestern Missouri and forty-seven or more in the hills of southern Illinois, reflecting local topographic effects. The Great Lakes influence thunderstorms, particularly in summer and fall. These large water bodies serve as a moisture source for many passing air masses, adding to the moisture available to support cloud and storm development. However, lake water temperatures in summer are often much lower than those of passing air masses. This stabilizes the lower atmosphere and reduces thunderstorm development or minimizes storm activity in passing storm systems downwind of the lakes.

Major urban areas produce sizable effects on the atmosphere (see discussion in chapter 22 of this volume). Under certain circumstances, these effects are sufficient to create thunderstorms and increase rainfall (Changnon 1978). Thunderstorm observations at large Midwestern cities reveal enhanced storm activity. For example, climatological studies of historical weather records in and around the Chicago metropolitan area discerned 10–20% increases in thunderstorm frequencies in and just east of the city. Isolines of constant average storm frequencies are shifted eastward at Chicago and St. Louis. Figure 20.2 shows the temporal distribution of major damaging thunderstorm events and associated losses during 1949–2003. Both distributions exhibited relatively low values during the early period (1949–1973), followed by an upward trend over time to peak in 1994–1998.

LIGHTNING

The annual average number of lightning flashes over a square mile in the Midwest varies from fifteen in southern Missouri to less than one flash per year in northern Minnesota and Upper Michigan (Changnon and Kunkel 2006). Increased lightning occurs downwind of Midwestern urban areas (Westcott 1995), and increased hail and rain occur at major Midwestern urban areas (Huff and Changnon 1973) due to elevated temperatures contributing to increased instability.

Lightning is dangerous, killing an average of 81 people annually in the Midwest, causing more deaths than any other form of severe storms, including tornadoes and floods. Lightning also injures an average of 158 persons each year and kills many farm animals in the Midwest. Lightning causes a variety of damages, including forest and building fires. It also causes local and large-scale outages in

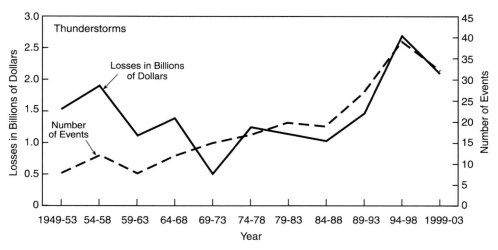

Figure 20.2. Temporal distribution of thunderstorm events in the Midwest creating more than $1 million in property losses, 1949–2003. The dashed line represents the number of events per five-year period, and the solid line losses in billions of dollars. Figure reproduced with permission of the American Meteorological Society from Changnon (2001a).

a power grid, and damages communication systems and electrical systems inside structures, including computers. Yet, lightning also produces benefits. It is an essential element in maintaining the earth-atmosphere electrical balance. Lightning also converts gaseous nitrogen in the atmosphere into nitrogen compounds that serve as a fertilizer for Midwestern soils. Lightning played a positive role in the history of the Midwest and is the reason that outstanding soils developed from a tall grass prairie for ten thousand years. Lightning-ignited fires in the region often developed into huge conflagrations that destroyed trees, but enhanced the growth of the prairie grasses that once existed over most of the Midwest (Changnon, Kunkel, and Winstanley 2003).

Only 3% of all thunderstorms in the Midwest produce damaging lightning, most commonly in June–August, when 80% of all such events occur. Damaging lightning occurs most often between 1 and 4 PM and is least prevalent between 10 PM and 9 AM (Changnon 2001b). Most lightning-related

damages in rural areas occur to farm buildings (82%) and rural schools and churches (9%). In cities with a 100,000 population or more, 40% of all damages occur to residences, 23% to commercial structures, and 22% to industrial buildings/facilities.

Hail

Hail occurrences decrease from west to the east across the Midwest (figure 20.3). Even in areas of high hail frequency, the number of hailfalls at a given location during a year often varies greatly because hail falls typically over only a few square miles. In any given year, a farm in western Iowa may experience five to six hailstreaks, whereas an adjacent farm may experience only one to two hailstreaks. These differences occur because the small size of hailstreaks results in a spotty distribution of hail. Hailstone sizes vary widely, with most between 0.2 and 3.0 inches (0.5–7.5 cm) in diameter. Most hailstones are small, and 85% of all hailstones have diameters of 0.5 inch or less.

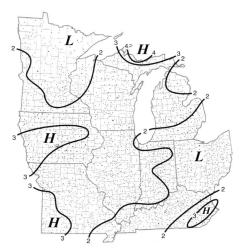

Figure 20.3. Annual average number of days with hail, 1948–2001. Dots indicate observing station locations. Reproduced with permission of the American Meteorological Society from Changnon (2002).

Hail diameters of an inch (2.5 cm) or larger occur in only 5% of all U.S. hailstorms (Flora 1956). Most hail falls in the afternoon, although it can occur at any time of the day. The average duration of hail at a point is five to six minutes, but durations vary from a few seconds up to fifteen minutes. Hail occurs in the warm season when convective activity peaks. Hail season typically begins

in early spring and peaks during summer, with infrequent hail in the fall and no hail in the winter. However, the effect of the relatively warm water in the Great Lakes during fall leads to hailstorms downwind of the lakes. October is the peak month of hail activity in parts of western Michigan, whereas April is the peak month elsewhere in the Midwest (Changnon 1966). Figure 20.4 depicts a time history of hail-caused losses to property and crops in the Midwest during 1949–2003. Property losses from hail were highest in 1959–1968 and lowest during 1969–1978 (Changnon 2004a). Crop hail losses had a somewhat similar distribution over time. Losses were highest during 1954–1968 with a secondary peak in 1994–1998, and like property losses exhibited a minimum during 1969–1978. Inadvertent urban modification of hail also occurs. Studies of hail activity at Chicago and St. Louis found that these cities influenced the atmosphere sufficiently to lead to local increases of 10–30% in the number of hailfalls (Changnon 1978).

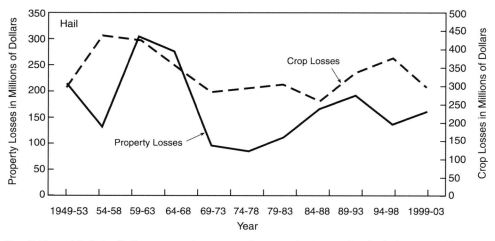

Figure 20.4. Temporal distribution of hail losses to crops and to property over five-year periods, 1949–2003. Reproduced with permission of the American Meteorological Society from Changnon (2002).

Tornadoes and High Winds

Tornadoes in the Midwestern USA occur most frequently in the spring, and during the afternoon and early evening. The peak occurs in April in the southern parts of the Midwest, and in May–June in the northern sections of the Midwest (Changnon and Kunkel 2006). However, tornadoes have occurred in all months of the year, at all hours of the day, and in all parts of the Midwest. Their average duration is twenty-two minutes, although a few long-track storms have lasted two to three hours. The average size of a tornado track in the Midwest is 11.8 miles long and 585 feet wide. A few record-setting tornadoes have had tracks covering 200 miles and widths of a mile, but 45% of all Midwestern tornadoes have tracks only half a mile long (Grazulis 1991). Forward speeds average 35–40 mph. The preferred direction of tornado movement is from southwest to northeast. The average annual number of tornadoes during 1901–1980, expressed as the number per 10,000 square miles (figure 20.5), shows the peak of

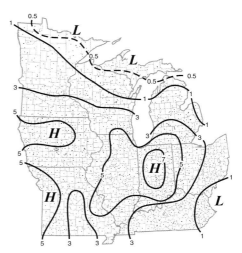

Figure 20.5. Average annual number of tornado occurrences per 10,000 square miles, 1901–1980. Figure reproduced with permission of the American Meteorological Society from Fujita (1981).

activity is in Indiana, Illinois, western Iowa, and Missouri. During 1953–1989, Iowa had 1,105 tornadoes, the most in the Midwest, and also ranked sixth nationally. Missouri had 996 tornadoes and ranked seventh nationally. Illinois had 960 and ranked eighth nationally.

The 1901–2000 temporal distribution of tornadoes that caused deaths in the Midwest (killer tornadoes) (figure 20.6) shows the

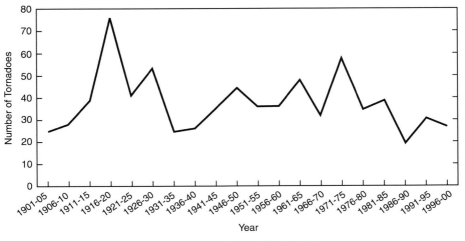

Figure 20.6. Temporal distribution of tornadoes that caused deaths in the Midwestern USA, 1901–2000.

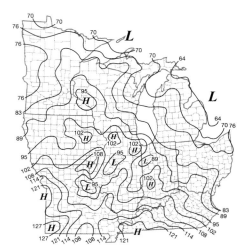

Figure 20.7. Point rainfall amounts (mm) in twenty-four hours expected at least once every five years (adapted from Huff and Angel, 1992).

ty-three storms), 1917 (thirty-two storms), and 1965 (twenty-nine storms). Values in more recent years, 1976–2000, have been relatively low. Low storm values occurred in 1931–1940 and 1986–1990, both periods of extreme droughts, fewer thunderstorms, and thus fewer strong tornadoes. Three years had no killer tornadoes: 1910, 1970, and 1989.

Heavy Rainfall

As discussed in chapters 9 and 13 of this volume, precipitation regimes over the Midwestern USA are strongly influenced by intense events, and exhibit strong spatial gradients across the Midwest. Figure 20.7 presents the pattern for heavy rainfall frequencies for the 24-hour, 5-year return period (i.e., a 24-hour accumulation expected to be reached or surpassed once in a 5-year time period). Values range from a low of 2.5 inches (64 mm) in Michigan up to 5 inches (120 mm) in Missouri. The area of highest rainfall values is oriented west-east from southwestern Missouri eastward across Il-

high inter-annual variability and the decline in deaths associated with improved tornado forecasting and detection capabilities after 1950. The 100-year distribution shows an early peak of seventy-six killer tornadoes in 1916–1920 and then fifty-four others in 1926–1930. Another peak of fifty-eight killer storms occurred during 1971–1975. Years with large tornado counts include 1974 (thir-

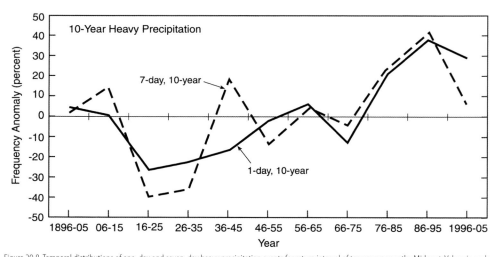

Figure 20.8. Temporal distributions of one-day and seven-day heavy precipitation events for return interval of ten years across the Midwest. Values in each decade are expressed as a % anomalies of long-term averages.

linois and Indiana. This pattern is somewhat reflective of the gradient of annual total precipitation described in chapter 9, which shows highest annual totals in the south of the region. The spatial pattern shown in figure 20.7 is also observed in analysis of precipitation totals for other return periods.

The frequency of heavy rain events exhibits variations with hour of the day, at synoptic timescales, seasonally, and inter-annually. Figure 20.8 shows frequencies for daily and weekly intervals, and for average frequencies of once per ten years (i.e., the maximum one-day or seven-day precipitation accumulation expected to be experienced once per decade). The most prominent feature of this graphic is the generally above-average frequency since 1976, with maximum values during 1986–1995. This tendency is consistent with expectations described in chapter 8 of this volume and temporal trends presented in chapter 9. The elevated frequency of heavy precipitation events during recent decades is one of the more notable trends in severe storm occurrences in the Midwest. This has led to increased flooding of streams (Changnon and Kunkel 1995) and severe losses from some flood events (Changnon 1996; Changnon and Kunkel 1999).

Snowstorms

Research on snowstorm impacts in Illinois and surrounding states found that ≥6 inches (15 cm) of snow in a day or less was the threshold for major damages and high recovery costs (Changnon 1969). The average point frequency of storms meeting this criterion (figure 20.9) reveals a considerable north-south gradient, from less than one storm per year at points in the southern

extremes of the Midwest to more than eight storms annually in Upper Michigan, where lake effects help create numerous snowstorms (Changnon, Changnon, and Karl 2006). Lake effects on storm activity are also obvious by the increased storm frequency downwind of the Great Lakes. Midwest storm snowfall amounts expected at least once in 10 years reveal values range from 10 inches (25 cm) for storms in southern sections, to 20 inches (50 cm) or more once every 10 years in the northern Midwest. The Midwestern USA experienced 856 snowstorms during 1950–2000, an average of 17 storms per year. The average storm size was 41,300 square miles. Most storms were elongated, and 74% of all snowstorms were oriented from the west-southwest toward the east-northeast.

Two other weather conditions also often occur with snowstorms and add to the level of impacts. A key one, freezing rain, occurs with many Midwestern snowstorms (Changnon 1969). Twelve of the eighteen storms in the 1977–1978 record storm winter in Illinois had freezing rain in and adjacent to the snow region. Study of the highly damaging snowstorms revealed 58% also had damaging freezing rain. High winds often occur with Midwestern snowstorms and add to the damages. Ninety percent of major snowstorms in the Midwest have high winds (>30 mph, i.e., >13.5 ms^{-1}) across large areas of the storm. A study of major damaging snowstorms found that 92% also had damaging high winds (Changnon and Changnon 2006). High winds (>30 mph) increase damages that heavy snow alone would create by 25%.

Figure 20.10 shows the temporal distribution of damaging Midwestern snowstorm events during 1949–2000. The fewest dam-

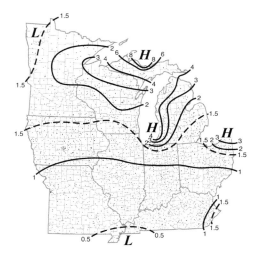

Figure 20.9. Average annual number of snowstorms, defined as events producing 6 inches (15 cm) of snow or more in twenty-four hours or less. A value of 0.5 indicates an average of five storms in ten years. Figure reproduced with permission of the American Meteorological Society from Changnon, Changnon, and Karl (2006).

aging events occurred in the first eight years, 1949–1956. The events had three peaks. The fifty-two-year trend of events is slightly upward over time. The time distribution of storm-produced losses is similar to the number of damaging events; the distribution increased with a major peak of losses in

1993–1996. Increased losses reflect increased storm intensity and society's greater vulnerability to storm damage.

Freezing Rain and Sleet

Freezing rain and sleet have long been conditions of concern because icing and ice storms are very damaging to property and the environment. For example, property losses from U.S. ice storms average $326 million annually (Changnon 2003). Major damages also occur to natural and planted environments. Tree damages from a January 1998 ice storm in Canada and New England were estimated at $2 billion. Additionally, ice and sleet storms also cause deaths and injuries, which are often a result of vehicular accidents (Changnon and Changnon 2004).

The pattern of the average annual number of days with freezing rain in the Midwestern USA (Figure 20.11) reveals that the highest frequencies, five days a year, occur in western Iowa and Minnesota, and along

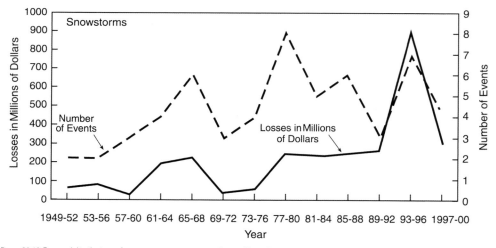

Figure 20.10. Temporal distributions of snowstorms causing property losses of $1 million or more and the amount of losses in the Midwest, 1949–2000. Figure reproduced with permission of Springer from Changnon and Changnon (2006).

a west-east zone from Missouri eastward to Ohio. Incidences are least in southern Kentucky and extreme northern Minnesota. Sleet, which is most frequent in the south-central section of the Midwest, averaging six to seven days a year, occurs on fewer than four days a year in northern Minnesota.

Sizes of ice storm and sleet areas in the Midwest vary widely. The smallest ice storm during 1950–2000 covered only 80 square miles, whereas the largest storm produced damage over 147,050 square miles. Half of the ice storms covered 8,400 square miles or less. Most storm areas are elongated, with length three times longer than width.

The first freezing rain and sleet events of the year occur over most of North America in the fall. The Midwest experiences its earliest freezing rains during November. The months of last occurrences of freezing rains and sleet have a latitudinal distribution. March is the last month for southern Kentucky, while April is the month with the last freezing rain and sleet occurrences over much of the Midwest. Freezing rains

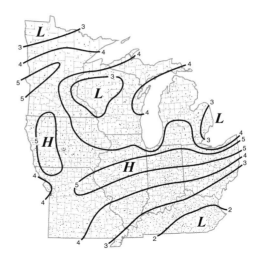

Figure 20.11. Annual average number of days with freezing rain, 1948–2000. Figure reproduced with permission of the American Meteorological Society from Changnon (2003).

occasionally occur in May in northern Minnesota.

The duration of freezing rain events at a point ranges from less than an hour up to 49 hours (Changnon 2003). Most events are short-lived: 38% last an hour or less, 51% last two hours or less, and 64% last three hours

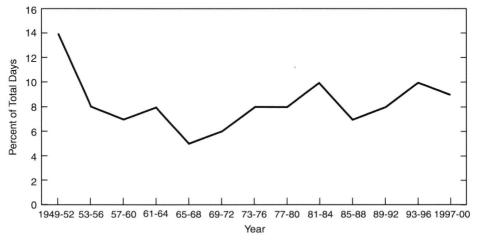

Figure 20.12. Temporal distribution of days with freezing rain in the Midwest. Values in each four-year period are expressed as a percent of the total, 1949–2000. Figure reproduced with permission from Changnon (2004b).

or less. Only 7% of the events exceed ten hours. Sleet, on average, occurs over two to three hours at a point. Freezing rain and sleet are most frequent in nocturnal hours, when temperatures are usually lowest. The greatest number of occurrences of freezing rain and sleet are from 1 AM until 10 AM. Freezing rain and sleet are least frequent in the afternoon, when the daily maximum temperature normally occurs, with lowest hourly values from noon until 6 PM.

Figure 20.12 shows a time history of days with freezing rain in the Midwestern USA during 1949–2000. Shown for each four-year period is the percentage of the total days that exhibited freezing rain. As an example, during 1949–1952 14% of all days exhibited freezing rain. The lowest frequency (11%) occurred in 1965–1972. The distribution suggests a decrease over time, with recent numbers of days less than in the early peak in 1949–1952.

Concluding Remarks

The Midwestern USA experiences a wide variety of severe storms in all seasons, in part as a result of frequent passages of different air masses and unstable atmospheric conditions. The period from March through November is characterized by: thunderstorms, lightning, hail, heavy rains, tornadoes, and high winds, all deadly and often damaging to the environment, crops, and property. The cold season, November–April, exhibits: snowstorms, ice storms, high winds, and sleet storms, also deadly and damaging to the environment, property, and transportation.

Thunderstorms are most frequent in the southern Midwest, occurring on an aver-

age of fifty-five days a year as compared to only twenty-five days or less in the extreme northern parts of Minnesota and Michigan. Thunderstorms and their products (hail, tornadoes, heavy rains, lightning, and high winds) cause an average loss to crops and property in the Midwest totaling $2.8 billion per year. The annual average loss of life due to lightning in the Midwest is eighty-one persons, with flash floods causing forty-five fatalities, and tornadoes averaging twenty-one deaths per year. However, thunderstorms also help the Midwest by providing between 40% (northern Midwest) and 60% (southern sections) of the total annual precipitation.

Severe snowstorms occur most frequently in the Michigan-Minnesota area, with an average of four to eight storms each year. In contrast, less than one snowstorm per year occurs in the southern Midwest. Ice storms are most frequent, averaging four to five days per year, in the central and northwestern Midwest, with less than two ice storm days occurring in the southern areas. The annual average number of deaths caused by winter storms is forty-three, and winter storms produce an average of $318 million in economic losses each year.

Midwestern severe storms exhibit different temporal distributions. The number of damaging thunderstorms, heavy rain events, and snowstorms show statistically significant temporal increases, with peaks in activity since 1990. In contrast, hailstorms, killer tornadoes, and freezing rain frequencies have decreased over time, by 10–20%.

The Great Lakes and the Midwest's large cities (Chicago, St. Louis, and Cleveland) affect the incidence of severe weather. The Great Lakes lead to more thunderstorms, more snowstorms, and record high hail incidences in the fall. The effect of the region's large cities on the atmosphere is the subject of a case study presented in chapter 22, and has led to increases in thunderstorms and hail in and immediately downwind of the cities, but also has led to fewer snowstorms and ice storms within the cities.

REFERENCES

Changnon, S. A. 1966. *Effect of Lake Michigan on Severe Weather*. Great Lakes Research Division Publication 15. Ann Arbor: University of Michigan.

———. 1969. *Climatology of Severe Winter Storms in Illinois*. Illinois State Water Survey Bulletin 53. Champaign, Ill.: Illinois State Water Survey.

———. 1978. "Urban Effects on Severe Local Storms." *Journal of Applied Meteorology* 17: 578–586.

———. 1996. "Defining the Flood: A Chronology of Key Events." In *The Great Flood of 1993*, ed. S. A. Changnon, 3–28. Boulder, Colo.: Westview Press.

———. 2001a. *Thunderstorms across the Nation*. Mahomet, Ill.: Changnon Climatologist.

———. 2001b. "Damaging Thunderstorm Activity in the United States." *Bulletin of the American Meteorological Society* 82: 597–608.

———. 2002. *Climatology of Hail Risk in the U.S.* Mahomet, Ill.: Changnon Climatologist.

———. 2003. "Characteristics of Ice Storms in the U.S." *Journal of Applied Meteorology* 42: 630–639.

———. 2004a. "Present and Future Economic Impacts of Climate Extremes in the U.S." *Environmental Hazards* 5: 47–50.

———. 2004b. *Climate Atlas: Freezing Rain and Ice Storms*. Mahomet, Ill.: Changnon Climatologist.

Changnon, S. A., and D. Changnon. 2006. "A Spatial and Temporal Analysis of Damaging Snowstorms in the U.S." *Natural Hazards* 37: 373–389.

Changnon, S. A., D. Changnon, and T. R. Karl. 2006. "Temporal and Spatial Characteristics of Snowstorms in the U.S." *Journal of Applied Meteorology and Climatology* 45: 1141–1155.

Changnon, S. A., and J. M. Changnon. 2004. "Major Ice Storms in the U.S., 1949–2000." *Environmental Hazards* 4: 105, 111.

Changnon, S. A., and K. E. Kunkel. 1995. "Climate-Related Fluctuations in Midwestern Floods during 1921–1985." *Journal of Water Resources Planning and Management, ASCE* 121: 326–334.

———. 1999. "Record Flood-Producing Rainstorms of 17–18 July, 1996 in the Chicago Metropolitan Area, Part 1: Synoptic and Mesoscale Features." *Journal of Applied Meteorology* 38: 257–265.

———. 2006. *Severe Storms in the Midwest*. Illinois State Water Survey Informational/Educational Material 2006-06. Champaign, Ill.: Illinois State Water Survey.

Changnon, S. A., K. E. Kunkel, and D. Winstanley. 2003. "Climate Factors That Caused the Unique Tall Grass Prairie in the Central U.S. *Physical Geography* 23: 259–280.

Flora, S. D. 1956. *Hailstorms of the United States*. Norman: University of Oklahoma Press.

Fujita, T. T. 1981. "Tornadoes and Downbursts in the Context of Generalized Planetary Scales." *Journal of Atmospheric Sciences* 38: 1511–1534.

Grazulis, T. P. 1991. *Significant Tornadoes 1880–1989*. St. Johnsbury, Vt.: Environmental Films.

Huff, F. A., and S. Changnon. 1973. "Precipitation Modification by Major Urban Areas." *Bulletin of the American Meteorological Society* 54: 1220–1232.

Huff, F. A., and J. R. Angel. 1992. *Rainfall Frequency Atlas of the Midwest*. Illinois State Water Survey Bulletin 71. Champaign, Ill.: Illinois State Water Survey.

Westcott, N. E. 1995. "Summertime Cloud-to-Ground Lightning Activity around Major Midwestern Urban Areas." *Journal of Applied Meteorology* 34: 1633–1642.

21. Climate Sensitivity of Great Lakes–Generated Weather Systems

D. A. R. KRISTOVICH

Introduction

The Laurentian Great Lakes of North America have important impacts on communities in both the Upper Midwest and Northeastern portions of the United States and southeastern Canada. Approximately 35 million people live within the Great Lakes basin, supporting important economic activities in agriculture, shipping, industry, and tourism. The commercial and sport fishing industry alone contributes $4 billion to the regional economy (GLERL 2007). Weather systems generated by the Great Lakes also have important societal consequences. For example, intense lake-effect snowstorms have considerable negative impacts (e.g., transportation delays, accidents, injuries) as well as positive implications (e.g., winter tourism) for the economy of near-shore communities (e.g., Kunkel, Westcott, and Kristovich 2002).

The Great Lakes can dramatically influence local weather by leading to intense winter weather or summer thunderstorms, or they can affect the regional climate in more subtle long-term ways. Much of the influence of the Great Lakes is due to the large thermal capacity of the lakes relative to the overlying air. The long-term average annual lake temperature variations tend to have a lower magnitude than, and are delayed by about a month from, the air temperature variations (e.g., Eichenlaub 1970; Kristovich 1988; Niziol, Snyder, and Waldstreicher 1995; Miner and Fritsch 1997). During the cold fall and early winter season, when the lakes tend to be warmer than the air, the lakes contribute to warming and moistening of the lower atmosphere as well as a regional decrease in surface pressure. Cold-air outbreaks accompanied by periods of strong upward sensible and latent heat fluxes (Laird and Kristovich 2002) can lead to rapid modification of the atmospheric boundary layer as cold air crosses the lakes (e.g., Kristovich, Laird, and Hjemmfelt 2003). Perhaps the most dramatic realization of these modifications are lake-effect snowstorms (figure 21.1; e.g., Eichenlaub 1970; Jiusto and Kaplan 1972), which in extreme cases can result in single-storm snowfall amounts measured in meters (e.g., Schmidlin and Kosarik 1999). During the late spring and summer, when the lakes tend to be cooler than the air crossing over them, generally downward-directed sensible and latent heat fluxes (Laird and Kristovich 2002) can lead to climatic decreases in air temperature, moisture, and precipitation over and very close to

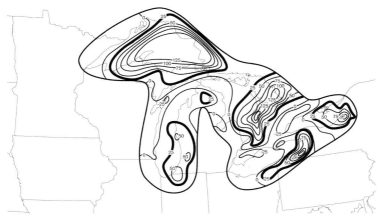

Figure 21.1. Lake-induced anomalies in average precipitation (mm) during winter. Outer curve represents the 80-km lake-effect boundary. Thick curves highlight the 25-mm boundaries. Modified from Scott and Huff (1996).

the lakes (e.g., Wilson 1977; Changnon and Jones 1972; Scott and Huff 1996).

Given the sensitivity of regional weather to the presence of the Great Lakes and their societal importance, it is fitting to examine the climatological variations in Great Lakes–generated mesoscale weather systems and their seasonal impacts. This chapter gives a review of the overall influences of the Great Lakes on local and regional atmospheric conditions, short-term mesoscale circulation patterns generated by the lakes, and climatological trends and variations in these mesoscale influences. The chapter is divided into two sections reflecting the different processes involved in lake influences on the atmosphere: (1) cold-season, when the lakes are typically warmer than the overlying air, and (2) warm-season, when the lakes are typically colder.

Cold-Season Weather Systems

IMPACTS OF THE GREAT LAKES ON COLD-SEASON CLIMATE: LONG-TERM

Discussion of the influences of the Great Lakes on weather conditions near their shores dates back at least to the nineteenth century in the meteorological scientific literature. For example, Hazen (1893) examined the influence of Lake Michigan on surface temperature, winds, and precipitation in the Chicago area. Hazen compared observed weather conditions for days when the winds were directed from the lake (easterly components) to those when the wind was toward the lake (westerly component). Hazen found the most dramatic influences of the lake were during the cold half of the year (October–March), when air temperatures averaged greater than 10°C warmer. In addition, Hazen observed more than a factor-of-two increase in the tendency of rain when the winds were from the lake. The influences of the lake on temperature and precipitation were less clear in the warm months. It must be noted, however, that no distinction was made whether the influences could be attributed directly to lake-generated weather systems.

Numerous studies have been conducted to quantify the influences of the Great Lakes on local surface climate conditions. A full discussion of all relevant research on this topic is beyond the scope of this chapter.

However, a review of the main themes of the research findings is beneficial for understanding Great Lakes influences.

Much of the previous research reported in the scientific literature emphasizes cold-season lake influences on precipitation. For example, research based on surface precipitation observations, radar reflectivity, and isotopic analysis has shown increases of 30% to more than 100% in wintertime total precipitation (or snow) close to the lakes over the amount received outside the lake-effect snowbelts (e.g. Eichenlaub 1970; Braham and Dungey 1984; Jiusto and Kaplan 1972; Ellis and Johnson 2004; Norton and Bolsenga 1993; Wilson 1977; Changnon et al. 1979; Burnett et al. 2003). A few studies specifically examined lake-effect rain events, rather than total precipitation or snowfall. Early cold-season lake-effect rain events were examined by Miner and Fritsch (1997), using surface precipitation observations in the eastern Great Lakes region. Such lake-influenced rain systems can be intense, as illustrated by a flash flood event southeast of Lake Erie (Nicosia et al. 1999).

One difficulty in examination of the cold-season climatic influences of the Great Lakes, and trends in these influences, is separating lake-effect precipitation from non-lake precipitation. A common method used for this analysis is to compare precipitation observations within several tens of kilometers from the lake shores (within the lake-effect "snowbelt") with those further away. The distances used are somewhat arbitrary. For example, Scott and Huff (1996) used a lake-effect/non-lake-effect boundary at 80 km from the shorelines of each of the lakes. Braham and Dungey (1984) defined an area within about 30–40 km from the shorelines as lake-effect,

and compared the areal-average precipitation to that in regions more than 80 km inland. Part of the difficulty in defining areas is that it is unclear how far inland the influences of the lakes extend. While precipitation observations taken over long time periods generally show large snowfall amounts close to the lake shores, it is not unusual to see lake-effect cloud and snow bands extending to distances over 80 km. Indeed, even if the precipitation does not extend inland more than a few tens of kilometers, the influences of atmospheric modification by the lakes (e.g., mesoscale wind circulations, temperature and humidity increases, etc.) can extend inland great distances (e.g., Sousounis and Fritsch 1994; Niziol, Snyder, and Waldstreicher 1995; Ballentine et al. 1998; Rodriguez, Kristovich, and Hjelmfelt 2007). Further research on the inland extent of the lake influences is needed.

In one of the most comprehensive studies of the influence of the Great Lakes on regional long-term climate conditions, Scott and Huff (1996) estimated how six climate variables near the lakes differed from those if the lakes were not present. In order to determine regional atmospheric conditions without the lake influences, spatial patterns in observed climatic variables were determined excluding observations within 80 km of any shoreline. Then, the interpolated non-lake conditions were subtracted from the observed variables within the 80-km limit to estimate the lake influences. The analyses were based on 1951–1980 observations in the United States and Canada, and were averaged by three-month meteorological seasons. It was found that in the winter months, the greatest increases in precipitation were east of Lake Superior (figure 21.1, increases of greater than 100%), with smaller increases to the east (about a 50% in-

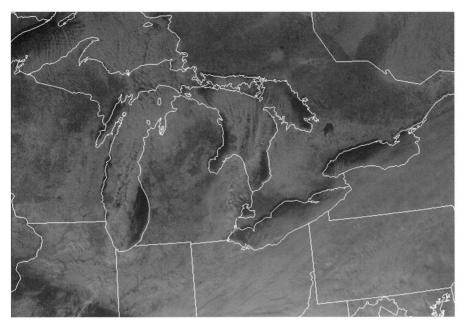

Figure 21.2. Visible satellite image of the Great Lakes region on 5 December 2007, 1815 UTC. Lake-effect and lake-enhanced clouds are seen over and downwind of all of the Great Lakes. Satellite imagery obtained from the Research Applications Program at the National Center for Atmospheric Research (http://www.rap.ucar.edu/weather/satellite) and the National Weather Service Aviation Digital Data Service (http://adds.aviationweather.gov).

crease downwind of Lakes Ontario and Erie) and south (35% southeast of Lake Michigan). In addition, there were important increases in minimum and maximum temperature, cloud cover, and water vapor pressure, but little change in average wind speed.

Influences of the Great Lakes on the wintertime atmosphere above the surface are not as well documented. On the large scale, the groundbreaking study by Petterssen and Calabrese (1959) showed that the large-scale atmosphere is altered by the presence of the lakes. During a six-day time period, they found that heating from the aggregate of all of the lakes resulted in generally lower pressure in the region, and concomitant changes in surface flow fields. These flow fields, in turn, affected precipitation distributions in the region. More recently, Angel and Isard (1997) found that the Great Lakes tended to increase the movement and rate of intensifi-

cation of cyclones propagating through the region during the cold season. In a series of papers, Sousounis and his colleagues quantified the influences of all of the Great Lakes on the region-scale circulation patterns (Sousounis and Shirer 1992; Sousounis and Fritsch 1994; Sousounis 1997; 1998; Weiss and Sousounis 1999; Sousounis and Mann 2000; Mann, Wagenmaker, and Sousounis 2002; Grover and Sousounis 2002; Chuang and Sousounis 2003). These studies describe how atmospheric circulations are significantly altered by heating from each of the Great Lakes and how the individual-lake influences combine into lake-aggregate changes. In addition, the studies examined how the lake-aggregate changes influence lake-effect precipitation intensity and location and detailed the physical processes causing these changes. This research not only highlighted lowering of surface atmospheric pressure, as pointed out by Petterssen and Calabrese (1959), but also showed how it caused increases in pressure and altered flows at mid-levels of the troposphere. The combination of surface and mid-level flows generated by the lakes, in turn, altered how the influences of the lake spread downwind. It was found that the influences of the Great Lakes can extend hundreds of kilometers or more downstream from the lakes, influencing climatic conditions in far-removed locations. Indeed, cloud bands observed in satellite imagery often extend long distances over downwind land areas (figure 21.2).

IMPACTS OF THE GREAT LAKES ON COLD-SEASON CLIMATE: SHORT-TERM AND LOCAL-SCALE

While large-scale atmospheric systems provide conditions needed for the development of lake-effect snowstorms (e.g., cold air, limited atmospheric stability, strong winds), the organization of the over-lake convection into different types of mesoscale structures determines snowfall location, intensity, and duration. Observed mesoscale convective structures can be categorized into three general types: widespread convection (consisting of rolls, cells, or random convection); banded convection parallel to the long axis of the lakes; and mesoscale vortices (Hjelmfelt 1990; Kristovich and Steve 1995; Niziol, Snyder, and Waldstreicher 1995; Laird, Kristovich, and Walsh 2002; Laird, Walsh, and Kristovich 2003; Laird and Kristovich 2004). Widespread convection (figure 21.3) often occurs when low-level winds are at large angles to the long axis of the lake (e.g., Kelly 1982; 1984; Kristovich 1993; Kristovich et al. 1999; Kristovich, Laird, and Hjelmfelt 2003;

Cooper et al. 2000; Liu and Moore 2004) or when excessive low-level directional wind shear prevents organization into a single band of convection (Niziol 1987). Widespread convection generally results in lighter precipitation, but over a more extensive area. Banded convection along the long axis of the lake (figure 21.3), often called shore-parallel or shoreline bands, typically occur when the large-scale winds are light or oriented along the long axis of the lake (e.g., Hjelmfelt 1990; Schoenberger 1984; Passarelli and Braham 1981; Grim, Laird, and Kristovich 2004). While these bands can vary greatly in the intensity of precipitation, depending on a large number of atmospheric and surface conditions, they can result in some of the most severe winter weather in the Great Lakes region (e.g., Niziol, Snyder, and Waldstreicher 1995). Mesoscale vortices generally occur when large-scale wind speeds are light and the lake surfaces are much warmer than the overlying air (e.g., Forbes and Merritt 1984; Laird 1999; Laird, Miller, and Kristovich 2001; Rodriguez, Kristovich, and Hjelmfelt 2007). Smaller-scale vortices are often observed along shore-parallel bands (e.g., Grim, Laird, and Kristovich 2004), but these are generally of much smaller scale than mesoscale vortices and may be caused by different processes. Mesoscale vortices generally produce only light snow, but some examples of intense vortices have been reported (Laird, Miller, and Kristovich 2001).

While much of what is known about the mesoscale structure of lake-effect snow bands is based on limited case studies or numerical modeling efforts, some attempts have been made to quantify how frequently these three mesoscale types occur. For example, Kelly (1986) examined the types of convective structures and accompanying snowfall for Lake Michigan using surface and satellite observations taken over a two-winter time period. He determined that widespread convection (called wind-parallel bands in this paper) occurred on more than half of the lake-effect cases, with shore-parallel bands occurring next most frequently. Kristovich and Steve (1995) and Rodriguez, Kristovich, and Hjelmfelt (2007) used daytime visible satellite imagery to determine the frequencies of the three types of lake-effect cloud bands over all of the Great Lakes based on different five-year time periods. In general, they found that widespread (wind-parallel bands) convection occurred most frequently over all of the lakes, but the relative frequency of shore-parallel bands increased from west to east. This increased tendency for shore-parallel bands over Lakes Erie and Ontario was thought to be largely due to typical wind directions during cold-air outbreaks over the region being more closely aligned with the long axes of these lakes. It should

be noted that while widespread lake-effect convection is more common, shore-parallel bands can often be much more intense (e.g., Laird and Kristovich 2004). Since the type of snow band that forms can often be associated with subtle changes in low-level wind directions, this has interesting implications for potential lake-effect responses to climatic variations in regional flow fields in the Great Lakes region.

TRENDS IN COLD-SEASON
LAKE-GENERATED PRECIPITATION

Lake-effect snowstorms develop under atmospheric conditions that allow boundary layer convection, clouds, and snow to develop. For a significant lake-effect snow event to develop, there must be adequate heating from the surface and limited low-level static stability, an over-lake fetch large enough for snow to develop, moderate wind speeds, and limited low-level wind shear (e.g., Rothrock 1969; Niziol 1987; Niziol, Snyder, and Wald-

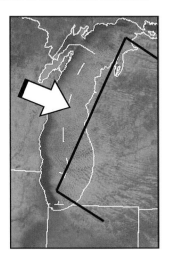
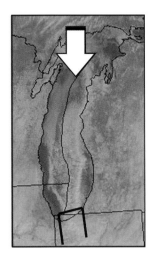

Figure 21.3. Examples of common types of lake-effect cloud patterns over Lake Michigan: (left) widespread convection on 13 January 1998 and (right) shore-parallel bands on 25 January 2000. Arrows represent the overall low-level synoptic wind flows. Satellite imagery obtained from the Research Applications Program at the National Center for Atmospheric Research (http://www.rap.ucar.edu/weather/satellite) and the National Weather Service Aviation Digital Data Service (http://adds.aviationweather.gov).

streicher 1995). A commonly cited stability threshold is that the difference between the lake surface and 850 hPa air temperature exceed 13°C, the approximate dry adiabatic lapse rate. Various minimum fetch distances have been cited; one of the most commonly cited is a minimum fetch of 80 km found by Rothrock (1969). Given these specific conditions, and the considerable inter-annual variability in lake-effect snows, it might be expected that these lake-generated systems are highly sensitive to regional climatic variations and trends.

Over the centuries, climatic and surface variations gave rise to long-term trends in lake-effect storms. Using oxygen-isotope analyses of Carbon 14–dated sediment cores from lakes in northern lower Michigan, Henne (2006) found that the Lake Michigan snowbelt likely developed between 8,500 and 5,500 calendar years before present, as the Laurentide ice sheet retreated and became less of a controlling factor on the region's climate and lake levels rose. After about 5,500 calendar years before present, a regional increase in snow was found, but the proportion due to lake-effect snows did not significantly change.

Increased interest in climate variability and change during the twentieth century, as well as questions about future climate, has led several authors to examine the lake-effect precipitation response to climate (figure 21.4). Most studies found increases in lake-effect snowfall during portions of the twentieth century based on surface observations (Ellis and Johnson 2004; Burnett et al. 2003; Braham and Dungey 1984), oxygen isotope analyses (Burnett et al. 2003), or identification of prevailing synoptic types (Ellis and Leathers 1996). Grover and Sou-

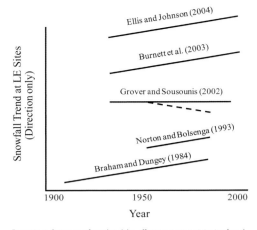

Figure 21.4. Summary of trends in lake-effect snow or precipitation found by some recent published studies. Lengths of the lines correspond to the approximate years over which the trends were derived. The tilts of the lines represent the direction of trends, but not the magnitude. See text for important additional information.

sounis (2002), using a synoptic climatology approach, found that for fall months, the differences in average precipitation from the 1950s to 1980s decreased in both number of lake-effect days and intensity. For a longer time period, comparing average precipitation from 1935–1965 to 1966–1995 showed a complicated picture, with decreases in lake-effect days, but increases in intensity. In either case, Grover and Sousounis (2002) found that changes in lake-effect occurrence had only a minor impact on the regional increase in Great Lakes fall precipitation.

Abrupt changes in lake-effect snowfall found by Braham and Dungey (1984) suggest that lake-effect storms might be quite sensitive to short-term changes in the regional climate. They examined seasonal snowfall observations at sites surrounding Lake Michigan over the period 1909 to 1981. For most sites outside of the Lake Michigan snowbelts (i.e., west of Lake Michigan or far inland over central lower Michigan), seasonal snowfall increased by a small amount, mainly during the 1970s. However, for sites close to the eastern or southern shores of the lake, very large increases in snowfall (more than a factor of three from the 1930s to the 1980s at Muskegon, Michigan) were observed. They argued that the increase in lake-effect precipitation was likely caused by the decrease in winter temperatures during that time period, leading to increased lake-effect snow and enhancements of snow from synoptic systems. This finding is consistent with other studies that identified much greater changes in lake-effect precipitation than regional precipitation (e.g., Ellis and Johnson 2004; Burnett et al. 2003; Norton and Bolsenga 1993). Taken together, these findings indicate that lake-effect precipitation varies over inter-annual and inter-decadal time periods and thus might be particularly sensitive to climate variability and change. Burnett et al. (2003) found that increases in snowfall downwind of the eastern Great Lakes were best explained by increases in lake surface temperatures and decreases in pack ice cover on the lakes. Based on this, they argued that "areas downwind of the Great Lakes may experience increased lake-effect snowfall for the foreseeable future." Kunkel, Westcott, and Kristovich (2002) identified surface and synoptic conditions favorable for heavy lake-effect snowstorms from the historical record. They then examined how the frequency of occurrence of favorable conditions changed in climate predictions by the HadCM2 and CGCM1 models. They also found that during the time period 2070–2099, these simulations showed decreases in the occurrence of favorable conditions by 50% and 90%, respectively. Since most of that decrease was the result of fewer days

with favorable surface air temperatures, they argued that lake-effect snow would be replaced by lake-effect rain in those scenarios.

Since lake effects are the result of interactions between lake and land surface processes, synoptic weather systems, and mesoscale circulations generated by the lake, a full understanding of the impacts of the lakes on regional climate is highly complicated. Spatio-temporal variations in pack ice cover and associated variations in surface heat and moisture fluxes (e.g., Burnett et al. 2003; Gerbush, Kristovich, and Laird 2007), surface temperature (e.g., Kristovich and Laird 1998), and wind fields can have large impacts on the lake-effect response to climate variability and change. For example, Gerbush, Kristovich, and Laird (2007) found that decreases in sensible and latent heat fluxes behave differently in response to changes in pack ice cover. Indeed, examples of lake-effect clouds and precipitation have been observed over Great Lakes that are more than 90% ice covered (see figure 21.5 and Cordeira and Laird 2008).

Warm-Season Weather Systems

IMPACTS OF THE GREAT LAKES ON WARM-SEASON CLIMATE

During the spring and summer months, the Great Lakes tend to have surface temperatures that are lower than the regional air temperatures, particularly during daytime hours. The lower lake surface temperatures lead to generally negative sensible and latent heat fluxes (Laird and Kristovich 2002), and thus a cooling and drying of the overlying air. In some cases, this cooling can help

Figure 21.5. False color composite image from two channels measured by the AVHRR satellite. Blue colors on the lakes represent older or thicker ice. Pink colors represent new or thinner ice. Blue-black streaks represent lake-effect clouds above the lakes. AVHRR imagery processed by Drew Pilant, U.S. Environmental Protection Agency. From http://www.geo.mtu.edu/great_lakes/ice/feb4_1996_lake_superior_freeze.html.

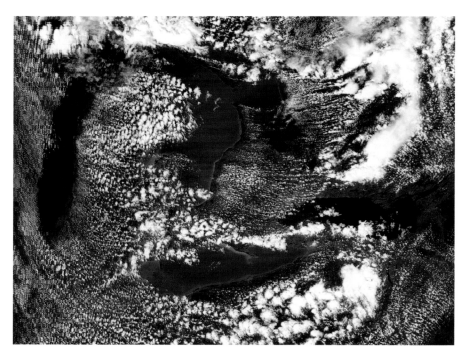

Figure 21.6. Aqua MODIS visible wavelength satellite image of Lakes Michigan, Huron, Erie, and Ontario at 1840 UTC 12 July 2007. Convective clouds, and in some locations thunderstorms, developing over land areas. Note that the clouds and storms are suppressed over the lakes. Obtained from http://rapidfire.sci. gsfc.nasa.gov/.

mitigate heat wave conditions close to the shoreline (as in the 1995 Chicago, Illinois, heat wave [Kunkel et al. 1996]). In addition, mesoscale lake-generated circulations have complex impacts on near-shore and inland precipitation development. During the daytime, slower warming of the air over the lakes than over land areas leads to surface pressure gradients that can drive lake breeze circulations (e.g., Lyons 1972). The resulting subsidence over the lakes, and rising air ahead of the lake breeze front, can lead to clear skies over the lakes and deep clouds around their edges (as seen in the example MODIS satellite imagery, figure 21.6). In some cases, regions of greater risk of severe weather have been identified in common locations of interacting lake breezes from different lakes (King et al. 2003).

Relative to wintertime lake-effect systems, less quantitative research has been conducted on the summertime influence of Great Lakes–generated weather systems and climatic impacts. The Great Lakes tend to have a more subtle impact on near-shore climate conditions in the warm months (e.g., Hazen 1893; Scott and Huff 1996) than the cold months. Scott and Huff (1996) provide a useful overview of the impacts of the Great Lakes on warm-season climatic variables, using thirty years of surface observations. As seen by previous studies (e.g., Wilson 1977; Changnon and Jones 1972; Changnon et al. 1979), precipitation tended to be decreased by up to about 25% over and near each of the Great Lakes (a less distinct minimum is seen over Lake Erie), relative to areas further than 80 km from the shores. Interestingly, the lake region tended to have warmer minimum temperatures when the air often is warmed by the underlying lake, and lower

maximum temperatures when the overlying air is cooled by the lake. Scott and Huff (1996) also found distinct decreases in surface vapor pressure near the lakes, likely due to mesoscale subsiding air and condensation onto the cooler lake surface.

VARIABILITY OF WARM-SEASON WEATHER SYSTEMS

While considerable research has been done on wintertime mesoscale circulations, much less work has been reported on the climatological frequencies and variations of warm-season lake-induced mesoscale wind flows (lake or land breezes) (Lyons 1972). To gain insight into the climatological variations in lake breeze occurrence, Laird et al. (2001) developed a method to use surface hourly meteorological observations near Lake Michigan to identify lake breeze occurrence, and applied it to those taken during the warm months of 1982 to 1996. They found that lake breeze occurrence was evident most frequently on the east side of Lake Michigan (Muskegon), peaking at 40% of the days in July and August. Lake breezes were evident on both shores simultaneously at about half that frequency. Interestingly, the peak in lake breeze occurrence did not coincide with the typical May to June maximum climatological differences between the air and lake temperatures. This was attributed to lower wind speeds during the late summer in the region, which would allow these weak lake breeze circulations to develop. Laird et al. (2001) also reported rather large inter-annual variability in the occurrence of lake breezes, which included multi-year periods of above- or below-average frequencies.

Care must be exercised in developing climatological analyses of mesoscale lake breeze occurrence. It is well-documented in the meteorological literature that local land cover variations (such as urban areas, agricultural practices) and topographical features (i.e., hills, inland water bodies, shoreline variations) have important impacts on lake breezes and their similar cousin, sea breezes.

Needed Future Work

Given the importance of the Great Lakes to the climate in the upper Midwest and northeastern United States and southeastern Canada, it is necessary to develop a better understanding of the variability of mesoscale systems in the cold and warm season and their impacts on other physical processes (e.g., hydrologic processes) and society. While much is known about the influences of the Great Lakes on regional weather and climate systems, there are a number of outstanding climatological issues that should be addressed:

• Climatological analyses of cold-season lake-effect clouds and precipitation are still in their infancy. Little research has been reported on the inter-annual and inter-decadal variations in the mesoscale organization of lake-effect storms. Since the organization of lake-effect convective systems is known to be sensitive to regional climatological conditions (such as low-level wind fields), understanding of their multi-year variations may be critical to predicting future trends.

• Several numerical modeling and observational studies point out that the cold-season influences of the Great Lakes may extend far downwind from the lakes. Mechanisms controlling the inland extent of this influence, as well as implications for storm development far removed from the Great Lakes region, should be explored.

• Analyses of mesoscale climatological warm-season influences of the Great Lakes are very limited in the literature. For example, further work is needed to understand the observed multi-year variations in lake breeze occurrence and how these variations may be manifested in lake-influenced climatic conditions (e.g., temperature, atmospheric water vapor, and precipitation). In addition, research is needed to understand the influences of land use and land cover changes on variations in warm-season mesoscale circulations. While considerable observational and numerical modeling work has been conducted on the influences of the Great Lakes on regional weather and climate conditions during the cold season, virtually nothing is known about the warm-season influences on those scales.

Concluding Remarks

The Great Lakes have important impacts both on communities within their basins and on the meteorological processes and climatology of the region. This chapter outlines the current understanding of the influence of the Great Lakes on the climatological conditions of the region and identifies possible foci for future research.

The most notable influences of the Great Lakes occur during the fall and winter months, when the lakes are climatologically warmer than the overlying air. During the cold-season, the lakes contribute to locally greater precipitation and snowfall, increase near-surface temperatures and humidity, and produce greater regional cloud cover.

Much of the influence from the lakes occurs during periodic lake-effect snow (or rain) storms, which can produce very severe winter weather conditions in near-shore communities. The communities affected by these storms and storm intensity are largely determined by the mesoscale organization of the lake-effect convection. Lake-effect clouds and precipitation tend to organize into several mesoscale types: widespread (often with wind-parallel bands), shore-parallel bands (i.e., bands oriented along the long axis of the lake), and mesoscale vortices. Sometimes subtle changes in environmental conditions, such as lake-air temperature differences and regional wind direction, can give rise to dramatic changes in mesoscale organization.

Most of the observational studies focusing on climatic variations and trends in lake-effect precipitation indicate an upward trend during the twentieth century. Several studies found that this is in sharp contrast to weaker or nonexistent upward trends in non-lake-effect precipitation (such as that due to synoptic weather systems). This, in combination with some studies indicating abrupt shifts in seasonal precipitation amounts within the lake-effect snow belts, suggests that lake-effect precipitation systems might be particularly sensitive to shifts in the regional climate. Limited research has been reported on potential future trends. One study indicated that lake-effect snow may continue to increase in the near future due to warming lake water temperatures. A second study found that some lake-effect snowstorms may eventually be replaced with lake-effect rainstorms during the twenty-first century.

Less quantitative work has been reported on the influences of the Great Lakes on summertime climatological conditions, particularly with regard to mesoscale circulations (i.e., lake and land breezes and associated phenomena). During the warm season, the lakes tend to decrease clouds and precipitation, particularly over and very close to the lakes, and decrease regional temperatures and vapor pressures. Lake breeze circulations, however, may lead to occasional local increases in precipitation and severe storms inland from the lakes. While long-term meso-climatologies of lake breeze occurrence are not available, one study found large multi-year variations in the frequency of lake breezes. Reasons for these variations are not currently understood.

ACKNOWLEDGMENTS
The author greatly appreciates review comments from Prof. Jill Coleman (Ball State University), Dr. Jim Angel, and Dr. Nancy Westcott (Illinois State Water Survey). This summary chapter was funded by the Illinois State Water Survey. Any opinions, findings, conclusions, or recommendations expressed in this material are those of the authors and do not necessarily reflect the views of the Illinois State Water Survey.

REFERENCES
Angel, J. R., and S. A. Isard. 1997. "An Observational Study of the Influence of the Great Lakes on the Speed and Intensity of Passing Cyclones." *Monthly Weather Review* 125: 2228–2237.

Ballentine, R. J., et al. 1998. "Mesoscale Model Simulation of the 4–5 January 1995 Lake-Effect Snowstorm." *Weather and Forecasting* 13: 893–920.

Braham, R. R., and M. J. Dungey. 1984. "Quantitative Estimates of the Effect of Lake Michigan on Snowfall." *Journal of Applied Meteorology* 23: 940–949.

Burnett, A. W., et al. 2003. "Increasing Great Lake–Effect Snowfall during the Twentieth Century: A Regional Response to Global Warming?" *Journal of Climate* 16: 3535–3542.

Changnon, S. A., Jr., and D. M. Jones. 1972. "Review of the Influences of the Great Lakes on Weather." *Water Resources Research* 8: 360–371.

Changnon, S. A., Jr., et al. 1979. *Studies of Urban and Lake Influences on Precipitation in the Chicago Area.* Champaign, Ill.: Illinois State Water Survey.

Chuang, H. Y., and P. J. Sousounis. 2003. "The Impact of the Prevailing Synoptic Situation on the Lake-Aggregate Effect." *Monthly Weather Review* 131: 990–1010.

Cooper, K. A., et al. 2000. "Numerical Simulations of Convective Rolls and Cells in Lake-Effect Snow Bands." *Monthly Weather Review* 128: 3283–3295.

Cordeira, J. M., and N. F. Laird. 2008. "The Influence of Ice Cover on Two Lake-Effect Snow Events over Lake Erie." *Monthly Weather Review* 136: 2747–2763.

Eichenlaub, V. L. 1970. "Lake Effect Snowfall to the Lee of the Great Lakes: Its Role in Michigan." *Bulletin of the American Meteorological Society* 51: 403–412.

Ellis, A. W., and D. J. Leathers. 1996. "A Synoptic Climatological Approach to the Analysis of Lake-Effect Snowfall: Potential Forecasting Applications." *Weather and Forecasting* 11: 216–229.

Ellis, A. W., and J. J. Johnson. 2004. "Hydroclimatic Analysis of Snowfall Trends Associated with the North American Great Lakes." *Journal of Hydrometeorology* 5: 471–486.

Forbes, G. S., and J. H. Merritt. 1984. "Mesoscale Vortices over the Great Lakes in Wintertime." *Monthly Weather Review* 112: 377–381.

Gerbush, M. R., D. A. R. Kristovich, and N. F. Laird. 2007. "Mesoscale Boundary Layer and Heat Flux Variations over Pack Ice–Covered Lake Erie." *Journal of Applied Meteorology and Climatology* 47: 668–682.

Great Lakes Environmental Research Laboratory (GLERL). 2007. *About Our Great Lakes: Great Lakes Basin Facts.* Available at: http://www.glerl.noaa.gov/pr/ourlakes/facts.html. Accessed December 2007.

Grim, J. A., N. F. Laird, and D. A. R. Kristovich. 2004. "Mesoscale Vortices Embedded within a Lake-Effect Shoreline Band." *Monthly Weather Review* 132: 2269–2274.

Grover, E. K., and P. J. Sousounis. 2002. "The Influence of Large-Scale Flow on Fall Precipitation Systems in the Great Lakes Basin." *Journal of Climate* 15: 1943–1956.

Hazen, H. A. 1893. *The Climate of Chicago.* Bulletin No. 10. U.S. Department of Agriculture, Weather Bureau.

Henne, P. D. 2006. "Spatial Variation and Millennial-Scale Change in Lake-Effect Snow Control Vegetational Distributions in the Great Lakes Region." Ph.D. thesis, University of Illinois.

Hjelmfelt, M. 1990. "Numerical Study of the Influence of Environmental Conditions on Lake-Effect Snowstorms on Lake Michigan." *Monthly Weather Review* 118: 138–150.

Jiusto, J. E., and M. L. Kaplan. 1972. "Snowfall from Lake-Effect Storms." *Monthly Weather Review* 100: 62–66.

Kelly, R. D. 1982. "A Single Doppler Radar Study of Horizontal-Roll Convection in a Lake-Effect Snow Storm." *Journal of the Atmospheric Sciences* 39: 1521–1531.

———. 1984. "Horizontal Roll and Boundary-Layer Interrelationships Observed over Lake Michigan." *Journal of the Atmospheric Sciences* 41: 1816–1826.

———. 1986. "Mesoscale Frequencies and Seasonal Snowfalls for Different Types of Lake Michigan Snow Storms." *Journal of Applied Meteorology* 25: 308–312.

King, P. W. S., et al. 2003. "Lake Breezes in Southern Ontario and Their Relation to Tornado Climatology." *Weather and Forecasting* 18: 795–807.

Kristovich, D. A. R. 1988. "Reflectivity Profiles and Core Characteristics along Horizontal Roll Convection in Lake-Effect Snowstorms." M.S. thesis, University of Chicago.

———. 1993. "Mean Circulations of Boundary-Layer Rolls in Lake-Effect Snow Storms." *Boundary-Layer Meteorology* 63: 293–315.

Kristovich, D. A. R., and N. F. Laird. 1998. "Observations of Widespread Lake-Effect Cloudiness: Influences of Lake Surface Temperature and Upwind Conditions." *Weather and Forecasting* 13: 811–821.

Kristovich, D. A. R., N. F. Laird, and M. R. Hjelmfelt. 2003. "Convective Evolution across Lake Michigan during a Widespread Lake-Effect Snow Event." *Monthly Weather Review* 131: 643–655.

Kristovich, D. A. R., and R. A. Steve III. 1995. "A Satellite Study of Cloud-Band Frequencies over the Great Lakes." *Journal of Applied Meteorology* 34: 2083–2090.

Kristovich, D. A. R., et al. 1999. "Transitions in Boundary Layer Meso-γ Convective Structures: An Observational Case Study." *Monthly Weather Review* 127: 2895–2909.

Kunkel, K. E., N. E. Westcott, and D. A. R. Kristovich. 2002. "Effects of Climate Change on Heavy Lake-Effect Snowstorms near Lake Erie." *Journal of Great Lakes Research* 28: 521–536.

Kunkel, K. E., et al. 1996. "The July 1995 Heat Wave in the Midwest: A Climatic Perspective and Critical Weather Factors." *Bulletin of the American Meteorological Society* 77: 1507–1518.

Laird, N. F. 1999. "Observation of Coexisting Mesoscale Lake-Effect Vortices over the Western Great Lakes." *Monthly Weather Review* 127: 1137–1141.

Laird, N. F., and D. A. R. Kristovich. 2002. "Variations of Sensible and Latent Heat Fluxes from a Great Lakes Buoy and Associated Synoptic Weather Patterns." *Journal of Hydrometeorology* 3: 3–12.

———. 2004. "Comparison of Observations with Idealized Model Results for a Method to Resolve Winter Lake-Effect Mesoscale Morphology." *Monthly Weather Review* 132: 1093–1103.

Laird, N. F., D. A. R. Kristovich, and J. E. Walsh. 2002. "Idealized Model Simulations Examining the Mesoscale Structure of Winter Lake-Effect Circulations." *Monthly Weather Review* 131: 206–221.

Laird, N. F., L. J. Miller, and D. A. R. Kristovich. 2001. "Synthetic Dual-Doppler Analysis of a Winter Mesoscale Vortex." *Monthly Weather Review* 129: 312–331.

Laird, N. F., J. E. Walsh, and D. A. R. Kristovich. 2003. "Model Simulations Examining the Relationship of Lake-Effect Morphology to Lake Shape, Wind Direction, and Wind Speed." *Monthly Weather Review* 131: 2101–2111.

Laird, N. F., et al. 2001. "Lake Michigan Lake Breezes: Climatology, Local Forcing, and Synoptic Environment." *Journal of Applied Meteorology* 40: 409–424.

Liu, A. Q., and G. W. K. Moore. 2004. "Lake-Effect Snowstorms over Southern Ontario, Canada, and Their Associated Synoptic-Scale Environment." *Monthly Weather Review* 132: 2595–2609.

Lyons, W. A. 1972. "The Climatology and Prediction of the Chicago Lake Breeze." *Journal of Applied Meteorology* 11: 1259–1270.

Mann, G. E., R. B. Wagenmaker, and P. J. Sousounis. 2002. "The Influence of Multiple Lake Interactions upon Lake-Effect Storms." *Monthly Weather Review* 130: 1510–1530.

Miner, T. J., and J. M. Fritsch. 1997. "Lake-Effect Rain Events." *Monthly Weather Review* 125: 3231–3248.

Nicosia, D. J., et al. 1999. "A Flash Flood from a Lake-Enhanced Rainband." *Weather and Forecasting* 14: 271–288.

Niziol, T. A. 1987. "Operational Forecasting of Lake Effect Snowfall in Western and Central New York." *Weather and Forecasting* 2: 310–321.

Niziol, T. A., W. R. Snyder, and J. S. Waldstreicher. 1995. "Winter Weather Forecasting throughout the Eastern United States, Part 4: Lake Effect Snow." *Weather and Forecasting* 10: 61–77.

Norton, D., and S. Bolsenga. 1993. "Spatiotemporal Trends in Lake Effect and Continental Snowfall in the Laurentian Great Lakes, 1951–1980." *Journal of Climate* 6: 1943–1956.

Passarelli, R. E., and R. R. Braham, Jr. 1981. "The Role of the Winter Land Breeze in the Formation of Great Lake Snow Storms." *Bulletin of the American Meteorological Society* 62: 482–492.

Petterssen, S., and P. A. Calabrese. 1959. "On Some Weather Influences due to Warming of the Air by the Great Lakes in Winter." *Journal of the Atmospheric Sciences* 16: 646–652.

Rodriguez, Y., D. A. R. Kristovich, and M. R.

Hjelmfelt. 2007. "Lake-to-Lake Cloud Bands: Frequencies and Locations." *Monthly Weather Review* 135: 4202–4213.

Rothrock, H. J. 1969. "An Aid in Forecasting Significant Lake Snows." Tech. Memo. WBTM CR-30. National Weather Service, Central Region.

Schmidlin, T. W., and J. Kosarik. 1999. "A Record Ohio Snowfall during 9–14 November 1996." *Bulletin of the American Meteorological Society* 80: 1107–1116.

Schoenberger, L. M. 1984. "Doppler Radar Observation of a Land-Breeze Cold Front." *Monthly Weather Review* 112: 2455–2465.

Scott, R. W., and F. A. Huff. 1996. "Impacts of the Great Lakes on Regional Climate Conditions." *Journal of Great Lakes Research* 22: 845–863.

Sousounis, P. J. 1997. "Lake-Aggregate Mesoscale Disturbances, Part 3: Description of a Mesoscale Aggregate Vortex." *Monthly Weather Review* 125: 1111–1134.

———. 1998. "Lake-Aggregate Mesoscale Disturbances, Part 4: Development of a Mesoscale Aggregate Vortex." *Monthly Weather Review* 126: 3169–3188.

Sousounis, P. J., and J. M. Fritsch. 1994. "Lake-Aggregate Mesoscale Disturbances, Part 2: A Case Study of the Effects on Regional and Synoptic-Scale Weather Systems." *Bulletin of the American Meteorological Society* 75: 1793–1811.

Sousounis, P. J., and G. E. Mann. 2000. "Lake-Aggregate Mesoscale Disturbances, Part 5: Impacts on Lake-Effect Precipitation." *Monthly Weather Review* 128: 728–745.

Sousounis, P. J., and H. N. Shirer. 1992. "Lake Aggregate Mesoscale Disturbances, Part 1: Linear Analysis." *Journal of the Atmospheric Sciences* 49: 80–96.

Weiss, C. C., and P. J. Sousounis. 1999. "A Climatology of Collective Lake Disturbances." *Monthly Weather Review* 127: 565–574.

Wilson, J. W. 1977. "Effect of Lake Ontario on Precipitation." *Monthly Weather Review* 105: 207–214.

22. Severe Weather Hazards in the Twin Cities Metropolitan Area

K. A. BLUMENFELD

Introduction

Severe convective storms are capable of producing tornadoes, large hail, damaging winds, and excessive rainfall, and they pose obvious threats to life and property (see chapter 20 of this volume). The greatest risks from these phenomena arise when they encounter large and/or dense populations, such as in major metropolitan areas. Forecasting severe storms in urban areas has been identified as a research need and priority (Dabberdt et al. 2000), and Edwards and Lemon (2002) have described the potentially catastrophic compound threats from significant severe weather events in urban areas, yet severe storms research has generally proceeded without much consideration of how urban areas influence the climatology of these hazardous weather systems.

One might hypothesize that large urban areas alter normal severe convective storm behavior, given that major cities have been shown to alter the spatial patterns of rainfall (Changnon 1969; Huff and Changnon 1973; Shepherd et al. 2002), cloud-to-ground lightning (Westcott 1995; Stallins, Bentley, and Rose 2006), storm initiation (Dixon and Mote 2003), and mesoscale surface features

(Loose and Bornstein 1977; Bornstein and Leroy 1990). Unfortunately, few authors have examined how severe convective storms interact with large cities. Limited findings from the METROMEX investigation (Changnon 1978; 1981) indicate the frequencies of hail, strong winds, and heavy rains were substantially higher over and downwind from St. Louis, Missouri, than on the upwind (west) side of the city. While that investigation provided the first in-depth analysis of the complex relationships between a major city and convective storms, it also raised numerous questions that have yet to be addressed thoroughly by the research community. For instance, will all cities intensify storms all the time, or are there certain conditions that favor urban storm intensification and others that do not?

This chapter uses severe weather reports, Doppler radar, and tropospheric sounding data to examine two hypotheses regarding the intensity and morphological characteristics of severe convective storms that interact with the urban core of the Minneapolis–St. Paul, Minnesota, metropolitan area (hereinafter Twin Cities Metropolitan Area or TCMA). A null hypothesis, H_o, and an alternative, H_1, will be examined. The null

Figure 22.1. The central TCMA (inner blue circle), the greater TCMA (medium black circle), and ten randomly generated control areas for comparing against central TCMA events. Control areas selected to fall within 100 mi. of MPX for use of MPX proximity soundings.

hypothesis postulates that the character (e.g., frequency, intensity) of severe thunderstorms over the urban core (the central TCMA) is identical to that from suburban and rural control regions, while the alternative postulates that severe thunderstorms, and especially tornadoes, are much less likely in the central TCMA than in other areas. Standard National Weather Service (NWS) guidelines defining severe weather are followed here (Moller 2001). "Severe weather" is any tornado, hailstone with diameter ≥ 2 cm (.75 in.), or wind gust ≥ 26 ms^{-1} (50 kts) associated with deep moist convection. Following from Grazulis (1993) and Doswell, Brooks, and Kay (2005), "significant severe" will denote tornadoes $\geq$F2, hail ≥ 5.1 cm, and wind gusts ≥ 33.5 ms^{-1}. A third threshold, "extreme," denotes F4–F5 tornadoes, hail ≥ 7.6 cm, and winds ≥ 41 ms^{-1}.

The TCMA lies on the northern periphery of the temporally variable "tornado alley" (Brooks, Doswell, and Kay 2003; see also chapter 20 of this volume), within a region of local wind damage maxima

(Doswell, Brooks, and Kay 2005), as well as within a high-frequency corridor of long-lived, extreme convective windstorms, known as derechos (Johns and Hirt 1987; Bentley and Mote 1998; Johns and Evans 2000; Coniglio and Stensrud 2004). In most instances, the "central TCMA" or alternatively the "urban core" is used here to imply an area within a ~16 km (or 10 mi.) radius of the Minneapolis–St. Paul border (figure 22.1). The "greater" or "outer" TCMA denotes the area extending to ~48 km from the central TCMA boundary. Ten randomly generated control regions equal in size to the central TCMA are used to compare severe weather frequencies and magnitudes, and for comparing differences in convective variables between storms in and out of the central TCMA (figure 22.1). The center of each circle is required to be within 100 km of the Chanhassen, Minnesota NWS (MPX) office for the random point generation procedure, and so that MPX soundings can be used to approximate the environments in which the severe storms developed.

Approach

Several measures are used to test both the null hypothesis (storms over the central TCMA are approximately identical to control area storms), and the alternative hypothesis (severe storms are less frequent and less intense in the central TCMA than in control regions). These include:

1. Raw severe weather frequencies in the central TCMA relative to control areas of equal size.

2. Severe weather day frequencies across a 140 km (north-south) by 160 km (east-west) area, centered on the TCMA.

3. Convective available potential energy (CAPE) and lifted index (LI) from MPX soundings for any storms occurring over TCMA or control areas within 2 hours of sounding times.

4. Mode of severe convection (e.g., super-cellular, linear, cluster, "pulse").

5. Behavior of storms over central TCMA (e.g., initiating, intensifying, diminishing, splitting).

The data include severe weather events from the Storm Prediction Center (SPC), tornado tracks from the SVRGIS project (Smith 2006), archived WSR-88D "NEXRAD" data from MPX and La Crosse (ARX), and archived radiosonde observations (RAOBs) (table 22.1). The hypotheses under consideration cannot be evaluated in the standard statistical sense, owing to sampling issues or data irregularities. Data evaluation techniques, the often subjective bases for hypotheses validation decisions, and other data-related issues are discussed below.

Severe Weather Frequencies

During 1950–2006, the central TCMA had greater raw severe weather frequencies than each of the ten control areas, and higher frequencies per unit area (table 22.2). This relationship was consistent for the total number of reports, for each type of event, and for all "significant" severe events. The "extreme" events were too few for statistical comparison, though the central TCMA did experience more high-end tornado and

Table 22.1. Data types and uses in the present investigation

Data type	Source	Description	Purpose
Storm events	Storm Prediction Center (SPC), through *svrplot2* (Hart 1993)	Point-based report data, with lat-lon, date, time, type (e.g., tornado, hail, wind), magnitude; 1950–2005	Locate individual reports, and map extent/severity of outbreaks
Tornado tracks	SVRGIS (Smith 2006)	Tornado touchdown/lift-off points (line geometry in GIS), 1950–2006	Map, analyze tornadoes as tracks rather than points
RAOBs	Univ. of Wyoming; SPC index of storm events[1,2]	Chanhassen (MPX) soundings, 1995–present (Univ. of WY); 2000–present (SPC)	Investigate CAPE (e.g., Blanchard 1998) and Lifted Index for "proximity events"
Radar data	NCDC HAS[3]	Archived level-II and -III data from MPX, ARX, 1995–present	Track changes in storm intensity and organization

1. http://www.weather.uwyo.edu/upperair/

2. http://www.spc.noaa.gov/exper/archive/events/

3. http://hurricane.ncdc.noaa.gov/pls/plhas/has.dsselect

Table 22.2. Comparison of number of tornado, wind, and hail events, partitioned by magnitude, between Central TCMA and 10 control areas. Density of reports are also compared between Central and Greater TCMA.

		Tornado				Wind			Hail		
	ALL	**Total**	**F2+**	**F4/F5**	**Total**	**≥33.5 ms⁻¹**	**≥33.5 m⁻¹**	**Total**	**≥5.1 cm**	**≥7.6 cm**
Central TCMA	270	9	7	2	144	14	2	117	7	2
Control avg	91.9	6.7	2.9	0.6	50	4.6	1	34.4	2.1	0.2
Control min	51	3	0	0	27	0	0	18	0	0
Control max	173	10	4	1	91	11	3	74	5	1
Central TCMA density (rpts/km²)	.34	.011	.009	.003	.182	.018	.003	.148	.009	.003
Outer TCMA density	.113	.005	.002	.0003	.060	.007	.001	.048	.004	.002

In the table header, the Wind subcolumns are labeled ≥33.5 ms⁻¹ and ≥33.5 m⁻¹; rendered as $\geq 33.5\ \text{ms}^{-1}$ and $\geq 33.5\ \text{m}^{-1}$.

hail reports than any control area, and more high-end wind events than all but one control area. At face value, these results would point to rejecting both the null and the alternative hypotheses because the central TCMA storms appear more frequent and somewhat more intense than control-area storms. These results, however, require careful interpretation because severe weather reports have many documented irregularities and many of the data values (e.g., hail size, wind speed) are estimated rather than measured (Doswell, Brooks, and Kay 2005; Trapp et al. 2006). First, the data misrepresent the areal nature of wind and hail events by treating those phenomena as points. Additionally, the locations in the database are rarely true to the actual location of the report, and induce random spatial error, often on the order of 10 km. Also, population changes and erratic reporting procedures over time and space have led to serious inhomogeneities within the database. For instance, from the 1950s through the 1990s, the number of severe weather reports increased exponentially, in response to population growth, better spotter training, and procedural changes within the NWS (Johns and

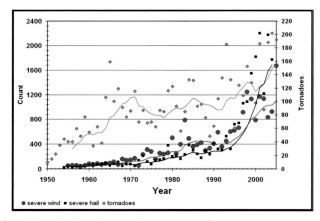

Figure 22.2. Reports of severe hail, wind, and tornadoes over study areas, 1950–2005, fit with seven-year moving averages. Note that tornadoes are scaled to secondary vertical axis.

Evans 2000; Doswell, Brooks, and Kay 2005) (figure 22.2). Also the density of severe weather reports crudely mimics population density across space, which leads to false maxima over cities.

SEVERE WEATHER DAYS

An alternative method for investigating frequencies is to use an "event-day" methodology. This approach reduces the influence of numerous reports clustered around a population center by turning individual analysis cells "on" or "off" depending on whether they contained a severe weather report on a given day. Brooks, Doswell, and Kay (2003), and Doswell, Brooks, and Kay (2005) used the event-day approach to map daily tornado and non-tornadic severe weather probabilities over the United States. They used a grid of 80 km by 80 km cells and Gaussian kernel density smoothers in time and space. Modified versions of their kernel density smoothers are applied to cells 20 km on a side, over a domain of 140 km by 160 km. The modified smoothers and smaller cells reflect the relatively localized nature

of this investigation. Any report in the 1200 UTC to 1159 UTC time frame is assigned to one severe weather day, consistent with SPC operations. For each cell, the value of m on any day is 0 or 1, depending on the presence of severe weather within that cell. Thus, over the study period, for any given day, the average value of m will range from 0 to 1. The following filter is then used:

$$f_n = \sum_{k=1}^{366} \frac{m}{\sqrt{2\pi\sigma_t}} \exp\left[-\frac{1}{2}\left(\frac{n-k}{\sigma_t} \right)^2 \right] \quad (22.1)$$

where f_n is the estimated frequency on the nth day of the year, σ_t is the temporal smoothing parameter, set to $t =10$ days, and k is an indexed value for the day of the year.

This filter, in essence, reduces the influence that the temporal abruptness of a particular event has on probabilistic estimates. For instance, in the study area, severe weather on 23 June is exactly twice as common as severe weather on 24 June. Obviously, this effect cannot be explained by real differences in the expected atmospheric conditions between those two dates (figure 22.3).

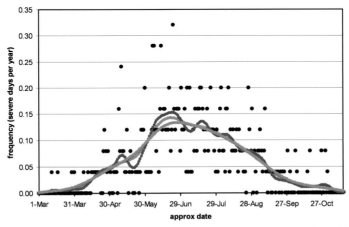

Figure 22.3. Annual cycle of severe weather days over 140-by-160 km area, fit with Gaussian kernel density smoothers with t = 5 (blue), 10 (red), and 15 (green) days.

After smoothing each grid cell series in time, a similar smoothing function is applied spatially:

$$p_{x,y,n} = \sum_{j=1}^{J}\sum_{i=1}^{I} \frac{f_n}{2\pi\sigma_x^2} \exp\left[-\frac{1}{2}\left(\frac{d_{i,j}}{\sigma_x}\right)^2\right] \quad (22.2)$$

where $p_{x,y,n}$ is the probability of severe weather at grid box (x,y) on day n, $d_{i,j}$ is the distance between the location of a grid box (i, j) and the analysis location (x, y), σ_x is the temporal smoothing parameter, set to x=1.5 grid boxes (30 km), and I and J represent the number of boxes in the east-west and north south directions, respectively.

The supposed strength of this approach is that it spreads the higher probabilities of the central TCMA to adjacent cells, and the lower probabilities of outer cells inward. The results, however, suggest that the number of severe weather days in the central TCMA vastly exceeds surrounding areas, and also that the effect is so large that no smoothing function can remedy it (figure 22.4). This result is consistent for all magnitudes of wind and hail days, but not for tornadoes. Population and reporting biases most likely explain

this spatial pattern (Doswell, Brooks, and Kay 2005): historically, the central TCMA has had the highest population density in the area, leaving more people to observe and report severe weather, and leading to the illusion of more severe weather days compared with surrounding areas. In this sense, the central TCMA may represent a "false spatial maximum" in severe weather day frequencies. The results nevertheless indicate that tornadoes and non-tornadic severe storms of all intensities do cross the central TCMA, so at a minimum, we would reject the popular hypothesis that the stronger storms avoid the urban core.

Convective Variables from Proximity Soundings

The alternative hypothesis does not suggest that the central TCMA never experiences severe weather; rather, it suggests that severe storms tend to cross the TCMA with a lower frequency and a lesser intensity than other areas. While the raw frequency and event-day analyses would argue against this notion, irregularities in the data make any true

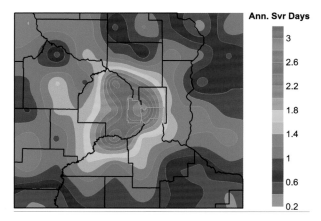

Figure 22.4. Spatial smooth of mean annual severe weather days over 140-by-160 km area. Even after smoothing, central TCMA has expected frequencies higher than surrounding areas.

determination almost impossible. Assuming the alternative hypothesis is accurate, the appropriate question to ask might be "why do some severe weather complexes survive the trip across the central TCMA, while others do not?" To address this question proximity soundings from MPX are used to evaluate the convective potential for each severe weather day in the central TCMA and the ten control regions used for the raw frequency analyses. As described by Brooks, Doswell, and Cooper (1994) and Evans and Doswell (2001), proximity soundings are meant to represent the convective environment of a given severe weather event at a given location. Thus, the event in question should occur within a certain distance and a certain amount of time of a 0000 UTC or 1200 UTC sounding. For instance, to discriminate tornadic from non-tornadic mesocyclones, Brooks, Doswell, and Cooper (1994) required that storms be within 160 km and 1 h of balloon launch, whereas Evans and Doswell (2001) used 167 km and 2 h in their analyses of derecho environments. Two hours and ~100 km are used in this investigation. Each control circle is therefore centered within 100 km of MPX. For qualifying events, it is assumed the severe weather arose in the environment sampled by the MPX sounding.

Convective-available potential energy (CAPE) for a surface-based parcel and the lifted index (LI) are analyzed for any event, 1995–2005, that crossed the central TCMA or any of the control areas within 2 h either side of balloon launch. CAPE is defined as the accumulated buoyant energy of a lifted parcel between the level of free convection (LFC) and the parcel equilibrium level (EL), expressed as:

$$CAPE = g \int_{z_{LFC}}^{z_{EL}} \left(\frac{T_{v_p} - T_{v_e}}{T_{v_e}} \right) dz \qquad (22.3)$$

where T_{v_p} and T_{v_e} are parcel and environmental virtual temperatures (Doswell and Rasmussen 1994), z is height, and g is gravity (Blanchard 1998).

Estimates and conditions vary, but many forecasters recognize 1,000 J kg^{-1} of CAPE as a critical threshold for sustained deep moist convection. The LI (derived using temperatures in °C at 500 hPa) is:

$$T_{v_e} - T_{v_p} \qquad (22.4)$$

LI is typically negative when CAPE is positive. Larger negative values indicate more instability, with values typically between 0 and –10 (°C). Of the seventy-three candidate events, fifty-six (77 %) useful profiles were available. Fourteen of the qualifying soundings had unusually low CAPE values (≤200 J kg^{-1}), but were maintained for supplemental analyses because of the possibility they formed in weakly unstable conditions.

In general, the central TCMA events arose in environments with greater instability than events that specifically did not affect the central TCMA (table 22.3). Inclusion of the low-CAPE soundings only strengthens this relationship, suggesting that perhaps the central TCMA environment is hostile to severe thunderstorms in low-instability conditions. For example, surface processes over the urban core may exert a negative buoyant force upon transient air masses. This finding needs to be reconciled with the earlier results of greater urban severe weather frequencies, which would suggest a nearly opposite effect regarding instability thresholds. On the other hand, perhaps the urban core

Table 22.3. Convective variables from proximity soundings, 1995–2005, from Central TCMA and control regions

Proximity events including:	n	Accepted proximity events CAPE (J kg⁻¹)	LI (°C)	n	All proximity events CAPE (J kg⁻¹)	LI (°C)
ALL	42	1791	-6.0	56	1366	-4.5
Central TCMA	8	1954	-6.7	9	1750	-5.85
Not Cent TCMA	34	1753	-5.9	47	1293	-4.3

does modulate the convective environment, but does so only up to a point, beyond which synoptic and mesoscale forcing for severe convection override local effects. Expanding the study period will certainly help clarify this issue, especially as the number of qualifying central TCMA events increases.

Storm Intensity, Mode, and Evolution

The alternative hypothesis of severe storm behavior in urban areas implies not only that the storms in urban areas are weaker than in surrounding areas, but also that the urban area exerts a destructive force upon storm systems entering the area. As suggested in previous sections, evaluating this hypothesis is difficult. Ascertaining the relative intensities of storms from point-based, population-dependent estimates poses one problem, while understanding the varying environments leading to particular convective scenarios poses yet another. This section uses storm reports and radar data to infer the "convective modes"—the organizational patterns of parent convective systems—and the "behavior" of individual events in the TCMA.

EVENT RANKING

In the period 1995–2005, the TCMA experienced 141 severe weather days, 51 (36%)

of which also affected the central TCMA. The extent and intensity of severe weather during these events ranged from a single, low-end wind gust report to widespread, regional outbreaks of significant tornadoes (T) and extreme winds (W) and hail (H) (figure 22.5). The following identification of intense events (inspired by Doswell et al. [2006]) was applied: The full spatial extent of each contiguous severe weather episode is retrieved, and the distributions of the number of total reports (n_t), maximum magnitude (M), number of significant reports (n_s) and number of extreme reports (n_e) are then calculated and standardized for each type of severe weather (T, W, H). After experimentation and some subjective judgment, the following weighted linear index is used for ranking TCMA severe weather events:

$$I = 2.5(n_t + M + n_s + n_e)_T + (1.5n_t + 3M + 3.5n_s + 3.5n_e)_W + (n_t + 2M + 3n_s + 4n_e)_H \quad (22.5)$$

In essence, the weighting scheme attempts to preserve important aspects of outbreaks while downplaying others. For instance, tornadoes will tend to have better chances of achieving a higher rating on the former F-scale when there are more property "targets." The number of reported tornadoes is, therefore, assumed to contain approximately as much information about the severity of a given outbreak as the reported number of significant or extreme tornadoes. For wind and hail reports, lower weights are assigned to n_t compared to the significant and extreme events. Different weighting schemes were examined, and while these did rearrange some of the rank orders, the top thirty or so events were generally stable.

For all 141 TCMA events, the average value of I was –10.6 with a standard deviation of 50. The negative mean I indicates that the distribution is dominated by relatively small events and is positive-skewed by a small number of large events. For reference, the event mapped in figure 22.5 is the fourth-ranking outbreak, with I=113.6 and a z-score of 2.48. The central TCMA events had an average I of 11.7, suggesting that events in the urban core tend to be larger than the background average. Although this result may be an artifact of the greater number of reports in the central TCMA, it does support the results from the proximity events, which also suggested that the central TCMA somehow selects for higher-end events.

EVENT MORPHOLOGY AND BEHAVIOR OVER THE TCMA

Storm morphology and evolution of the thirty largest TCMA events was analyzed by assigning the spatial tracks of the convective systems associated with major axes of storm reports to one of three classes of severe convection: (1) weakly organized "pulse" storms, (2) multicell linear or clustered storms, or (3) supercell storms. These classes reflect a crude thunderstorm organization-intensity spectrum, similar to the one employed in SKYWARN spotter training programs (see NOAA PA92055: *Advanced Spotters' Field Guide*). In essence, the storm organization or convective mode is a "proxy" for storm intensity: supercell thunderstorms are considered to be most severe, followed by multicell, and lastly pulse storms. Supercell thunderstorms have three known varieties—*high-precipitation* (HP), *low-precipitation* (LP), and *classic*—each with its own preferred severe weather hazards (Moller et al. 1994). Even multicellular storms have many

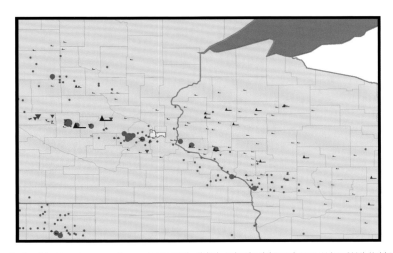

Figure 22.5. Example of spatial pattern of severe weather reports associated with high-end outbreak (event shown is 11 June 2001). Hail (circles) and wind (50-kt flags) have been scaled by magnitude (severe, significant, extreme), while tornado symbols (triangles) are scaled by F-rating (largest shown is F2).

known varieties, including, among others, bow echoes, line-echo-wave-patterns, and mesoscale convective complexes (Fritsch and Forbes 2001), so the classes used here are by no means definitive. In this section, the focus on the thirty largest storms systematically excludes pulse severe storms from the analyses.

Initially, the basic mode of convection is determined from the number, visual geographic dispersion, and type of reports within an event. Multicell and supercell storms exhibit systematic spatial organization, from which the appropriate designation is determined, depending on the nature of the reports. For example, the tornadoes present in figure 22.5 suggest that the event contained at least some supercells, while the progressive wind damage east of the TCMA indicates evolution toward linear, "bowing" structures (Wakimoto 2001).

To supplement the structural "clues" about the storm mode within the storm reports, archived data from the Weather Surveillance Radar 1988 Doppler (WSR-88D) NEXRAD radar network are also used, particularly from Chanhassen, Minnesota (MPX), but occasionally from La Crosse, Wisconsin (ARX). Unfortunately, Chanhassen is quite near to the central TCMA, and

storms over Minneapolis are often only just exiting the radar's blind region, or "cone of silence" (Howard, Gourley, and Maddox 1997). Indeed, tracking a typical storm from west-to-east over the area will generally bring it right through the MPX cone of silence. The opposite problem arises with the ARX radar, as storms over the Twin Cities are often on the edge of its beam domain. The ARX radar nevertheless can "see" some storms that challenge the MPX radar, and together, the two radar sites provide a rather complete picture of convective morphology (Figure 22.6). The MPX/ARX radar data and the report magnitudes are then used to determine whether storms weakened, intensified, or maintained their strength over the central TCMA.

Twenty of the thirty top events originated as supercells, and nine events were supercellular upon entering the central TCMA (table 22.4). Twenty-nine of the events were producing severe weather as they entered the central TCMA, and of those, twelve weakened appreciably upon crossing the urban core. Eleven of the weakening storms were multicellular, but only one supercellular event exhibited weakening characteristics over the central TCMA; the remaining supercells either maintained intensity or

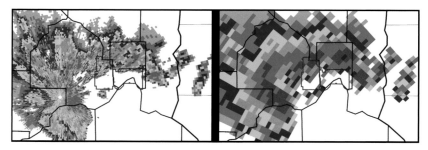

Figure 22.6. Base reflectivity perspective of same storm system entering central TCMA, from MPX radar (left) and ARX radar (right). While both reflectivity scans are at 0.5°, ARX is actually sampling much higher elevations within the storm than MPX.

Table 22.4. Primary storm mode for top 30 TCMA events at initiation and over Central TCMA, and event behavior over Central TCMA

| Primary storm mode | Number events exhibiting storm mode | | Observed behavior of events | | |
	At event initiation	In Central TCMA	Weakening over Central TCMA	Intensifying over Central TCMA	Maintaining intensity over Central TCMA
Multicellular (all linear)	10	21	11 (52.4%)	4 (19.0%)	6 (28.6%)
Supercellular	20	8	1 (12.5%)	3 (37.5%)	4 (50%)

strengthened. These results lend further support to the earlier speculation that the most intense thunderstorms maintain severity over the central TCMA. Indeed, it appears that supercell thunderstorms are relatively unfazed by interaction with the urban core. Results from this section suggest rejecting both the null and the alternative hypotheses.

Concluding Remarks

Conservative treatment of the data supports the following conclusions: (1) all manner of severe weather at all intensity levels have affected the central TCMA; (2) severe weather events in the central TCMA are associated with stronger instability than control area events; (3) "survival rates" of storms entering the central TCMA improve as storm severity—approximated by organizational mode (e.g., multicell, supercell)—increases; and (4) supercell thunderstorms are unlikely to weaken over the central TCMA. These conclusions support conditional rejection of the null hypothesis that severe storms in and out of the central TCMA are approximately identical in frequency and intensity. More importantly, though, the results support categorical rejection of the alternative hypothesis that the central TCMA is less susceptible to severe weather than other parts of the area (table 22.5).

A brief review of local history further supports the notion that tornadic and extreme non-tornadic storms constitute real and serious hazards within the local urban core. Minneapolis and St. Paul proper have been hit by tornadoes more than a dozen

Table 22.5. Hypothesis validation decisions based on data and analyses in this investigation

| Data/analysis parameter | Hypotheses and Decisions | |
	H₀: central TCMA storms equivalent to control storms	H₁: severe storms less common, weaker than control storms; storms break apart or weaken over central TCMA
Severe wx reports	Reject conditionally: Central TCMA frequency maxima, but serious data problems	Reject categorically: evidence of at least some significant and extreme events
Severe wx days	Same as above	Same as above
Proximity events	Reject: Central TCMA storms higher instability than control storms	Reject conditionally: urban core may deter some storms, but *not* strongest ones
Storm mode	Fail to reject; all storm modes observed throughout TCMA	Reject: supercells, common in Central TCMA
Storm evolution	Reject for weaker storms; fail to reject for well-organized storms	Reject: supercellular events maintain or increase intensity over TCMA

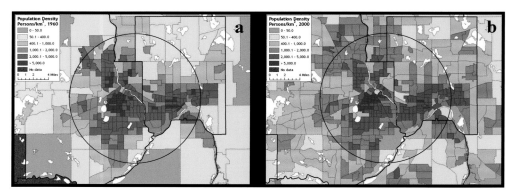

Figure 22.7. Census-tract population density in 1960 (a), and 2000 (b), for the central TCMA (shown as circle) and surrounding areas. The most significant growth, sometimes by a factor of fifty, occurred outside the urban core, though the central TCMA did grow by approximately 25%. Images modified from: J. Schroeder, 2005: "Unpublished maps," University of Minnesota, Minneapolis, Minn.

times since the late nineteenth century (Grazulis 1993). One tornado family in 1952 had a cumulative path length in excess of seventy miles, and crossed the heart of the TCMA. In 1965, a violent tornado outbreak, consisting of multiple F-4 killer tornadoes, struck the western and northern suburbs of Minneapolis during a three-hour period. F-3 tornadoes then struck very near downtown Minneapolis in 1981, and again in 1984. In addition, non-tornadic convective winds damaged trees and structures inside Minneapolis on five separate occasions in 2007 alone, and have caused extensive structural damage throughout the central TCMA over a dozen times since 1983. Many of these systems have been classified as derechos, which are on the "extreme" end of the non-tornadic storm organization/intensity spectrum (Johns and Hirt 1987; Bentley and Mote 1998; Coniglio and Stensrud 2004). Thus, the intense, well-organized storms that are most likely to cause damage also appear least likely to be perturbed by crossing the TCMA.

Two points pertaining to the climatology and history of severe weather in the TCMA warrant discussion. First, while this investigation focused on that which has already happened, all indications are that the central TCMA will continue to experience a wide variety of severe convective weather, as long as the conditions favorable for such systems are present. Future investigations could use Regional Climate Models to investigate potential changes in the annual cycle of favorable severe convective storm parameters identified by Brooks et al. (2007). Second, the entire TCMA has become more heavily and densely populated since the early part of this study period (figure 22.7). The area-wide population is now approximately three million. The at-risk population has thus increased, and a future tornado outbreak comparable to that of 1952 or 1965 would affect substantially more people, perhaps by an order of magnitude. This sort of research, combined with research into the effects of urban surface roughness and the urban energy balance on ambient convective processes, would certainly add to the little we currently know about severe storm interactions with urban areas.

REFERENCES

Bentley, M. L., and T. L. Mote. 1998. "A Climatology of Derecho-Producing Mesoscale Convective Systems in the Central and Eastern United States, 1986–95, Part 1: Temporal and Spatial Distribution." *Bulletin of the American Meteorological Society* 79: 2527–2540.

Blanchard, D. O. 1998. "Assessing the Vertical Distribution of Convective Available Potential Energy." *Weather and Forecasting* 13: 870–877.

Bornstein, R., and M. Leroy. 1990. "Urban Barrier Effects on Convective and Frontal Thunderstorms." In *Extended Abstracts, Fourth Conference on Mesoscale Processes,* 120–121. Boulder, Colo.: American Meteorological Society.

Brooks, H. E., C. A. Doswell, and J. Cooper. 1994. "On the Environments of Tornadic and Nontornadic Mesocyclones." *Weather and Forecasting* 9: 606–618.

Brooks, H. E., C. A. Doswell III, and M. P. Kay. 2003. "Climatological Estimates of Local Daily Tornado Probability for the United States." *Weather and Forecasting* 18: 626–640.

Brooks, H. E., et al. 2007. "Climatological Aspects of Convective Parameters from the NCAR/NCEP Reanalysis." *Atmospheric Research* 83: 294–305.

Changnon, S. A., Jr. 1969. "Recent Studies of Urban Effects on Precipitation in United States." *Bulletin of the American Meteorological Society* 50: 411–419.

———. 1978. "Urban Effects on Severe Local Storms at St. Louis." *Journal of Applied Meteorology* 17: 578–586.

———, ed. 1981. *METROMEX: A Review and Summary.* Meteorological Monographs, no. 40. Boston: American Meteorological Society.

Coniglio, M. C., and D. J. Stensrud. 2004. "Interpreting the Climatology of Derechos." *Weather and Forecasting* 19: 595–605.

Dabberdt, W. F., et al. 2000. "Forecast Issues in the Urban Zone: Report of the 10th Prospectus Development Team of the U.S. Weather Research Program." *Bulletin of the American Meteorological Society* 81: 2047–2064.

Dixon, P. G., and T. L. Mote. 2003. "Patterns and Causes of Atlanta's Urban Heat Island–Initiated Precipitation." *Journal of Applied Meteorology* 42: 1273–1284.

Doswell, C. A., and E. N. Rasmussen. 1994. "The Effect of Neglecting the Virtual Temperature Correction on CAPE Calculations." *Weather and Forecasting* 9: 625–629.

Doswell, C. A., III, H. E. Brooks, and M. P. Kay. 2005. "Climatological Estimates of Daily Local Nontornadic Severe Thunderstorm Probability for the United States." *Weather and Forecasting* 20: 577–595.

Doswell, C. A., III, et al. 2006. "A Simple and Flexible Method for Ranking Severe Weather Events." *Weather and Forecasting* 21: 939–951.

Edwards, R., and L. R. Lemon. 2002. "Proactive or Reactive? The Severe Storm Threat to Large Event Venues." Preprints, *21st Conf. Severe Local Storms.* San Antonio, Tex.: American Meteorological Society.

Evans, J. S., and C. A. Doswell. 2001. "Examination of Derecho Environments using Proximity Soundings." *Weather and Forecasting* 16: 329–342.

Fritsch, J. M., and G. S. Forbes. 2001. "Mesoscale Convective Systems." In *Severe Convective Storms.* Meteorological Monographs, no. 50, ed. C. A. Doswell III, 323–357. Boston: American Meteorological Society.

Grazulis, T. P. 1993. *Significant Tornadoes, 1680–1991.* St. Johnsbury, Vt.: Environmental Films.

Hart, J. A. 1993. "SVRPLOT: A New Method of Accessing and Manipulating the NSSFC Severe Weather Data Base." Preprints, *17th Conference on Severe Local Storms,* 40–41. St. Louis, Mo.: American Meteorological Society.

Howard, K. W., J. J. Gourley, and R. A. Maddox. 1997. "Uncertainties in WSR-88D Measurements and Their Impacts on Monitoring Life Cycles." *Weather and Forecasting* 12: 166–174.

Huff, F. A., and S. A. Changnon Jr. 1973. "Precipitation Modification by Major Urban Areas." *Bulletin of the American Meteorological Society* 54: 1220–1232.

Johns, R. H., and W. D. Hirt. 1987. "Derechos: Widespread Convectively Induced Windstorms." *Weather and Forecasting* 2: 32–49.

Johns, R. H., and J. S. Evans. 2000. Comments on "A Climatology of Derecho-Producing Mesoscale Convective Systems in the Central and Eastern United States, 1986–95, Part 1: Temporal and Spatial Distribution." *Bulletin of the American Meteorological Society* 81: 1049–1054.

Loose, T., and R. Bornstein. 1977. "Observations of Mesoscale Effects on Frontal Movement through an Urban Area." *Monthly Weather Review* 105: 563–571.

Moller, A. R. 2001. "Severe Local Storms Forecasting." In *Severe Convective Storms*. Meteorological Monographs, no. 50, ed. C. A. Doswell III, 433–480. Boston: American Meteorological Society.

Moller, A. R., et al. 1994. "The Operational Recognition of Supercell Thunderstorm Environments and Storm Structures." *Weather and Forecasting* 9: 327–347.

Shepherd, J. M., H. Pierce, and A. J. Negri. 2002. "Rainfall Modification by Major Urban Areas: Observation from Spaceborne Rain Radar on the TRMM Satellite." *Journal of Applied Meteorology* 41: 689–701.

Smith, B.T. 2006. "SVRGIS: Geographic Information System (GIS) Graphical Database of Tornado, Large Hail, and Damaging Wind Reports in the United States (1950–2005)." Preprints, *23rd Conference on Severe Local Storms*. St. Louis, Mo.: American Meteorological Society.

Stallins, J. A., M. Bentley, and L. Rose. 2006. "Cloud-to-Ground Flash Patterns for Atlanta, Georgia (USA) from 1992 to 2003." *Climate Research* 30: 99–112.

Trapp, R. J., et al. 2006. "Buyer Beware: Some Words of Caution on the Use of Severe Wind Reports in Postevent Assessment and Research." *Weather and Forecasting* 21: 408–415.

Wakimoto, R. M. 2001. "Convectively Driven High Wind Events." In *Severe Convective Storms*. Meteorological Monographs, no. 50, ed. C. A. Doswell III, 255–298. Boston: American Meteorological Society.

Westcott, N. 1995. "Summertime Cloud-to-Ground Lightning Activity around Major Midwestern Urban Areas." *Journal of Applied Meteorology* 34: 1633–1642.

23. Where Is Climate Science in the Midwestern USA Going?

E. S. TAKLE AND S. C. PRYOR

Introduction: National and International Research Initiatives

There are multiple ongoing international and national research efforts focused on improving the state-of-the-art with respect to climate change science, increasing confidence in impact analyses, and developing appropriate adaptation and mitigation strategies, many of which are referenced in the Fourth Assessment Reports of the Intergovernmental Panel on Climate Change (IPCC 4AR). An additional key area of increasing activity is improving climate science policy by improved articulation and dissemination of climate science research. Despite near-universal consensus in the climate science community that the recent rise in global temperatures is unequivocal and primarily caused by human modification of the climate system, 56% of Americans "believe there is a lot of disagreement among scientists about whether global warming is even occurring" (Hassol 2008).

In this chapter we:

1. Briefly review a small fraction of the international climate science research efforts coordinated by the World Climate Research Programme (WCRP) (http://www.wmo.ch/pages/prog/wcrp/) and summarize one key climate change initiative that is focused on the United States—the North American Regional Climate Change Assessment Program.

2. Use these national and international research efforts, and the results of analyses presented herein, to identify key research themes within climate science that could or should be pursued with a critical focus on the Midwestern USA.

EXAMPLES OF INTERNATIONAL RESEARCH EFFORTS

The WCRP "identifies gaps in scientific knowledge of climate change and variability; through observations and modeling of the Earth system it enables policy-relevant climate predictions" (http://www.wmo.ch/pages/prog/wcrp/). Among the programs organized by WCRP that have specific relevance for the Midwestern USA are:

• GEWEX (Global Energy and Water Experiment) (http://www.gewex.org/). GEWEX is focused on the hydrologic cycle and the exchange of energy and water at the atmosphere-surface interface. As the first of five continental-scale experiments (CSEs), the GEWEX Continental Scale International Project (GCIP) was established to quantitatively assess the hydrologic cycle and energy fluxes of the Mississippi River basin. GCIP focuses on "understanding the annual, interannual, and spatial variability of hydrology and climate within the Mississippi River basin; the development and evaluation of regional coupled hydrologic/atmospheric models; the development of data assimilation schemes; and the development of accessible, comprehensive databases" (http://www.gewex.org/).

The CSEs have allowed GEWEX hydrological research results to be intercompared across continents and hence different climatic regimes. The Transferability Work Group of the GEWEX Hydrometeorology Panel was developed to build on and extend the Regional Climate Model (RCM) intercomparison projects (MIPs) to "improve understanding and prediction of water and energy cycle processes through systematic intercomparisons of ensembles of regional climate simulations on several continents with coordinated continental-scale observations and analyses" (Takle et al. 2007). One such extension of these intercomparisons is the current North American Regional Climate Change Assessment Program discussed below.

• CLIVAR (Climate Variability and Predictability) (http://www.clivar.org/). This program is focused on climate variability and predictability, with a particular focus on the role of ocean-atmosphere interactions in climate. Specific objectives of CLIVAR include: (1) description of the processes responsible for climate variability and predictability on seasonal, inter-annual, decadal, and centennial

time scales, (2) extension of the paleoclimatic record, (3) extension of the range and accuracy seasonal to inter-annual climate prediction, and (4) improved prediction of the climate system response to increases of radiatively active gases and aerosols and comparison of these predictions to the observed climate record in order to detect the anthropogenic modification of the natural climate signal.

THE NORTH AMERICAN REGIONAL CLIMATE CHANGE ASSESSMENT PROGRAM

The North American Regional Climate Change Assessment Program (NARCCAP) is an international, multi-agency program to produce multiple realizations of future scenario climates at regional scales by use of coupled Atmosphere-Ocean General Circulation Models (AOGCMs) and Regional Climate Models (RCMs) (Mearns et al. 2005). Results from four AOGCMs for contemporary and future scenario climates created for the IPCC AR4 are used as input for six RCMs to produce dynamically downscaled climate information at 50-km grid spacing for most of North America (figure 23.1). In addition, two high-resolution atmospheric Global Climate Models (GCMs) are producing time-slice climate information at the same resolution for comparison with RCM results. Output from this modeling program will be available for assessing the impacts of climate change at regional scales for the United States, Canada, and northern Mexico, and for climate science studies.

The goals of NARCCAP are to (1) quantify the multiple uncertainties of AOGCM-RCM regional projections of future climate,

(2) develop multiple high-resolution regional climate-change scenarios for impact and risk assessments, (3) evaluate regional model performance over North America by nesting RCMs in reanalysis products such as those discussed in chapters 14 and 15 of this volume, (4) understand critical regional climate-change processes, (5) create greater collaboration among U.S., Canadian, and European climate modeling groups, and (6) derive added value from diverse, ongoing regional and global modeling projects. In Phase I, results of RCM simulations driven by reanalysis data ("observed") boundary conditions from 1979–2004 are being used to quantify uncertainty introduced by RCM and tracking uncertainty through impacts models. In Phase IIa, the RCMs are being run with lateral boundary conditions for

1971–2000 produced by the "Climate of the 20th Century" runs of the AOGCMs. This period was chosen to overlap with the period of the reanalysis-driven simulations. In Phase IIb, RCMs will use lateral boundary conditions from AOGCM output from the emission scenario SRES A2 (nominally 2041–2070). Results of Phases IIa and IIb will be available for studying impacts of climate change at regional scales. Output from the regional models and time-slice models will be available in standard data format similar to that used for AOGCM results produced for the IPCC Fourth Assessment Report. Phase I of this program has been completed, and Phase II is well under way as of the middle of 2008. More details are available at the NARCCAP website (http://www.narccap.ucar.edu/).

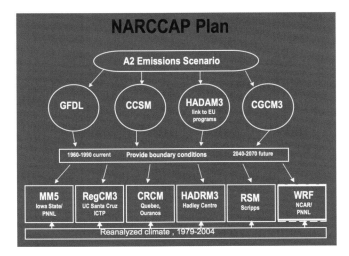

Figure 23.1. Matrix of NARCCAP experiments to use output of four GCM/AOGCMs driven by the A2 SRES emissions scenario as input to six RCMs for producing scenarios of future regional climate. GFDL = Geophysical Fluid Dynamics Laboratory model, CCSM = NCAR model, HadAM3 = Hadley Centre model, CGCM3 = Canadian model, MM5 = NCAR/Penn State model used by Iowa State University, RegCM3 = Regional Climate Model 3 used by UC–Santa Cruz, CRCM = Canadian Regional Climate Model, HadRM3 = regional model of the Hadley Centre, RSM= Regional Spectral Model of the Scripps Oceanographic Institution at UC–San Diego, and WRF is the Weather Research and Forecast Model used by the Pacific Northwest National Laboratory. The NARCCAP plan also includes regional simulations of time slices for the indicated periods by high-resolution GCM run by Lawrence Livermore National Laboratory and the Geophysical Fluid Dynamics Laboratory.

Identification of Key Research Avenues: Climate Science

The last decade has seen tremendous strides in our abilities to quantify and assign causality to changes in the climate system, and national and international efforts such as those described above will continue to make progress in further specifying likely future climate states and reasonable adaptation and mitigation approaches. However, our understanding remains imperfect, and further research is required in order that we can address with confidence the question: how will global climate change be manifest in the Midwestern USA? Some of the science research priorities that are required in order to respond to this question are given below.

RECONCILE DIFFERENCES IN HISTORICAL CLIMATE TRAJECTORIES DERIVED USING DIFFERING DATA SETS

A critical context for development of projected changes in climate regimes requires analysis of historical data to (1) quantify natural variability and possibly (2) detect a climate change signal. This, in turn, is critically reliant on data quality and homogeneity, which remains problematic for some climate parameters, and may generate divergent climate trajectories or very high uncertainty in historical trends of characterization of the past climate (e.g., chapter 15). There is need for further efforts to generate robust data sets and reconciliation of discrepancies between data sets and reanalysis products.

QUANTIFICATION OF THE ROLE OF TELECONNECTIONS IN DICTATING HISTORICAL CLIMATE VARIABILITY AND POSSIBLE FUTURE CLIMATE STATES

As discussed in chapter 1 and multiple chapters herein, inter-annual and intra-annual variations in the intensity and tracking of synoptic-scale systems and several surface parameters (e.g., temperature and precipitation) are explicable, at least in part, by a number of teleconnection indices. Hence, there is a need for improved quantification of these linkages, and description of the physical mechanisms by which they are expressed. These teleconnection linkages also provide a clear mechanism for evaluation and validation of AOGCMs (see chapter 17) and for identification of aspects of the models that require further refinement.

ENHANCEMENT OF OBSERVATIONAL NETWORKS

Regional cooperation should be encouraged to extend the list of routine measurements taken throughout the region to enable more rigorous decision tools to be developed. As just one example, soil moisture is a key factor relating to decision making for several issues listed in table 23.1. Promotion of a region-wide network of soil moisture observations would provide a valuable data resource for seasonal to inter-annual planning for these issues.

Table 23.1. Issues in the Midwest that would benefit from inter-annual to decadal projections of climate variables

Crop & horticulture production	Soil erosion
Conservation practices	Water supplies
Stream flow	Water quality
Beef and pork daily gains	Livestock breeding success
Milk and egg production	Crop and livestock pests and pathogens
Biofuel production projection	Power demand for heating & cooling
Planning for droughts & floods	Agricultural tile drainage systems
Natural ecosystem species distributions	Human health (heat waves, influenza)
Building designs	Recreation opportunities
River navigation	Pavement performance (roads)
Corrosion rates (bridges)	Carbon sequestration/loss by soil
Forest productivity	Shipping limitations (Great Lakes)
Wind power resources	Winter maintenance costs (roads & bridges)
Air quality	

FURTHER EVALUATION AND DEVELOPMENT OF DOWNSCALING TOOLS FOR REGIONAL CLIMATE PROJECTIONS

Regional climate scientists should actively participate in national activities to create new climate-change information at regional scales and develop methods for using and assessing the value of such information for region-specific applications. Methods for extracting higher resolution projections from AOGCMs can be broadly classified into two categories of downscaling: dynamical (application of Regional Climate Models) and statistical (development of empirical transfer functions between aspects of the large-scale climate and local climate variables). These approaches are briefly described below:

• Regional climate modeling is yet in its infancy and will require many years of refinement to reduce the uncertainty introduced to regional climate projections by these models.

Although RCMs resolve terrain and coastlines better than AOGCMs, they are dependent on the quality of lateral boundary information provided by the global models in which they are imbedded (e.g., Pryor, Barthelmie, and Kjellström 2005). They do produce higher spatial variability over complex topographic features, but such detail is not always an assurance of higher-quality simulations. An area where RCMs should be expected to add value to AOGCM output is where high spatial variability results from fine-scale dynamical processes, such as mesoscale circulations. An example is the interaction of mesoscale convergence and divergence with the low-level jet that leads to enhanced summertime precipitation in the western Midwest (see chapters 3 and 8 of this volume). Anderson et al. (2003) showed that some, but not all, RCMs were capable of simulating the nocturnal maximum of precipitation in the western Midwest, which is a unique signature of the fine-scale dynamical processes in this region.

• Statistical downscaling. Statistical downscaling is based on development of transfer functions between descriptors of the large-scale climate and the local variable of interest (see chapters 4 and 6 of this volume). Traditionally a major challenge to statistical downscaling has been underestimation of variability. Two approaches have been proposed to address this issue: inflation, where the simulated variability is increased by multiplication by a specified factor (in some instances using techniques which ensure the correct resultant variability), and randomization, where noise (often derived from the synoptic scale) is added to increase the variability. However, new techniques are being developed that do not require this type of adjustment (Pryor, Schoof, and Barthelmie 2005b), thereby greatly enhancing the range of applications for which this type of approach is applicable. A further major advantage of statistical downscaling is that the associated computational costs are generally modest, allowing for downscaling of multiple AOGCMs (see chapter 6 of this volume).

Dynamical downscaling is theoretically preferable and can be conducted for any location, regardless of the availability of observations of the surface variable. However, even dynamical downscaling is not completely based on first principles, but employs "parameterizations" to represent unresolvable or unknown physical processes relating to convective motions, heterogeneous and non-neutral conditions in the boundary layer, and cloud microscale processes. Empirically (at best) derived constants are used in these parameterizations based on observations in domains (climate regimes) in which they were developed. Transfer-

ability experiments (Takle et al. 2007) are designed to intercompare models on many climate regimes as a means of improving RCM performance on all currently available climates, with a goal of adding confidence in their ability to simulate future (unknown) climates. Such experiments have shown that RCMs may have a "home domain" advantage (Takle et al. 2007); that is, they perform better in the region for which they were developed than in distinctly different climate regimes. This raises the question, if climate changes, will our RCMs be able to simulate the new climate?

Deriving realistic scenarios from empirical tools relies on strong and stationary relationships between predictors and predictands, while RCMs do not require specific training and can respond in physically consistent ways to external conditions not realized in the training period. However, statistical downscaling may be undertaken without requiring additional data such as surface orography and roughness maps, but requires in situ data for transfer function development. Statistical downscaling based on known spatial variability may be a more appropriate method for determining topographically anchored patterns from AOGCM results.

There is a continued need for technique development, refinement, and intercomparison of downscaling methods and results. The Midwestern USA may be an ideal location for such analyses, given it has sufficient uniformity over large regions to allow spatial coherence uninterrupted by topographic features to be evaluated.

An important consideration in using of all climate science products as input to climate change impacts models is the propagation of errors and quantification of uncertainties. Output of a modeling sequence that employs global, regional, and impacts models to assess climate change impacts contains errors from each of these component models. Pan et al. (2001) and Jha et al. (2004) discuss strategies for evaluating such errors as measures of uncertainty in climate change impacts assessments. Jha et al. (2004) outline a modeling strategy that allows at least an estimate of the contributions to total modeling error from each of the component models. For their application, the combined errors introduced by the GCM, the RCM, and the hydrological (streamflow) model were less that the magnitude of change in streamflow attributed to climate change arising from increases in greenhouse gas concentrations. In a combined statistical and dynamical downscaling analysis for wind climates, an analysis of uncertainties in wind climate projections resolved that uncertainties derived from variations in the AOGCM simulations of the climate downscaling predictors had a greater impact on wind climate uncertainties than either stochastic effects from individual AOGCMs or emission scenarios (Pryor, Barthelmie, and Kjellström 2005; Pryor, Schoof, and Barthelmie 2005a; 2006). Such work can only enhance the utility of climate projections and indeed is a cornerstone of the NARCCAP project.

Further work on development of tools to quantify error propagation and uncertainty reduction in climate scenarios should be a focus of future research initiatives. Success in quantifying uncertainty will add credibility and usefulness of future-climate information conveyed to scientists studying impacts of climate change.

Identification of Key Research Avenues: Climate Change Impacts, Mitigation, and Adaptation

So where is climate science going in the Midwestern USA in terms of "end-users"? Societal needs demand high-quality information on future "normals" (or at least some kind of "points of departure") (Livezey et al. 2007). Armed with the results of research such as that presented in this volume that has quantified trends in Midwestern USA climate variables observed over the last century (and more specifically the last thirty years) and projections into the future of these variables (and associated uncertainty) for this region, we have the tools to provide advice on regional climate change. In some cases of regional vulnerability (table 23.1, e.g., pavement performance, milk production, streamflow, crop yield), models already exist that can be used to ingest conditions representing a future climate (see for example chapters 10 and 11 of this volume). In other cases risk assessment models or decision tools need to be developed. What would such advice be used for? And what activities should be undertaken to optimize the ways in which to direct climate science to address these key impacts? An intense dialog is needed between climate-impacts scientists and the climate-modeling community to ensure that the appropriate climate model information is being generated and that this information is being accessed and used properly by impacts scientists.

In the following sections we highlight just a few research avenues that might provide useful foci for impact assessment and mitigation and adaptation activities.

EXPLOITATION OF RENEWABLE ENERGY RESOURCES

Emissions of greenhouse gases (GHGs) covered by the Kyoto Protocol increased by about 70% between 1970 and 2004 (by 24% during 1990–2004) (Barker et al. 2007). In 2005 atmospheric carbon dioxide (CO_2) concentrations reached 379 ppm, with "mean annual growth rates in the 2000–2005 period higher than in the 1990s," and total CO_2-equivalent (CO_2-eq) concentration of all long-lived GHGs reached 455 ppm CO_2-eq (Barker et al. 2007). On a global basis, in 2004, energy supply accounted for about 26% of GHG emissions, with a further 19% from industry, 17% from land-use change and forestry, 14% from agriculture, 13% from transport, 8% from residential, commercial, and service sectors, and 3% from waste (Barker et al. 2007). Working Group III of the IPCC 4AR resolved that if there is "no change in energy policies, the energy mix supplied to run the global economy in the 2025–30 timeframe will essentially remain unchanged, with more than 80% of energy supply based on fossil fuels. On this basis, the projected emissions of energy-related CO_2 in 2030 are 40–110% higher than in 2000" (Barker et al. 2007). As discussed in chapter 1 of this volume, the need to reduce GHG emissions can, in part,

be met by greater exploitation of renewable energy sources. The Midwestern USA is well positioned to play a role in this endeavor. However, there is a need for further research to quantify what the climate commitment we have already undertaken and have yet to undertake means in terms of implications for changing:

• Thermal regimes—for example, changes in frost-free period, growing-season length (chapter 4), plant hardiness (chapter 7), and diurnal temperature range (chapter 6)—and specifically how will these changes impact biodiesel or ethanol production?

• Precipitation regimes—for example, changes in extreme precipitation and precipitation frequency (chapter 9 and 13), surface water balance components (chapter 11), streamflow (chapter 10), and flooding (chapter 13)—and specifically how will these changes impact bio-diesel or ethanol production, and/or the feasibility of solar and/or hydroelectric deployment? Will there be sufficient water supply to meet needs for agriculture and cooling for fossil fuel power plants?

• Wind regimes (chapters 15 and 16), and specifically how may changes in near-surface wind speeds influence the viability of the wind energy industry and/or agricultural activities due to soil erosion, evapo-transpiration, and plant-atmosphere heat and carbon dioxide exchange?

DEVELOPING ENERGY DEMAND PROJECTIONS

As discussed in chapter 1, the climate of the Midwestern USA is such that there is relatively high energy demand for both heating and cooling purposes. This leads to both relatively large regional GHG emissions and also a need for accurate energy demand projections. Climate science has already made a significant contribution to the latter, by demonstrating the need to move away from use of thirty-year normals to trend-based or other referencing methods (Livezey et al. 2007). For instance, the natural gas industry in northern Illinois has replaced the thirty-year (1971–2000) National Weather Service normals for temperature with the mean of the most recent ten years as a "point of departure" for estimating natural gas demands for space heating in the upcoming heating season. Statistics of past trends alone were not sufficient to convince the Illinois Commerce Commission (ICC) to abandon its time-honored reliance on the conventional NWS normals, but when those were combined with projections of climate models of future cold-season conditions in the northern Illinois region, the ICC agreed to the switch to the ten-year base period. However, there is a need for further research to quantify what the climate commitment we have already undertaken and have yet to undertake means in terms of likely future energy demand on a range of time scales from intra-annual to inter-annual as a consequence of climate change (i.e., alteration of the need for electricity due to shifts in the number of heat degree days and cooling degree days, chapter 1) and other societal evolution (Ruth and Lin 2006). Making such projections may require refinement of tools for energy demand modeling in the face of climate change and uncertainty.

ENSURING SECURITY OF FOOD SUPPLY

As discussed in chapter 1, the Midwestern USA plays a critical role in food production. However, competition for cropland with the energy sector and increasing world and national population may lead to increased demand for increased agricultural yields, which may not be possible under a modified climate. Hence there is a need for further research to quantify what the climate commitment we have already undertaken and have yet to undertake means in terms of implications for changing:

- Thermal regimes—for example, changes in frost-free period, growing season length (chapter 4), plant hardiness (chapter 7), and diurnal temperature range (chapter 6)—and specifically how will these changes impact the agricultural sector?
- Precipitation regimes—for example, changes in extreme precipitation and precipitation frequency (chapter 9 and 13), changes in surface water balance components (chapter 11), streamflow (chapter 10), and flooding (chapter 13)—and specifically how will these changes impact the agricultural sector?
- Wind regimes (chapters 15 and 16), and specifically how may changes in near-surface wind speeds influence the viability of agricultural activities due to soil erosion, evapo-transpiration, and plant-atmosphere heat and carbon dioxide exchange?

PROTECTING HUMAN HEALTH AND WELL-BEING

The world's urban population is growing by sixty million people a year, about three times the increase in the rural population (Cohen 2003). The movement of people toward cities has accelerated in the past forty years, and the fraction of the world's population that dwell in urban areas has increased from one-third in 1960 to over 50% in 2008. By 2030, it is expected that over 60% of the global population will live in cities. This increasing urbanization may lead to increased vulnerability to climate extremes and severe weather (De Sherbinin, Schiller, and Pulsipher 2007), and associated enhancement of urban heat island effects, coupled with increased temperatures, may potentially prompt a greater frequency of heat waves and associated mortality and morbidity (Meehl and Tebaldi 2005; Ebi et al. 2006).

Currently no definitive statement can be made regarding the total impact of natural hazards and particularly severe/extreme weather on socioeconomic systems in the Midwest, and development of future projections is even more problematic. However, analyses presented in chapters 19, 20, and 22 emphasize the need for atmospheric scientists to engage in continued research to develop improved understanding of these phenomena. There is a further need for climate scientists to engage in development of hazard management strategies and climate change adaptation measures to minimize the risk posed and reduce the vulnerability of the Midwestern USA to climate variability and change.

There is a need for further research to quantify what climate commitment we have already undertaken and have yet to undertake means in terms of implications for changes in:

• The frequency with which heat waves may be observed in the Midwest, and a key confounding influence on the occurrence and human health impacts of heat waves: the interaction of temperature and atmospheric moisture (see chapter 5).

• Air quality (Leung and Gustafson 2005). Several counties in the state of Indiana currently fail National Ambient Air Quality Standards (Pryor and Spaulding 2009), and changes in synoptic regimes (chapters 14, 17, and 18) may have profound influences in air pollution–related mortality and morbidity in the region.

• The frequency with which climate hazards are observed (chapters 19, 20, 22).

Concluding Remarks

Improving confidence in past climate evolution, developing possible future projections, and providing decision makers with reliable information for planning in climate-sensitive areas present climate scientists with a host of challenging tasks. Nevertheless, as documented in this volume, much knowledge has been gained about the Midwestern USA climate in terms of the degree of variability, predictability, and past and possible future changes. Additionally, tools are being developed to provide ever better climate projections and impact assessments. It is the intent of this collection of papers, and the workshop that led us to this point, to provide a first step in developing region-wide coordination among climate scientists to enhance the state of climate science in the Midwestern USA and better serve the data needs of our region in this era of climate change.

ACKNOWLEDGMENTS

Financial support was supplied by grants from the National Science Foundation (0618823, 0533567, 0618364, and 0647868), and the U.S. Department of Agriculture NRI (20063561516724).

REFERENCES

Anderson, C. J., et al. 2003. "Hydrological Processes in Regional Climate Model Simulations of the Central United States Flood of June–July 1993." *Journal of Hydrometeorology* 4: 584–598.

Barker, T., et al. 2007. "Technical Summary." In *Climate Change 2007: Mitigation.* Contribution of Working Group III to the Fourth Assessment Report of the Intergovernmental Panel on Climate Change, ed. B. Metz, et al., 70. New York: Cambridge University Press.

Cohen, J. 2003. "Human Population: The Next Half Century." *Science* 302: 1172–1175.

De Sherbinin, A., A. Schiller, and A. Pulsipher. 2007. "The Vulnerability of Global Cities to Climate Hazards." *Environment and Urbanization* 19: 39–64.

Ebi, K. L., et al. 2006. "Climate Change and Human Health Impacts in the United States: An Update on the Results of the U.S. National Assessment." *Environmental Health Perspectives* 114: 1318–1324.

Hassol, S. J. 2008. "Improving How Scientists Communicate about Climate Change." *EOS* 89: 106–107.

Jha, M., et al. 2004. "Impacts of Climate Change on Stream Flow in the Upper Mississippi River Basin: A Regional Climate Model Perspective." *Journal of Geophysical Research* 109, doi:10.1029/2003/JD003686.

Leung, L. R., and W. I. Gustafson Jr. 2005. "Potential Regional Climate Change and Implications to U.S. Air Quality." *Geophysical Research Letters* 32, doi:10.1029/2005GL022911.

Livezey, R. E., et al. 2007. "Estimation and Extrapolation of Climate Normals and Climatic Trends." *Journal of Applied Meteorology and Climatology* 46: 1759–1776.

Mearns, L. O., et al. 2005. "NARCCAP, North American Regional Climate Change Assessment Program: A Multiple AOGCM and RCM Climate Scenario Project over North America." In *16th Conference on Climate Variability and Change.* San Diego: American Meteorological Society.

Meehl, G. A., and C. Tebaldi. 2005. "More Intense, More Frequent, and Longer Lasting Heat Waves in the 21st Century." *Science* 305: 994–997.

Pan, Z., et al. 2001. "Evaluation of Uncertainties in Regional Climate Change Simulations." *Journal of Geophysical Research* 106: 17735–17751.

Pryor, S. C., R. J. Barthelmie, and E. Kjellström. 2005. "Analyses of the Potential Climate Change Impact on Wind Energy Resources in Northern Europe Using Output from a Regional Climate Model." *Climate Dynamics* 25: 815–835.

Pryor, S. C., J. T. Schoof, and R. J. Barthelmie. 2005a. "Climate Change Impacts on Wind Speeds and Wind Energy Density in Northern Europe: Results from Empirical Downscaling of Multiple AOGCMs." *Climate Research* 29: 183–198.

———. 2005b. "Empirical Downscaling of Wind Speed Probability Distributions." *Journal of Geophysical Research* 110, doi: 10.1029/2005JD005899.

———. 2006. "Winds of Change? Projections of Near-Surface Winds under Climate Change Scenarios." *Geophysical Research Letters* 33, doi:10.1029/2006GL026000.

Pryor, S. C., and A. M. Spaulding. 2009. "Air Quality in Indiana." In *Indiana's Weather and Climate,* ed. J. Oliver, 57–62. Bloomington: Indiana University Press.

Ruth, M., and A. C. Lin. 2006. "Regional Energy Demand and Adaptations to Climate Change: Methodology and Application to the State of Maryland, U.S.A." *Energy Policy* 34: 2820–2833.

Takle, E. S., et al. 2007. "Transferability Intercomparison: An Opportunity for New Insight on the Global Water Cycle and Energy Budget." *Bulletin of the American Meteorological Society* 88: 375–384.

Further Reading

Achtemei, G. 1967. "Trajectories of Thunderstorm and Accompanying Severe Weather over Midwest on 25 August 1965." *Bulletin of the American Meteorological Society* 48: 498–507.

Alfaro, E. J., A. Gershunov, and D. Cayan. 2006. "Prediction of Summer Maximum and Minimum Temperature over the Central and Western United States: The Roles of Soil Moisture and Sea Surface Temperature." *Journal of Climate* 19: 1407–1421.

Altwicker, E. R., and A. H. Johannes. 1987. "Spatial and Historical Trends in Acidic Deposition: A Graphical Intersite Comparison." *Atmospheric Environment* 21: 129–135.

Anderson, C. J., and R. W. Arritt. 2001. "Mesoscale Convective Systems over the United States during the 1997–98 El Niño." *Monthly Weather Review* 129: 2443–2457.

Anderson, C. J., et al. 2003. "Hydrological Processes in Regional Climate Model Simulations of the Central United States Flood of June–July 1993." *Journal of Hydrometeorology* 4: 584–598.

Anderson, R. C. 1998. "Overview of Midwestern Oak Savanna." *Transactions of the Wisconsin Academy of Sciences Arts and Letters* 86: 1–18.

Andrews, J. F. 1969. "Weather and Circulation of April 1969—a Warm Month Accompanied by Severe Flooding in Upper Midwest and Increased Westerlies." *Monthly Weather Review* 97: 523–526.

Angel, J. R., and F. A. Huff. 1995. "Seasonal Distribution of Heavy Rainfall Events in Midwest." *Journal of Water Resources Planning and Management-ASCE* 121: 110–115.

———. 1997. "Changes in Heavy Rainfall in Midwestern United States." *Journal of Water Resources Planning and Management-ASCE* 123: 246–249.

Angel, J. R., and S. A. Isard. 1998. "The Frequency and Intensity of Great Lake Cyclones." *Journal of Climate* 11: 61–71.

Assel, R. A. 1980. "Maximum Freezing Degree-Days as a Winter Severity Index for the Great Lakes, 1897–1977." *Monthly Weather Review* 108: 1440–1445.

Assel, R. A. 1992. "Great Lakes Winter-Weather 700-hPa PNA Teleconnections." *Monthly Weather Review* 120: 2156–2163.

———. 2005. "Classification of Annual Great Lakes Ice Cycles: Winters of 1973–2002." *Journal of Climate* 18: 4895–4905.

Assel, R. A., K. Cronk, and D. Norton. 2003. "Recent Trends in Laurentian Great Lakes Ice Cover." *Climatic Change* 57: 185–204.

Bae, D. H., and K. P. Georgakakos. 1994. "Climatic Variability of Soil-Water in the American Midwest, Part 1, Hydrologic Modeling." *Journal of Hydrology* 162: 355–377.

Baker, R. G. 1998. "Late Quaternary Environmental Changes in the Midwestern United States." In *Status and Conservation of Midwestern Amphibians,* ed. M. J. Lannoo. Iowa City: University of Iowa Press.

Baldwin, C. K., and U. Lall. 1999. "Seasonality of Streamflow: The Upper Mississippi River." *Water Resources Research* 35: 1143–1154.

Barlage, M. J., et al. 2002. "Impacts of Climate

Change and Land Use Change on Runoff from a Great Lakes Watershed." *Journal of Great Lakes Research* 28: 568–582.

Barnston, A. G., R. E. Livezey, and M. S. Halpert. 1991. "Modulation of Southern Oscillation Northern-Hemisphere Midwinter Climate Relationships by the QBO." *Journal of Climate* 4: 203–217.

Bartlein, P. J., T. Webb, and E. Fleri. 1984. "Holocene Climatic-Change in the Northern Midwest: Pollen-Derived Estimates." *Quaternary Research* 22: 361–374.

Bell, G. D., and J. E. Janowiak. 1995. "Atmospheric Circulation Associated with the Midwest Floods of 1993." *Bulletin of the American Meteorological Society* 76: 681–695.

Bentley, M. L., and T. L. Mote. 1998. "A Climatology of Derecho-Producing Mesoscale Convective Systems in the Central and Eastern U.S., 1986–1995, Part 1, Temporal and Spatial Distribution." *Bulletin American Meteorological Society* 79: 2527–2540.

Berger, C. L., et al. 2002. "A Climatology of Northwest Missouri Snowfall Events: Long-term Trends and Interannual Variability." *Physical Geography* 23: 427–448.

Bernstein, B. C., C. A. Wolff, and F. McDonough. 2007. "An Inferred Climatology of Icing Conditions Aloft, including Supercooled Large Drops, Part 1, Canada and the Continental United States." *Journal of Applied Meteorology and Climatology* 46: 1857–1878.

Bonan, G. B. 2001. "Observational Evidence for Reduction of Daily Maximum Temperature by Croplands in the Midwest United States." *Journal of Climate* 14: 2430–2442.

Booth, R. K., J. E. Kutzbach, S. C. Hotchkiss, and R. A. Bryson. 2006. "A Reanalysis of the Relationship between Strong Westerlies and Precipitation in the Great Plains and Midwest Regions of North America." *Climatic Change* 76: 427–441.

Booth, R. K., M. Notaro, S. T. Jackson, and J. E. Kutzbach. 2006. "Widespread Drought Episodes in the Western Great Lakes Region during the Past 2000 Years: Geographic Extent and Potential Mechanisms." *Earth and Planetary Science Letters* 242: 415–427.

Bowen, G. J. 2008. "Spatial Analysis of the Intra-Annual Variation of Precipitation Isotope Ratios and Its Climatological Corollaries." *Journal of Geophysical Research* 113, doi:10-1029/2007JD009295.

Bowes, M. D., and R. A. Sedjo. 1993. "Impacts and Responses to Climate-Change in Forests of the MINK Region." *Climatic Change* 24: 63–82.

Braham, R. R. 1983. "The Midwest Snow Storm of 8–11 December 1977." *Monthly Weather Review* 111: 253–272.

Braham, R. R., and M. J. Dungey. 1984. "Quantitative Estimates of the Effect of Lake Michigan on Snowfall." *Journal of Applied Meteorology* 23: 940–949.

Brikowski, T. H. 2008. "Doomed Reservoirs in Kansas, USA? Climate Change and Groundwater Mining on the Great Plains Lead to Unsustainable Surface Water Storage." *Journal of Hydrology* 354: 90–101.

Brinkmann, W. A. R. 1983. "Variability of Temperature in Wisconsin." *Monthly Weather Review* 111: 172–180.

——— 1999. "Within-Type Variability of 700 hPa Winter Circulation Patterns over the Lake Superior Basin." *International Journal of Climatology* 19: 41–58.

Brown, B. G., and R. W. Katz. 1995. "Regional-Analysis of Temperature Extremes: Spatial Analog for Climate-Change. *Journal of Climate* 8: 108–119.

Brown, M. E., and D. L. Arnold. 1998. "Land-Surface-Atmosphere Interactions Associated with Deep Convection in Illinois." *International Journal of Climatology* 18: 1637–1653.

Budikova, D. 2005. "Impact of the Pacific Decadal Oscillation on Relationships between Temperature and the Arctic Oscillation in the USA in Winter." *Climate Research* 29: 199–208.

Burnett, A. W., et al. 2003. "Increasing Great Lake–Effect Snowfall during the Twentieth Century: A Regional Response to Global Warming?" *Journal of Climate* 16: 3535–3542.

Butler, T. J., et al. 2008. "Regional Precipitation Mercury Trends in the Eastern USA, 1998–2005: Declines in the Northeast and Midwest, no Trend in the Southeast." *Atmospheric Environment* 42: 1582–1592.

Carleton, A. M., et al. 1994. "Climatic-Scale Vegetation: Cloud Interactions during Drought using Satellite Data." *International Journal of Climatology* 14: 593–623.

Carleton, A. M., et al. 2001. "Summer Season Land Cover: Convective Cloud Associations for the Midwest US 'Corn Belt.'" *Geophysical Research Letters* 28: 1679–1682.

Carlson, R. E. 1990. "Heat-Stress, Plant-Available Soil-Moisture, and Corn Yields in Iowa: A Short-Term and Long-Term View." *Journal of Production Agriculture* 3: 293–297.

Carlson, R. E., D. P. Todey, and S. E. Taylor. 1996. "Midwestern Corn Yield and Weather in Relation to Extremes of the Southern Oscillation." *Journal of Production Agriculture* 9: 347–352.

Changnon, D. 1996. "Changing Temporal and Spatial Characteristics of Midwestern Hydrologic Droughts." *Physical Geography* 17: 29–46.

Changnon, S. A. 1979. "Illinois-Climate-Center." *Bulletin of the American Meteorological Society* 60: 1157–1164.

———. 1981. "Midwestern Cloud, Sunshine and Temperature Trends since 1901: Possible Evidence of Jet Contrail Effects." *Journal of Applied Meteorology* 20: 496–508.

———. 1982. "Views of Climate Change: A Delphi Experiment in the Midwest." *Bulletin of the American Meteorological Society* 63: 1160–1161.

———. 1984. "Temporal and Spatial Variations in Hail in the Upper Great Plains and Midwest." *Journal of Climate and Applied Meteorology* 23: 1531–1541.

———. 1985. "Climate Fluctuations and Impacts: The Illinois Case." *Bulletin of the American Meteorological Society* 66: 142–151.

———. 1987. "Climatic Fluctuations and Record-High Levels of Lake Michigan." *Bulletin of the American Meteorological Society* 68: 1394–1402.

———. 2001. "Damaging Thunderstorm Activity in the United States." *Bulletin of the American Meteorological Society* 82: 597–608.

———. 2001. "Thunderstorm Rainfall in the Conterminous United States." *Bulletin of the American Meteorological Society* 82: 1925–1940.

———. 2002. "Impacts of the Midwestern Drought Forecasts of 2000." *Journal of Applied Meteorology* 41: 1042–1052.

———. 2003. "Characteristics of Ice Storms in the U.S." *Journal of Applied Meteorology* 42: 630–639.

Changnon, S. A., and D. Changnon. 2000. "Long-Term Fluctuations in Hail Incidences in the United States." *Journal of Climate* 13: 658–664.

———. 2001. "Long-Term Fluctuations in Thunderstorm Activity in the United States." *Climatic Change* 50: 489–503.

Changnon, S. A., D. Changnon, and T. R. Karl. 2006. "Temporal and Spatial Characteristics of Snowstorms in the U.S." *Journal of Applied Meteorology and Climatology* 45: 1141–1155.

Changnon, S. A., and J. Changnon. 2003. "Major Ice Storms in the U.S., 1949–2000." *Natural Hazards* 4: 105–111.

Changnon, S. A., and F. A. Huff. 1991. "Potential Effects of Changed Climates on Heavy Rainfall Frequencies in the Midwest." *Water Resources Bulletin* 27: 753–759.

Changnon, S. A., F. A. Huff, and C. F. Hsu. 1988. "Relations between Precipitation and Shallow Groundwater in Illinois." *Journal of Climate* 1: 1239–1250.

Changnon, S. A., and T. R. Karl. 2003. "Temporal and Spatial Variations of Freezing Rain in the Contiguous United States: 1948–2000." *Journal of Applied Meteorology* 42: 1302–1315.

Changnon, S. A., and K. E. Kunkel. 1992. "Assessing Impacts of a Climatologically Unique Year (1990) in the Midwest." *Physical Geography* 13: 180–190.

———. 1995. "Climate-Related Fluctuations in Midwestern Floods during 1921–1985." *Journal of Water Resources Planning and Management-ASCE* 121: 326–334.

Changnon, D., M. Sandstrom, and M. Bentley. 2006. "Midwestern High Dew Point Events 1960–2000." *Physical Geography* 27: 494–504.

Chumbley, C. A., R. G. Baker, and E. A. Bettis. 1990. "Midwestern Holocene Paleoenvironments Revealed by Floodplain Deposits in Northeastern Iowa." *Science* 249: 272–274.

Clark, J. S., B. J. Stocks, and P. J. H. Richard. 1996. "Climate Implications of Biomass Burning since the 19th Century in Eastern North America." *Global Change Biology* 2: 433–442.

Coleman, J. S. M., and J. C. Rogers. 2003. "Ohio River Valley Winter Moisture Conditions Associated with the Pacific/North American Teleconnection Pattern." *Journal of Climate* 16: 969–981.

———. 2007. "A Synoptic Climatology of the Central United States and Associations with Pacific Teleconnection Pattern Frequency." *Journal of Climate* 20: 3485–3497.

Colle, B. A., and C. Mass. 1995. "The Structure and Evolution of Cold Surges East of the Rocky Mountains." *Monthly Weather Review* 123: 2577–2610.

Cooter, E. J., J. Swall, and R. Gilliam. 2007. "Comparison of 700-hPa NCEP-R1 and AMIP-R2 Wind Patterns over the Continental United States using Cluster Analysis." *Journal of Applied Meteorology and Climatology* 46: 1744–1758.

Cruse, R., et al. 2006. "Daily Estimates of Rainfall, Water Runoff, and Soil Erosion in Iowa." *Journal of Soil and Water Conservation* 61: 191–199.

Dai, A., I. Y. Fung, and A. D. DelGenio. 1997. "Surface Observed Global Land Precipitation Variations during 1900–88." *Journal of Climate* 10: 2943–2962.

Davis, P. M., N. Brenes, and L. L. Allee. 1996. "Temperature Dependent Models to Predict Regional Differences in Corn Rootworm (Coleoptera: Chrysomelidae) Phenology." *Environmental Entomology* 25: 767–775.

Dixon, B., and K. Segerson. 1997. "Midwest Farmer Adaptation to Climate Change: Impacts of Changes in Mean and Variation of Climate." *American Journal of Agricultural Economics* 79: 1708–1718.

Douglas, E. M., R. M. Vogel, and C. N. Kroll. 2000. "Trends in Floods and Low Flows in the United States: Impact of Spatial Correlation." *Journal of Hydrology* 240: 90–105.

Eaton, J. G., and R. M. Scheller. 1996. "Effects of Climate Warming on Fish Thermal Habitat in Streams of the United States." *Limnology and Oceanography* 41: 1109–1115.

Ellis, A. W., and J. J. Johnson. 2004. "Hydroclimatic Analysis of Snowfall Trends Associated with the North American Great Lakes." *Journal of Hydrometeorology* 5: 471–486.

Ellis, A. W., and D. J. Leathers. 1996. "A Synoptic Climatological Approach to the Analysis of Lake-Effect Snowfall: Potential Forecasting Applications." *Weather Forecasting* 11: 216–229.

Enfield, D. B., A. M. Mestas-Nunez, and P. J. Trimble. 2001. "The Atlantic Multidecadal Oscillation and Its Relation to Rainfall and River Flows in the Continental U.S." *Geophysical Research Letters* 28: 2077–2080.

Feng, S., et al. 2008. "Atlantic and Pacific SST Influences on Medieval Drought in North America Simulated by the Community Atmospheric Model." *Journal of Geophysical Research* 113, doi:10.1029/2007JD009347.

Forbes, G. S., R. A. Anthes, and D. W. Thomson. 1987. "Synoptic and Mesoscale Aspects of an Appalachian Ice Storm Associated with Cold-Air Damming." *Monthly Weather Review* 115: 564–591.

Fox-Rabinovitz, M. S., et al. 2005. "A Multiyear Ensemble Simulation of the U.S. Climate with a Stretched-Grid GCM." *Monthly Weather Review* 133: 2505–2525.

Fritsch, J. M., R. J. Kane, and C. R. Chelius. 1986. "The Contribution of Mesoscale Convective Weather Systems to the Warm Season Precipitation in the United States." *Journal of Climate and Applied Meteorology* 25: 1333–1345.

Gallus, W. A., N. A. Snook, and E. V. Johnson. 2008. "Spring and Summer Severe Weather Reports over the Midwest as a Function of Convective Mode: A Preliminary Study." *Weather and Forecasting* 23: 101–113.

Georgakakos, K. P., and D. H. Bae. 1994. "Climatic Variability of Soil-Water in the American Midwest, Part 2, Spatiotemporal Analysis." *Journal of Hydrology* 162: 379–390.

Georgakakos, A. P., et al. 1998. "Impacts of Climate Variability on the Operational Forecast and Management of the Upper Des Moines River Basin." *Water Resources Research* 34: 799–821.

Gershunov, A. 1998. "ENSO Influence on Intraseasonal Extreme Rainfall and Temperature Frequencies in the Contiguous United States: Implications for Long-Range Predictability." *Journal of Climate* 11: 3192–3203.

Giorgi, F., et al. 1996. "A Regional Model Study of the Importance of Local versus Remote Controls of the 1988 Drought and the 1993 Flood over the Central United States." *Journal of Climate* 9: 1150–1162.

Groisman, P. Y., and R. D. Easterling. 1994. "Variability and Trends of Precipitation and Snowfall over the United States and Canada." *Journal of Climate* 7: 184–205.

Grover, E. K., and P. J. Sousounis. 2002. "The Influence of Large-Scale Flow on Fall Precipitation Systems in the Great Lakes Basin." *Journal of Climate* 15: 1943–1956.

Gutowski, W. J., et al. 2004. "Diagnosis and Attribution of a Seasonal Precipitation Deficit in a U.S. Regional Climate Simulation." *Journal of Hydrometeorology* 5: 230–242.

Gutowski, W. J., Jr., et al. 2007. "A Possible Constraint on Regional Precipitation Intensity Changes under Global Warming." *Journal of Hydrometeorology* 8: 1382–1396.

Guttman, N. B., et al. 1994. "The 1993 Midwest Extreme Precipitation in Historical and Probabilistic Perspective." *Bulletin of the American Meteorological Society* 75: 1785–1792.

Guyette, R., et al. 1980. "A Climate History of Boone County, Missouri, from Tree-Ring Analysis of Eastern Red-Cedar." *Wood and Fiber* 12: 17–28.

Hanson, H. P., C. S. Hanson, and B. H. Yoo. 1992. "Recent Great-Lakes Ice Trends." *Bulletin of the American Meteorological Society* 73: 577–584.

Hilgendorf, E. R., and R. H. Johnson. 1998. "A Study of the Evolution of Mesoscale Convective Systems using WSR-88D Data." *Weather and Forecasting* 13: 437–452.

Hostetler, S. W., and E. E. Small. 1999. "Response of North American Freshwater Lakes to Simulated Future Climates." *Journal of the American Water Resources Association* 35: 1625–1637.

Huff, F. A. 1995. "Characteristics and Contributing Causes of an Abnormal Frequency of Flood-Producing Rainstorms at Chicago." *Water Resources Bulletin* 31: 703–714.

Jha, M., et al. 2004. "Impact of Climate Change on Stream Flow in the Upper Mississippi River Basin: A Regional Climate Model Perspective." *Journal of Geophysical Research* 109, doi:10.1029/2003JD003686.

Jha, M., et al. 2006. "Climate Change Sensitivity Assessment on Upper Mississippi River Basin Streamflows using SWAT." *Journal of the American Water Resources Association* (August): 997–1015.

Johnson, S. L., and H. G. Stefan. 2006. "Indicators of Climate Warming in Minnesota: Lake Ice Covers and Snowmelt Runoff." *Climatic Change* 75: 421–453.

Junker, N. W., R. S. Schneider, and S. L. Fauver. 1999. "A Study of Heavy Rainfall Events during the Great Midwest Flood of 1993." *Weather and Forecasting* 14: 701–712.

Karl, T. R., and R. W. Knight. 1997. "The 1995 Chicago Heat Wave: How Likely Is a Recurrence?" *Bulletin of the American Meteorological Society* 78: 1107–1119.

Keppenne, C. L. 1995. "An ENSO Signal in Soybean Futures Prices." *Journal of Climate* 8: 1685–1689.

Kim, D. H., R. D. Slack, and F. Chavez-Ramirez. 2008. "Impacts of El Niño—Southern Oscillation Events on the Distribution of Wintering Raptors." *Journal of Wildlife Management* 72: 231–239.

Klippel, W. E., G. Celmer, and J. R. Purdue. 1978. "The Holocene Naiad Record at Rodgers Shelter in the Western Ozark Highland of Missouri USA." *Plains Anthropologist* 23: 257–271.

Klink, K. 2002. "Trends and Interannual Variability of Wind Speed Distributions in Minnesota." *Journal of Climate* 15: 3311–3317.

———. 2007. "Atmospheric Circulation Effects on Wind Speed Variability at Turbine Height." *Journal of Applied Meteorology and Climatology* 46: 445–456.

Klugman, M. R. 1978. "Drought in Upper Midwest, 1931–1969." *Journal of Applied Meteorology* 17: 1425–1431.

Knox, J. C. 2000. "Sensitivity of Modern and Holocene Floods to Climate Change." *Quaternary Science Reviews* 19: 439–457.

Kongoli, C. E., and W. L. Bland. 2000. "Long-Term Snow Depth Simulations using a Modified Atmosphere-Land Exchange Model."

Agricultural and Forest Meteorology 104: 273–287.

Konrad, C. E., and S. J. Colucci. 1989. "An Examination of Extreme Cold Air Outbreaks over Eastern North America." *Monthly Weather Review* 117: 2687–2700.

Kothavala, Z. 1997. "Extreme Precipitation Events and the Applicability of Global Climate Models to the Study of Floods and Droughts." *Mathematics and Computers in Simulation* 43: 261–268.

Kristovich, D. A., and R. A. Steve. 1995. "A Satellite Study of Cloud-Band Frequencies over the Great Lakes." *Journal of Applied Meteorology* 34: 2083–2090.

Kristovich, D. A. R., et al. 2000. "The Lake-Induced Convection Experiment and the Snowband Dynamics Project." *Bulletin of the American Meteorological Society* 81: 519–542.

Kunkel, K. E. 1989. "A Surface Energy Budget View of the 1988 Midwestern United States Drought." *Boundary Layer Meteorology* 48: 217–225.

Kunkel, K. E., K. Andsager, and D. R. Easterling. 1999. "Long-Term Trends in Heavy Precipitation Events over North America." *Journal of Climate* 12: 2515–2527.

Kunkel, K. E., S. A. Changnon, and J. R. Angel. 1994. "Climatic Aspects of the 1993 Upper Mississippi River Basin Flood." *Bulletin of the American Meteorological Society* 75: 811–822.

Kunkel, K. E., S. A. Changnon, and R. T. Shealy. 1993. "Temporal and Spatial Characteristics of Heavy-Precipitation Events in the Midwest." *Monthly Weather Review* 121: 858–866.

Kunkel, K. E., and X.-L. Liang. 2005. "CMIP Simulations of the Climate in the Central United States." *Journal of Climate* 18: 1016–1031.

Kunkel, K. E., N. E. Westcott, and D. A. R. Kristovich. 2002. "Effects of Climate Change on Heavy Lake-Effect Snowstorms near Lake Erie." *Journal of Great Lakes Research* 28: 521–536.

Kunkel, K. E., et al. 1990. "A Real-Time Climate Information-System for the Midwestern United States." *Bulletin of the American Meteorological Society* 71: 1601–1609.

Kunkel, K. E., et al. 1996. "The July 1995 Heat Wave in the Midwest: A Climatic Perspective and Critical Weather Factors." *Bulletin of the American Meteorological Society* 77: 1507–1518.

Kunkel, K. E., et al. 1998. "An Expanded Digital Daily Data Set for Climatic Resources Applications in the Midwestern United States." *Bulletin of the American Meteorological Society* 79: 1357–1366.

Kunkel, K. E., et al. 2006. "Can CGCMs Simulate the Twentieth Century Warming Hole in the Central United States?" *Journal of Climate* 19: 4137–4153.

Kutzbach, J. E., J. Williams, and S. Vavrus. 2005. "Simulated 21st Century Changes in Regional Water Balance of the Great Lakes Region and Links to Changes in Global Temperature and Poleward Moisture Transport." *Geophysical Research Letters* 32, doi:10.1029/2005GL023506.

Laird, N. F., et al. 2001. "Lake Michigan Lake Breezes: Climatology, Local Forcing, and Synoptic Environment." *Journal of Applied Meteorology* 40: 409–424.

Lakshmi, V., and K. Schaaf. 2001. "Analysis of the 1993 Midwestern Flood Using Satellite and Ground Data." *IEEE Transactions on Geoscience and Remote Sensing* 39: 1736–1743.

Lau, K. M., K. M. Kim, and S. S. P. Shen. 2002. "Potential Predictability of Seasonal Precipitation over the United States from Canonical Ensemble Correlation Predictions." *Geophysical Research Letters* 29, doi:10.1029/2001GL014263.

Legler, D. M., K. J. Bryant, and J. J. O'Brien. 1999. "Impact of ENSO-Related Climate Anomalies on Crop Yields in the U.S." *Climatic Change* 42: 351–375.

Lenters, J. D., and T. E. Twine. 2000. "Sensitivity of Upper Midwest Hydrology to Changes in Climate: Results from an Improved Land Surface Model." *IAGLR Conference Program and Abstracts* 43: A-88.

Leung, L. R., and W. I. Gustafson. 2005. "Potential Regional Climate Change and Implications to U.S. Air Quality." *Geophysical Research Letters* 32, doi:10.1029/2005GL022911.

Liang, X. Z., K. E. Kunkel, and A. N. Samel. 2001. "Development of a Regional Climate Model for U.S. Midwest Applications, Part 1, Sensitivity to Buffer Zone Treatment." *Journal of Climate* 14: 4363–4378.

Liang, X. Z., et al. 2004. "Regional Climate Model Simulation of U.S. Precipitation during 1982–2002, Part 1, Annual Cycle." *Journal of Climate* 17: 3510–3529.

Lofgren, B. M., et al. 2002. "Evaluation of Potential Impacts on Great Lakes Water Resources Based on Climate Scenarios of Two GCMs." *Journal of Great Lakes Research* 28: 537–554.

Lu, Q. Q., R. Lund, and L. Seymour. 2005. "An Update of U.S. Temperature Trends." *Journal of Climate* 18: 4906–4914.

Matthews, W. J. 1986. "Geographic-Variation in Thermal Tolerance of a Widespread Minnow Notropis-Lutrensis of the North-American Mid-West." *Journal of Fish Biology* 28: 407–417.

Mauget, S. A. 2004. "Low Frequency Streamflow Regimes over the Central United States: 1939–1998." *Climatic Change* 63: 121–144.

Mauget, S. A., and D. R. Upchurch. 1999. "El Niño and La Niña Related Climate and Agricultural Impacts over the Great Plains and Midwest." *Journal of Production Agriculture* 12: 203–215.

McCormick, M. J., and G. L. Fahnenstiel. 1999. "Recent Climatic Trends in Nearshore Water Temperatures in the St. Lawrence Great Lakes." *Limnology and Oceanography* 44: 530–540.

McGregor, K. M. 1986. "Drought during the 1930s and 1950s in the Central United States." *Physical Geography* 6: 289–301.

———. 1987. "Anomalous Wet and Dry Spells in the Central United States." *Physical Geography* 8: 225–240.

Mo, K.-C., J. Nogues-Paegle, and J. Paegle. 1995. "Physical Mechanisms of the 1993 Summer Floods." *Journal of Atmospheric Science* 52: 879–895.

Mo, K. C., J. Nogues-Paegle, and R. W. Higgins. 1997. "Atmospheric Processes Associated with Summer Floods and Droughts in the Central United States." *Journal of Climate* 10: 3028–3046.

Moldenhauer, O., and M. K. B. Ludeke. 2002. "Climate Sensitivity of Global Terrestrial Net Primary Production NPP: Calculated Using the Reduced-Form Model NNN." *Climate Research* 21: 43–57.

Montroy, D. L. 1997. "Linear Relation to Central and Eastern North American Precipitation to Tropical Pacific Sea Surface Temperature Anomalies." *Journal of Climate* 10: 541–558.

Montroy, D. L., M. B. Richman, and P. J. Lamb. 1998. "Observed Nonlinearities of Monthly Teleconnections between Tropical Pacific Sea Surface Temperature Anomalies and Central and Eastern North American Precipitation." *Journal of Climate* 11: 1812–1835.

Negri, A. J., and R. F. Adler. 1981. "Relation of Satellite-Based Thunderstorm Intensity to Radar-Estimated Rainfall." *Journal of Applied Meteorology* 20: 288–300.

Norton, D., and S. Bolsenga. 1993. "Spatiotemporal Trends in Lake Effect and Continental Snowfall in the Laurentian Great Lakes, 1951–1980." *Journal of Climate* 6: 1943–1956.

Notaro, M., Z. Liu, and J. Williams. 2006. "Observed Vegetation-Climate Feedbacks in the United States." *Journal of Climate* 19: 763–786.

O'Neal, M. 1996. "Interactions between Land Cover and Convective Cloud Cover over Midwestern North America Detected from GOES Satellite Data." *International Journal of Remote Sensing* 17: 1149–1181.

O'Neal, M. R., et al. 2005. "Climate Change Impacts on Soil Erosion in Midwest United States with Changes in Crop Management." *Catena* 61: 165–184.

Orville, R. E., and G. R. Huffines. 2001. "Cloud-to-Ground Lightning in the United States: NLDN Results in the First Decade, 1989–98." *Monthly Weather Review* 129: 1179–1193.

Over, T. M., et al. 2005. "Soil Moisture and Precipitation Feedback in the U.S. Midwest: An Empirical Study." *Transactions of the Illinois State Academy of Science* 98: 35.

Palecki, M. A., S. A. Changnon, and K. E. Kunkel. 2001. "The Nature and Impacts of the

July 1999 Heat Wave in the Midwestern U.S.: Learning from the Lessons of 1995." *Bulletin of the American Meteorological Society* 82: 1353–1367.

Pan, Z. T., et al. 1999. "Long Simulation of Regional Climate as a Sequence of Short Segments." *Monthly Weather Review* 127: 308–321.

Pan, Z., et al. 2004. "Altered Hydrologic Feedback in a Warming Climate Introduces a Warming Hole." *Geophysical Research Letters* 31, doi:10.1029/2004GL020528.

Peterjohn, B. G. 1989. "The Winter Season December 1 1988–February 28 1989 Middle-Western Prairie Region USA." *American Birds* 43: 319–323.

———. 1990. "The Spring Season March 1 to May 31 1990 Middle-Western Prairie Region USA." *American Birds* 44: 432–439.

Polderman, N. J., and S. C. Pryor. 2004. "Linking Synoptic-Scale Climate Phenomena to Lake-Level Variability in the Lake Michigan–Huron Basin." *Journal of Great Lakes Research* 30: 419–434.

Potter, C., et al. 2008. "Terrestrial Vegetation Dynamics and Global Climate Controls in North America: 2001–05." *Earth Interactions* 12, doi:10.1175/2008EI1249.1171.

Primack, A. G. B. 2000. "Simulation of Climate-Change Effects on Riparian Vegetation in the Pere Marquette River, Michigan." *Wetlands* 20: 538–547.

Pryor, S. C., J. A. Howe, and K. E. Kunkel. 2009. "How Spatially Coherent and Statistically Robust Are Temporal Changes in Extreme Precipitation across the Contiguous USA?" *International Journal of Climatology* 29: 31–45.

Pryor, S. C., and J. T. Schoof. 2008. "Changes in Precipitation Seasonality over the Contiguous USA." *Journal of Geophysical Research* 113, doi:10.1029/2008JD010251.

Purcell, L. C., T. R. Sinclair, and R. W. McNew. 2003. "Drought Avoidance Assessment for Summer Annual Crops using Long-Term Weather Data." *Agronomy Journal* 95: 1566–1576.

Rajagopalan, B., et al. 2000. "Spatiotemporal Variability of ENSO and SST Teleconnections to Summer Drought over the United States during the Twentieth Century." *Journal of Climate* 13: 4244–4255.

Rhemtulla, J. M., D. J. Mladenoff, and M. K. Clayton. 2007. "Regional Land-Cover Conversion in the U.S. Upper Midwest: Magnitude of Change and Limited Recovery (1850–1935–1993)." *Landscape Ecology* 22: 57–75.

Richman, M. B., and W. E. Easterling. 1988. "Procrustes Target Analysis—a Multivariate Tool for Identification of Climate Fluctuations." *Journal of Geophysical Research-Atmospheres* 93: 10989–11003.

Ritchie, J. T., and B. Basso. 2008. "Water Use Efficiency Is Not Constant When Crop Water Supply Is Adequate or Fixed: The Role of Agronomic Management." *European Journal of Agronomy* 28: 273–281.

Robeson, S. M. 2002. "Increasing Growing-Season Length in Illinois during the 20th Century." *Climatic Change* 52: 219–238.

Robeson, S. M., and K. A. Shein. 1997. "Spatial Coherence and Decay of Wind Speed and Power in the North-Central United States." *Physical Geography* 18: 479–495.

Rodionov, S., and R. Assel. 2000. "Atmospheric Teleconnection Patterns and Severity of Winters in the Laurentian Great Lakes Basin." *Atmosphere-Ocean* 38: 601–635.

———. 2003. "Winter Severity in the Great Lakes Region: A Tale of Two Oscillations." *Climate Research* 24: 19–31.

Rodriguez, Y., D. A. R. Kristovich, and M. R. Hjelmfelt. 2007. "Lake-to-Lake Cloud Bands: Frequencies and Locations." *Monthly Weather Review* 135: 4202–4213.

Rogers, J. C. 1993. "Climatological Aspects of Drought in Ohio." *Ohio Journal of Science* 93: 51–59.

Rogers, J. C., and J. S. M. Coleman. 2003. "Interactions between the Atlantic Multidecadal Oscillation, El Niño/La Niña, and the PNA in Winter Mississippi Valley Stream Flow." *Geophysical Research Letters* 30, doi:10.1029/2003GL017216.

———. 2004. "Ohio Winter Precipitation and Stream Flow Associations to Pacific Atmospheric and Oceanic Teleconnection Patterns." *Ohio Journal of Science* 104: 51–59.

Rogers, J. C., S.-H. Wang, and J. S. M. Coleman. 2007. "Evaluation of a Long-Term (1882–2005) Equivalent Temperature Time Series." *Journal of Climate* 20: 4476–4485.

Rogers, J. C., and A. Yersavich. 1988. "Daily Air Temperature Variability Associated with Climatic Variability at Columbus, Ohio." *Physical Geography* 9: 120–138.

Rolf Olsen, J., et al. 1999. "Climate Variability and Flood Frequency Estimation for the Upper Mississippi and Lower Missouri Rivers." *Journal of the American Water Resources Association* 35: 1509–1523.

Roy, S. B., et al. 2003. "Impact of Historical Land Cover Change on the July Climate of the United States." *Journal of Geophysical Research* 108, doi:10.1029/2003JD003565.

Sandstrom, M. A., R. G. Lauritsen, and D. Changnon. 2004. "A Central-U.S. Summer Extreme Dew-Point Climatology (1949–2000)." *Physical Geography* 25: 191–207.

Scheller, R. M., and D. J. Mladenoff. 2008. "Simulated Effects of Climate Change, Fragmentation, and Inter-specific Competition on Tree Species Migration in Northern Wisconsin, USA." *Climate Research* 36: 191–202.

Schmidlin, T. W. 1989. "The Urban Heat Island at Toledo, Ohio." *Ohio Journal of Science* 89: 38–41.

Schmidlin, T. W., D. J. Edgell, and M. A. Delaney. 1992. "Design Ground Snow Loads for Ohio." *Journal of Applied Meteorology* 31: 622–627.

Schmidlin, T. W., et al. 1995. "Automated Quality Control Procedure for the Water Equivalent of Snow on the Ground Measurement." *Journal of Applied Meteorology* 34: 143–151.

Schoof, J. T., and S. C. Pryor. 2001. "Downscaling Temperature and Precipitation: A Comparison of Regression-Based Methods and Artificial Neural Networks." *International Journal of Climatology* 21: 773–790.

———. 2003. "Evaluation of the NCEP/NCAR Reanalysis in Terms of Synoptic Scale Phenomena: A Case Study from the Midwestern USA." *International Journal of Climatology* 23: 1725–1741.

———. 2006. "An Evaluation of Two GCMs: Simulation of North American Tele-connection Indices and Synoptic Phenomena." *International Journal of Climatology* 26: 267–282.

———. 2008. "On the Proper Order of Markov Chain Model for Daily Precipitation Occurrence in the Contiguous United States." *Journal of Applied Meteorology and Climatology* 47: 2477–2486.

Schoof, J. T., S. C. Pryor, and S. M. Robeson. 2007. "Downscaling Daily Maximum and Minimum Air Temperatures in the Midwestern USA: A Hybrid Empirical Approach." *International Journal of Climatology* 27: 439–454.

Scott, R. W., and F. A. Huff. 1996. "Impacts of the Great Lakes on Regional Climate Conditions." *Journal of Great Lakes Research* 22: 845–863.

Schubert, S. D., et al. 1998. "Subseasonal Variation in Warm-Season Moisture Transport and Precipitation over the Central and Eastern United States." *Journal of Climate* 11: 2530–2555.

Schubert, S. D., et al. 2008. "ENSO and Wintertime Extreme Precipitation Events over the Contiguous United States." *Journal of Climate* 21: 22–39.

Schwartz, M. D. 1991. "An Integrated Approach to Air Mass Classification in the North Central United States." *Professional Geographer* 43: 77–91.

Seneviratne, S. I., et al. 2002. "Summer Dryness in a Warmer Climate: A Process Study with a Regional Climate Model." *Climate Dynamics* 20: 69–85.

Serreze, M. C., et al. 1998. "Characteristics of Snowfall over the Eastern Half of the United States and Relationships with Principal Modes of Low-Frequency Atmospheric Variability." *Journal of Climate* 11: 234–250.

Skaggs, R. H., and D. G. Baker. 1985. "Fluctuations in the Length of the Growing Season in Minnesota." *Climate Change* 7: 403–414.

———. 1986. "The Contribution of Months and Seasons to Fluctuations of the Mean Annual Temperature in Eastern Minnesota." *Theoretical and Applied Climatology* 37: 158–165.

Smith, S. R., and J. J. O'Brien. 2001. "Regional Snowfall Distributions Associated with ENSO: Implications for Seasonal Forecasting."

Bulletin of the American Meteorological Society 82: 1179–1191.

Sobolowski, S., and A. Frei. 2007. "Lagged Relationships between North American Snow Mass and Atmospheric Teleconnection Indices." *International Journal of Climatology* 27: 221–231.

Sonka, S. T., et al. 1982. "Can Climate Forecasts for the Growing Season Be Valuable to Crop Producers? Some General Considerations and an Illinois Pilot-Study." *Journal of Applied Meteorology* 21: 471–476.

Southworth, J., et al. 2000. "Consequences of Future Climate Change and Changing Climate Variability on Maize Yields in the Midwestern United States." *Agriculture, Ecosystems, and Environment* 82: 139–158.

Sousounis, P. J., and E. K. Grover. 2002. "Potential Future Weather Patterns over the Great Lakes Region." *Journal of Great Lakes Research* 28: 496–520.

Sparks, J., D. Chagnon, and J. Starke. 2002. "Changes in Frequency of Extreme Warm-Season Surface Dewpoints in Northeastern Illinois: Implications for Cooling-System Design and Operation." *Journal of Applied Meteorology* 41: 890–898.

Stenger, P. J., and P. J. Michaels. 1992. "Climatic-Change in Mixed-Layer Trajectories over Large Regions." *Theoretical and Applied Climatology* 45: 167–175.

Sud, Y. C., et al. 2003. "Simulating the Midwestern U.S. Drought of 1988 with a GCM." *Journal of Climate* 16: 3946–3965.

Sun, W. Y., J. D. Chern, and M. Bosilovich. 2004. "Numerical Study of the 1988 Drought in the United States." *Journal of the Meteorological Society of Japan* 82: 1667–1678.

Takle, E. S., M. Jha, and C. J. Anderson. 2005. "Hydrological Cycle in the Upper Mississippi River Basin: 20th Century Simulations by Multiple GCMs." *Geophysical Research Letters* 32, L18407 10.1029/2005GL023630.

Takle, E. S., and L. O. Mearns. 1995a. "Midwest Temperature Means, Extremes, and Variability: Analysis of Observed Data." In *Progress in Biometeorology: Preparing for Global Change: A Midwestern Perspective,* ed. G. R. Carmichael, G. E. Folk, and J. L. Schnoor, 127–132. Proceedings of the Second Symposium on Global Change II: A Midwest Perspective, April 7–8, 1994, the University of Iowa, Iowa City, Iowa. Amsterdam, Neth.: SPB Academic Publishers.

———. 1995b. " Midwest Temperature Means, Extremes, and Variability: Analysis of Results from Climate Models." In *Progress in Biometeorology: Preparing for Global Change: A Midwestern Perspective,* ed. G. R. Carmichael, G. E. Folk, and J. L. Schnoor, 133–140. Proceedings of the Second Symposium on Global Change II: A Midwest Perspective, April 7–8, 1994, the University of Iowa, Iowa City, Iowa. Amsterdam, Neth.: SPB Academic Publishers.

Takle, E. S., et al. 1999. "Project to Intercompare Regional Climate Simulations (PIRCS): Description and Initial Results." *Journal of Geophysical Research* 104: 19,443–19,462.

Takle, E. S., et al. 2006. "Upper Mississippi River Basin Modeling Systems, Part 4, Climate Change Impacts on Flow and Water Quality." In *Coastal Hydrology and Processes,* ed. V. P. Singh and Y. J. Xu. Highlands Ranch, Colo.: Water Resources Publications LLC, 2006.

Takle, E. S., et al. 2007. "Transferability Intercomparison: An Opportunity for New Insight on the Global Water Cycle and Energy Budget." *Bulletin of the American Meteorological Society* 88: 375–384.

Tao, Z. N., et al. 2007. "Sensitivity of U.S. Surface Ozone to Future Emissions and Climate Changes." *Geophysical Research Letters* 34, doi:10.1029/2007GL029455.

Thompson, L. M. 1990. "Impact of Global Warming and Cooling on Midwestern Agriculture." *Journal of the Iowa Academy of Science* 97: 88–90.

Thomson, A. M., et al. 2005. "Climate Change Impacts for the Conterminous USA: An Integrated Assessment, Part 4, Water Resources." *Climatic Change* 69: 67–88.

Trenberth, K. E., and G. W. Branstator. 1992. "Issues in Establishing Causes of the 1988 Drought over North America." *Journal of Climate* 5: 159–172.

Trenberth, K. E., G. W. Branstator, and P. A. Arkin. 1988. "Origins of the 1988 North American Drought." *Science* 242: 1640–1645.

Trenberth, K. E., and C. J. Guillemot. 1996. "Physical Processes Involved in the 1988 Drought and 1993 Floods in North America." *Journal of Climate* 9: 1288–1298.

Undersander, D. J., and L. J. Greub. 2007. "Summer-Fall Seeding Dates for Six Cool-Season Grasses in the Midwest United States." *Agronomy Journal* 99: 1579–1586.

Van den Broeke, M. S., et al. 2005. "Cloud-to-Ground Lightning Production in Strongly Forced, Low-Instability Convective Lines Associated with Damaging Wind." *Weather and Forecasting* 20: 517–530.

Vavrus, S., et al. 2006. "The Behavior of Extreme Cold-Air Outbreaks under Greenhouse Warming." *International Journal of Climatology* 26: 1133–1147.

Vidon, P., and A. P. Smith. 2007. "Upland Controls on the Hydrological Functioning of Riparian Zones in Glacial Till Valleys of the Midwest." *Journal of the American Water Resources Association* 43: 1524–1539.

Wang, H., M. F. Ting, and M. Ji. 1999. "Prediction of Seasonal Mean United States Precipitation Based on El Niño Sea Surface Temperatures." *Geophysical Research Letters* 26: 1341–1344.

Warrach, K., H. T. Mengelkamp, and E. Raschke. 2001. "Treatment of Frozen Soil and Snow Cover in the Land Surface Model SEWAB." *Theoretical and Applied Climatology* 69: 23–37.

Weismueller, J. L., and S. M. Zubrick. 1998. "Evaluation and Application of Conditional Symmetric Instability, Equivalent Potential Vorticity, and Frontogenetic Forcing in an Operational Forecast Environment." *Weather and Forecasting* 13: 84–101.

Wendland, W. M. 1987. "Prominent November Coldwaves in the North Central United States since 1901." *Bulletin of the American Meteorological Society* 68: 616–619.

Wendland, W. M., et al. 1983. "Review of the Unusual Winter of 1982–83 in the Upper Midwest." *Bulletin of the American Meteorological Society* 64: 1346–1350.

Wendland, W. M., et al. 1984. "A Climatic Review of Summer 1983 in the Upper Midwest." *Bulletin of the American Meteorological Society* 65: 1068–1072.

Westcott, N. E. 2007. "Some Aspects of Dense Fog in the Midwestern United States." *Weather and Forecasting* 22: 457–465.

White, W. B., A. Gershunov, and J. Annis. 2008. "Climatic Influences on Midwest Drought during the Twentieth Century." *Journal of Climate* 21: 517–531.

Wilson, L. L., and E. Foufoulageorgiou. 1990. "Regional Rainfall Frequency-Analysis via Stochastic Storm Transposition." *Journal of Hydraulic Engineering-ASCE* 116: 859–880.

Winkler, M. G. 1994. "Sensing Plant Community and Climate Change by Charcoal-Carbon Isotope Analysis." *Ecoscience* 1: 340–345.

Winkler, M. G., A. M. Swain, and J. E. Kutzbach. 1986. "Middle Holocene Dry Period in the Northern Midwestern United-States—Lake Levels and Pollen Stratigraphy." *Quaternary Research* 25: 235–250.

Winstanley, D., and S. A. Changnon. 1999. "Long-Term Variations in Seasonal Weather Conditions Important to Water Resources in Illinois." *Journal of the American Water Resources Association* 35: 1421–1427.

Woodhouse, C. A., and J. T. Overpeck. 1998. "2000 Years of Drought Variability in the Central United States." *Bulletin of the American Meteorological Society* 79: 2693–2714.

Xie, H., J. W. Eheart, and H. An. 2008. "Hydrologic and Economic Implications of Climate Change for Typical River Basins of the Agricultural Midwestern United States." *Journal of Water Resources Planning and Management-ASCE* 134: 205–213.

Xin-Zhong, L., K. E. Kunkel, and A. N. Samel. 2001. "Development of a Regional Climate Model for U.S. Midwest Applications, Part 1, Sensitivity to Buffer Zone Treatment." *Journal of Climate* 14: 4363–4378.

Yang, S., et al. 2007. "Response of Seasonal Simulations of a Regional Climate Model to High-

Frequency Variability of Soil Moisture during the Summers of 1988 and 1993." *Journal of Hydrometeorology* 8: 738–757.

You, J. S., and K. G. Hubbard. 2006. "Quality Control of Weather Data during Extreme Events." *Journal of Atmospheric and Oceanic Technology* 23: 184–197.

Zhang, D. L., W. Z. Zheng, and Y. K. Xue. 2003. "A Numerical Study of Early Summer Regional Climate and Weather over LSA-East, Part 1: Model Implementation and Verification." *Monthly Weather Review* 131: 1895–1909.

Zhu, J. H., and X. Z. Liang. 2007. "Regional Climate Model Simulations of US Precipitation and Surface Air Temperature during 1982–2002: Interannual Variation." *Journal of Climate* 20: 218–232.

Contributors

S. C. Pryor is Professor of Atmospheric Science in the Department of Geography at Indiana University in Bloomington, Indiana. She has published over eighty journal articles in the field of global atmospheric change and is currently chair of the Midwest Assessment Group for Investigations of Climate (MAGIC). Author of chapters 1, 8, 9, 15, 17, 23, and editor.

E. S. Takle is Professor of Atmospheric Science in the Department of Geological and Atmospheric Sciences, Professor of Agricultural Meteorology in the Department of Agronomy at Iowa State University, and Director of the ISU Climate Science Initiative. He was contributing author to the IPCC Third Assessment Report and is a participant in the North American Regional Climate Change Assessment Program (NARCCAP). Author of chapters 1, 10, 23.

J. T. Schoof is an Assistant Professor in the Department of Geography and Environmental Resources at Southern Illinois University in Carbondale, Illinois. He has authored or coauthored more than twenty journal articles, focused primarily on downscaling techniques and statistical applications in climatology. Author of chapters 2, 4, 9, 17.

Z. Pan is Associate Professor in the Department of Earth and Atmospheric Sciences at St. Louis University in St. Louis, Missouri. He has published over fifty journal articles, principally in the field of regional climate change and land-surface interaction. Author of chapters 3, 8.

M. Segal is Scientist in the Department of Agronomy at Iowa State University in Ames, Iowa. He has published over sixty peer-reviewed journal articles in mesoscale meteorology and related fields. Author of chapter 3.

X. Li is a Researcher in the Institute of Earth Environments, Chinese Academy of Science, Xian, China. His research interests are mainly in climate change and ecosystem in arid areas. Author of chapter 3.

B. Zib is an undergraduate student in the Department of Earth and Atmospheric Sciences at St. Louis University in St. Louis, Missouri. He is focusing on weather and climate variability in the central United States of America. Author of chapter 3.

J. C. Rogers is Professor of Atmospheric Science in the Department of Geography, The Ohio State University in Columbus, Ohio. He is state climatologist for the state of Ohio and publishes in the field of mid- and high-latitude climate variability. Author of chapters 5, 18.

S.-H. Wang is a Research Associate in the Polar Meteorology Group at the Byrd Polar Research Center at The Ohio State University in Columbus, Ohio. He specializes in global and Arctic regional model simulations and re-analyses. Author of chapter 5.

J. S. M. Coleman is an Assistant Professor in the Department of Geography at Ball State University in Muncie, Indiana. Her research focuses on applied climatology, atmospheric teleconnections, and biometeorology. Author of chapters 5, 14, 18.

G. S. Guentchev is a Postdoctoral Fellow with UCAR CLIVAR Climate Prediction and Applications Postdoctoral Program at Boulder, Colorado. She has published in the area of climate variability and change. Author of chapter 6.

K. Piromsopa is a Lecturer in the Department of Computer Engineering at Chulalongkorn University in Thailand. He has two patents pending on hardware security and is an active researcher in the areas of computer architecture and security. Author of chapter 6.

J. A. Winkler is a Professor in the Department of Geography at Michigan State University in East Lansing, Michigan. She has published extensively in the areas of synoptic climatology and climate variability and change. Author of chapter 6.

D. J. Lorenz is an Assistant Scientist in the Center for Climatic Research at the University of Wisconsin–Madison. His research interests are in the hydrologic cycle and in atmospheric dynamics. Author of chapters 7, 12.

S. J. Vavrus is an Associate Scientist in the Center for Climatic Research at the University of Wisconsin–Madison. His research interests are global and polar climate change, cloud and cryospheric processes, extreme events, and paleoclimate. Author of chapters 7, 12.

D. J. Vimont is an Assistant Professor in the Department of Atmospheric and Oceanic Sciences at the University of Wisconsin–Madison. His research focuses on the dynamics of large-scale ocean-atmosphere interactions and on impacts of climate variations and climate change. Author of chapters 7, 12.

J. W. Williams is Associate Professor in the Department of Geography and Center for Climatic Research at the University of Wisconsin–Madison and is the Bryson Professor of Climate, People, and the Environment. His research focuses on plant species and community responses to climate change, and crosses from the recent geological record to twenty-first-century climate change. Author of chapters 7, 12.

M. Notaro is an Associate Scientist in the Center for Climatic Research at the University of Wisconsin–Madison. He is an author of sixteen journal articles, with focus on vegetation-climate interactions and climate change. Author of chapters 7, 12.

J. A. Young is Director of the Wisconsin State Climatology Office and Professor Emeritus in the Department of Atmospheric and Oceanic Sciences. His research focuses on global atmospheric dynamics, boundary layers, dynamics, and climate processes. Author of chapters 7, 12.

E. T. DeWeaver is an Assistant Professor in the Department of Atmospheric and Oceanic Sciences at the University of Wisconsin–Madison. His research interests include large-scale atmospheric dynamics, data assimilation, sea ice dynamics and thermodynamics, the global hydrological cycle, and the impact of sea ice decline on polar bears. Author of chapters 7, 12.

E. J. Hopkins is the Assistant Wisconsin State Climatologist, a Lecturer in the Department of Atmospheric and Oceanic Sciences at the University of Wisconsin–Madison, and a weather/climate education consultant with the American Meteorological Society's Education Program. He coauthored *Wisconsin's Weather and Climate* with Joseph Moran. Author of chapters 7, 12.

K. E. Kunkel was formerly Interim Chief of the Illinois Water Survey in Champaign, Illinois, and now serves as Executive Director of the Division of Atmospheric Sciences of the Desert Research Institute in Reno, Nevada. He was contributing author to the IPCC Third and Fourth Assessment Reports and a lead author on two synthesis and assessment reports of the U.S. Climate Change Science Program: "Weather and Climate Extremes in a Changing Climate" and "Climate Models: An Assessment of Strengths and Limitations." Author of chapters 9, 19, 20.

M. Jha is an Assistant Scientist in the Center for Agricultural and Rural Development (CARD) at Iowa State University. His research focuses primarily in modeling watershed hydrology and water quality. Author of chapter 10.

E. Lu is a Postdoctoral Research Associate in the Department of Agronomy at Iowa State University. His specialization is the analysis of regional climate information from observations and output of regional climate models, with applications to regional impacts of global climate change. Author of chapter 10.

J. A. Andresen is an Associate Professor in the Department of Geography at Michigan State University in East Lansing, Michigan. He serves as state climatologist for the state of Michigan and has published more than fifty articles in the atmospheric and agricultural sciences literature. Author of chapter 11.

W. J. Northcott is an Extension Specialist in the Department of Biosystems and Agricultural Engineering at Michigan State University and former Director of Sustainable Agriculture for the Michigan Chapter of the Nature Conservancy. He is responsible for outreach and research programs related to agricultural irrigation and water use in the Great Lakes Region. Author of chapter 11.

H. Prawiranata is an environmental engineer with professional experience in groundwater hydrology, bioremediation of contaminated water resources, and numerical modeling. He currently serves as an analyst for the Tri-County Regional Planning Commission in Lansing, Michigan. Author of chapter 11.

S. A. Miller is a hydrologist and Visiting Instructor with the Department of Biosystems and Agricultural Engineering at Michigan State University. He has research experience with groundwater modeling, irrigation, and use of geographic information systems for hydrological applications and was responsible for the development of a statewide groundwater inventory and mapping project in Michigan. Author of chapter 11.

K. A. Blumenfeld is a Visiting Assistant Professor in the Department of Geography at the University of Minnesota in Minneapolis, Minnesota. His research emphasizes the climatology of extreme and hazardous weather, as well as modern regional climatic change. Author of chapters 13, 22.

K. Klink is an Associate Professor in the Department of Geography at the University of Minnesota in Minneapolis, Minnesota. Her current research investigates wind climatology and variability within the United States, with a particular focus on the upper Midwest. Author of chapters 14, 16.

R. J. Barthelmie is a Professor of Atmospheric Science and Sustainability in the Department of Geography at Indiana University in Bloomington, Indiana, the Ewert Farvis Chair of Energy Systems in the School of Engineering and Electronics at the University of Edinburgh in Scotland, and editor of the journal *Wind Energy.* She has published over eighty journal articles, principally in the fields of wind resource estimation, wind-turbine wake effects, and regional climate change impacts on renewable energy. Author of chapter 15.

D. A. R. Kristovich is Senior Scientist and Head of the Center for Atmospheric Sciences at the Illinois Water Survey, Institute for Natural Resources Sustainability, and Adjunct Associate Professor of the Department of Atmospheric Science, at the University of Illinois in Champaign-Urbana. He has published over twenty-five articles in scientific journals and is serving as an Editor of the *Journal of Applied Meteorology and Climatology.* Author of chapters 19, 21.

S. A. Changnon is Chief Emeritus of the Illinois State Water Survey and Adjunct Professor of the Department of Geography at the University of Illinois in Champagne-Urbana. He has published approximately four hundred journal articles on a wide range of climate topics, including extremes, impacts, and hydroclimatology. Author of chapter 20.

Index